A GUIDE TO MODELING
COASTAL MORPHOLOGY

ADVANCES IN
COASTAL AND OCEAN ENGINEERING
VOLUME 12

A GUIDE TO MODELING COASTAL MORPHOLOGY

Dano Roelvink
UNESCO-IHE and Deltares, The Netherlands

Ad Reniers
University of Miami, USA

World Scientific

NEW JERSEY · LONDON · SINGAPORE · BEIJING · SHANGHAI · HONG KONG · TAIPEI · CHENNAI

Published by

World Scientific Publishing Co. Pte. Ltd.
5 Toh Tuck Link, Singapore 596224
USA office: 27 Warren Street, Suite 401-402, Hackensack, NJ 07601
UK office: 57 Shelton Street, Covent Garden, London WC2H 9HE

British Library Cataloguing-in-Publication Data
A catalogue record for this book is available from the British Library.

Advances in Coastal and Ocean Engineering — Vol. 12
A GUIDE TO MODELING COASTAL MORPHOLOGY

Copyright © 2012 by World Scientific Publishing Co. Pte. Ltd.

All rights reserved. This book, or parts thereof, may not be reproduced in any form or by any means, electronic or mechanical, including photocopying, recording or any information storage and retrieval system now known or to be invented, without written permission from the Publisher.

For photocopying of material in this volume, please pay a copying fee through the Copyright Clearance Center, Inc., 222 Rosewood Drive, Danvers, MA 01923, USA. In this case permission to photocopy is not required from the publisher.

ISBN-13 978-981-4304-25-2
ISBN-10 981-4304-25-5

Printed in Singapore by Mainland Press Pte Ltd.

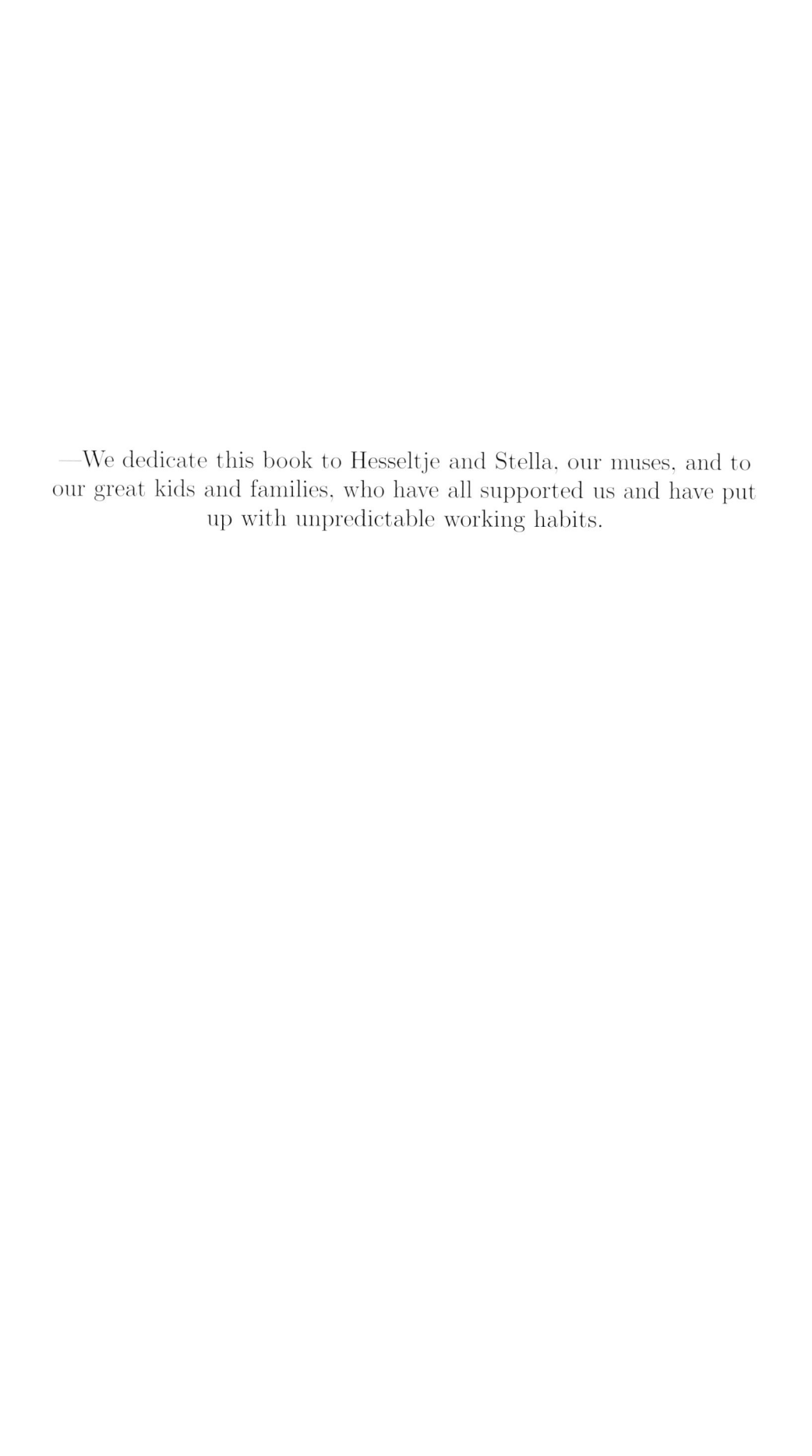

—We dedicate this book to Hesseltje and Stella, our muses, and to our great kids and families, who have all supported us and have put up with unpredictable working habits.

Preface

Models are formalized representations of reality. A clay model represents a car or a person without pretending to be it. A hydraulic scale model can simulate the flow around a harbor but it is not a harbor. Computer models of coastal morphology contain ideas about hydraulics, waves, sediment transport and sediment conservation that are captured in formulations. They produce interesting pictures and flashy animations of coastal behaviour but they cannot claim to represent reality in all its complexity. Rather, they quantify concepts we have in our heads and combine processes that are too difficult for us to combine just by reasoning. So far so good. The problem arises when the people who put the concepts into the model are different from the people who use the model. Then model concepts may be used out of context and results produced that are clearly wrong.

This guide to coastal morphodynamic modeling is meant to give coastal engineers and scientists an insight into some important processes in a relatively narrow strip of some kilometres from the coastline where most of the human activity takes place, where most of the sand transport and morphological changes occur and a large part of the transport of nutrients, pollutants and fine sediments are located.

This kind of understanding will help us to make quick assessments of important processes. It will make it easier to select a numerical model for a particular situation, and to assess whether the assumptions in it are acceptable. Finally, it will help us set up such models and to interpret and check the outcomes. Rather than being innocent victims of computer models we can master them and use them intelligently.

Dano Roelvink and Ad Reniers

Acknowledgment

Dano started working on this book many years ago during a visit at USGS in Woods Hole. The many discussions with colleagues there made him question everything he thought was evident and led to much delay but hopefully a better book. Ad took this time to collect his ideas and to present them in an organized fashion. We have had the great opportunity to work in very inspiring and different environments, at Delft University, Delft Hydraulics, Deltares, Naval Postgraduate School, University of Miami and UNESCO-IHE, surrounded by great colleagues and students. We would also like to acknowledge the support of and collaboration with the Office of Naval Research, Rijkswaterstaat, the National Science Foundation, the US Geological Survey and the many EU projects that allowed us to develop these ideas and tools.

Contents

Preface   vii

Acknowledgment   ix

1. Introduction   1

2. Wind Waves   3
   - 2.1 Introduction 3
   - 2.2 Wave generation, propagation and dissipation 4
   - 2.3 Spectral description 5
   - 2.4 Wave conditions 8
   - 2.5 Wave modeling 9
   - 2.6 Governing equations 10
     - 2.6.1 Wave action balance 10
     - 2.6.2 Wave energy balance 12
     - 2.6.3 Roller energy balance 15
   - 2.7 Propagation of wave groups 16
   - 2.8 Wave propagation over complex bathymetry 18
   - 2.9 Wave blocking 18

3. Currents   23
   - 3.1 Introduction 23
   - 3.2 Governing equations 23
     - 3.2.1 3D Shallow water equations 23
     - 3.2.2 Depth-averaged shallow water equations 27
     - 3.2.3 Stokes drift 28
     - 3.2.4 Wave forcing 28
     - 3.2.5 Bottom shear stress 33
     - 3.2.6 Turbulent eddy viscosity 36
   - 3.3 Tidal currents 37

	3.3.1	Open coasts	37
	3.3.2	Propagation into estuaries	42
	3.3.3	Resonance	43
	3.3.4	Funnelling effect	44
	3.3.5	Short, wide basins	45
	3.3.6	Tidal currents around structures	46
	3.3.7	Flow patterns around a realistic inlet	48
	3.3.8	Current pattern across a trench	49
3.4	Wind-driven longshore current and set-up on an alongshore uniform coast		51
	3.4.1	Wind-driven longshore current	51
	3.4.2	Wind-driven set-up	53
3.5	Wave-driven longshore current and set-up on an uniform coast ..		53
	3.5.1	Wave-driven longshore current	53
	3.5.2	Wave-driven set-up	54
	3.5.3	Numerical evaluation	55
	3.5.4	Shear instabilities	58
3.6	Wave-group driven motions		62
	3.6.1	Introduction	62
	3.6.2	Wave group induced bound long waves	63
	3.6.3	Leaky waves and trapped waves	66
	3.6.4	Edge wave resonance	69
	3.6.5	Very Low Frequency motions	72
3.7	Vertical structure of the current		74
	3.7.1	Tide (or slope) driven current profile	74
	3.7.2	Wind-driven current profile	76
	3.7.3	Wind driven longshore current profile	78
	3.7.4	Wind-driven cross-shore current profile	79
	3.7.5	Wave driven current profile	81
3.8	3D Wave-driven currents on a non-uniform coast		84

4. Sediment transport   89

4.1	Introduction	89
4.2	Suspended transport	90
	4.2.1 3D Advection-diffusion equation for sediment	90
	4.2.2 2DH Advection-diffusion equation for sediment	92
4.3	Bed load and total load transport formulations	94
	4.3.1 Current-only situation	94
	4.3.2 Waves plus current	95
4.4	Wave-driven transport	97
	4.4.1 Wave skewness and asymmetry	97
	4.4.2 Lagrangian drift	103

		4.4.3	Streaming	103
		4.4.4	Wave group induced bound long waves	104
	4.5	Return flow		107
		4.5.1	Breaker delay	109
	4.6	Rip circulation cells		109

5. Morphological Processes — 111

- 5.1 Introduction 111
- 5.2 Some principles 111
  - 5.2.1 Propagation of bed forms 111
  - 5.2.2 Equilibrium depth 113
- 5.3 Open coasts 115
  - 5.3.1 Cross-shore profile behavior 115
  - 5.3.2 Bed-slope related transport 117
  - 5.3.3 Dune erosion and overwash 120
  - 5.3.4 Rip channel dynamics 122
  - 5.3.5 Plan shape evolution 125
- 5.4 Tidal inlets and Estuaries 133
  - 5.4.1 Ebb and flood tidal delta formation 133
  - 5.4.2 Equilibrium relations 135
  - 5.4.3 Discussion 142

6. Modeling Approaches — 145

- 6.1 Coastal profile, coastline and area models 145
- 6.2 Scales of application 146
  - 6.2.1 Coastal profile models 146
  - 6.2.2 Coastline models 147
  - 6.2.3 Coastal area models 147
- 6.3 Input schematisation 150
  - 6.3.1 Input parameters 150
  - 6.3.2 General principle of schematisation 151
  - 6.3.3 Tidal schematisation 152
  - 6.3.4 Schematisation of wind/wave climate 159

7. Coastal Profile Models — 169

- 7.1 Introduction 169
  - 7.1.1 Principles and Approach 169
  - 7.1.2 Profile modeling 170
- 7.2 Short-term event modeling 173
- 7.3 Long-term evolution of barred profiles 175
  - 7.3.1 Including longshore transport gradients 176
- 7.4 Nourishments 176

8. Coastline Models — 179

- 8.1 Principles … 179
- 8.2 Existing models … 179
- 8.3 A simple Matlab version … 180
  - 8.3.1 Profile model to generate S-phi curves … 181
  - 8.3.2 Coastline computation … 181
  - 8.3.3 The basic version based on S-phi curves … 181
  - 8.3.4 Including large-scale variations in wave climate … 182
  - 8.3.5 Representing small-scale features … 184
- 8.4 Case study of IJmuiden, the Netherlands … 184

9. Coastal Area Models — 187

- 9.1 Introduction … 187
- 9.2 Wave drivers … 188
  - 9.2.1 Wave-averaged … 188
  - 9.2.2 Short wave averaged … 189
  - 9.2.3 Short wave resolving … 189
- 9.3 2DH, Q3D and 3D … 190
  - 9.3.1 Flow model … 190
  - 9.3.2 Sediment transport … 190
  - 9.3.3 Bottom … 191
- 9.4 Grids and numerical aspects … 192
  - 9.4.1 Overview of model components in some morphodynamic model systems … 193
- 9.5 Boundary conditions for coastal area models … 193
  - 9.5.1 Flow model … 193
  - 9.5.2 Waves … 199
  - 9.5.3 Sediment transport … 200
  - 9.5.4 Bed level … 200
- 9.6 Modeling strategies for wave-current interaction … 200
- 9.7 Strategies for morphodynamic updating … 204
  - 9.7.1 Tide-averaging approach … 205
  - 9.7.2 Continuity correction … 207
  - 9.7.3 RAM approach … 208
  - 9.7.4 Online approach with morphological factor … 209
  - 9.7.5 Tide-averaged approach vs. morphological factor … 211
  - 9.7.6 Parallel online approach … 213
  - 9.7.7 Efficiency of the methods … 214
- 9.8 Strategies for longer-term simulations … 215
  - 9.8.1 Beach profile extension … 215
  - 9.8.2 Representation of subgrid features … 217
  - 9.8.3 Representation of dredging … 217

		9.8.4	Beach nourishments	217
10.	Case Studies			219
	10.1	Toy models of small coastal problems		219
		10.1.1	Introduction	219
		10.1.2	Model setup	219
		10.1.3	Wave height patterns	221
		10.1.4	Current patterns	222
		10.1.5	Effect on bathymetry	223
		10.1.6	Relative erosion/sedimentation patterns	224
		10.1.7	Discussion	224
	10.2	Long-term modeling of tidal inlets, estuaries and deltas		225
		10.2.1	Introduction	225
		10.2.2	How far can upscaling lead us?	225
		10.2.3	Necessary model improvements for long-term modeling	226
		10.2.4	How much of the morphology of an estuary is forced by its boundaries?	227
		10.2.5	Effect of sediment sorting	230
		10.2.6	Some sample simulations	232
	10.3	Dune erosion		235
	10.4	Overwash		237
	10.5	Sand bars and rip channels		239
11.	Modeling Procedure			245
	11.1	Introduction		245
	11.2	Data collection and analysis		245
		11.2.1	Bathymetry data	245
		11.2.2	Wave and wind data	245
		11.2.3	Tidal data	246
		11.2.4	Longshore current data	246
		11.2.5	Sediment transport data	246
	11.3	Conceptual model		246
	11.4	Setting up modeling strategy		246
	11.5	Setting up model grid and bathymetry		247
		11.5.1	Flow and morphology grid	247
		11.5.2	Wave grids	247
		11.5.3	Bathymetry	247
	11.6	Boundary conditions		247
		11.6.1	Wave schematisation	247
		11.6.2	Representative tide	248
		11.6.3	Sediment transport	248
		11.6.4	Bottom change	248

11.7	Calibration	248
11.8	Validation	248
11.9	Preparing scenarios	249
11.10	Defining output	251
11.11	Running and postprocessing	251
11.12	Interpretation	252
11.13	Reporting	252
11.14	Archiving	253

12. Modeling Philosophy — 255

12.1	Virtual reality or realistic analogue?	255
12.2	Process-based or data-driven?	257
12.3	Top-down or bottom-up?	257
12.4	More physics, better model?	258
12.5	How to judge model skill?	258
12.6	Absolute vs. relative skill	259

Bibliography — 261

Chapter 1

Introduction

This guide is roughly divided into three parts.

We start with a description of the most important processes governing the morphological evolution of our coasts at some relevant time and space scales: from individual storms to long-term evolution and from tens of metres to large tidal basins. Chapters 2, 3, 4 and 5 are devoted respectively to waves, currents, sediment transport and morphology. The processes considered are wave propagation and dissipation, currents driven by tides, waves and winds. Here, we are not concerned with where and how these waves or winds or tides were generated, but with what happens in the coastal area only. We shall thus assume that there is ample information on tidal water levels, large-scale wind fields, wave conditions and river discharges at the landward boundaries and at, say, twenty metres depth, from monitoring stations, buoys or regional and global models. Our main objective here will be to understand some typical flow situations and to be able to estimate current patterns and strengths by analysing the balance of the most relevant terms in the governing equations.

The next chapters discuss modeling approaches in general (Chapter 6), after which we have separate discussions on coastal profile models (Chapter 7), coastline models (Chapter 8) and 2DH/3D coastal area models (Chapter 9). By models we mean model concepts rather than software systems, proprietary or otherwise, and we will try to keep the discussions generic by using easy-to-understand Matlab programs where possible, as in the case of profile and coastline models. For the more complex 2DH/3D models we will stay away from particularities of the different systems but rather describe relevant concepts, boundary conditions, typical schematization procedures and then the interesting morphological phenomena that can result. When we want to describe a particular phenomenon, we will try to find the simplest model that can generate it and explain how it works. Each situation can be seen as a subset of all the processes that can happen in complex models used in practice. By isolating the processes responsible for a given phenomenon we learn what is the minimum set required to represent it.

In Chapter 10 we discuss some case studies grouped around common coastal problems. Again, we want to stimulate thinking about the coastal processes and how they can be effectively modeled and we do that using different model concepts for different purposes.

In Chapter 11 we provide an attempt at a systematic procedure for modeling studies, like an extended checklist, based on our own experience and studies we have seen carried out around us. This is followed by a discussion on modeling philosophy in Chapter 12 focusing on the given mismatch between the model representation and reality and still be able to use these complex morphodynamic models to make useful predictions.

All Matlab scripts and functions referred to in this book are available through OpenEarth.nl, an open-source initiative by Deltares. After a simple registration procedure you can download, among many other tools, our code under Tools, directory matlab/applications/CoastalMorphologyModeling.

Chapter 2

Wind Waves

2.1 Introduction

The modeling of surface gravity waves is an essential component in the prediction of coastal morphodynamics as outlined throughout this book. Here we consider wind-generated surface gravity waves, i.e. sea and swell, only.

Out at sea wind waves typically radiate away from the depression centre, thereby transporting energy, mass and momentum. At deeper water these waves are only weakly non-linear and as a consequence the wave field can characterized as a summation of a large number of independent wave components resulting in a Gaussian sea state creating an irregular wave field. As the waves propagate from deep to shallow water they start interacting with the bottom, with increasing non-linearities, decreasing wave propagation speed and energy loss due to bed friction. In the case of obliquely incident waves the latter results in refraction due to the difference in propagation speed along the wave crests. At further decreasing water depths, shoaling becomes evident, up to a point where the waves become unstable and breaking occurs. A broken wave can be easily recognized by the aerated water at the front, known as a roller, which results from the inclusion of air-bubbles by the overtopping wave crest.

Waves play an important role in the coastal response. The near-bed orbital motion is known to stir up sediment, which is transported by (ambient) currents, consequently the sediment transport rate significantly increases in the presence of waves (see Section 4.3.2). On the other hand, the interaction between wave orbital motions and current velocities within the wave boundary layer increases the bottom shear stress for the ambient currents, slowing the current down (see Section 3.2.5). Within the surfzone, the dissipation of short wave energy forces the wave-driven nearshore currents, both alongshore (in the case of obliquely incident waves) and cross-shore (see Sections 3.5 and 3.6). In addition, set-up gradients induced by spatial variation in wave energy can generate strong alongshore and cross-shore flows such as a rip-current circulations (Section 3.8). These littoral flows are known to be capable of transporting large amounts of sediment [Komar (1976)]. Obstruction of the littoral drift results in an area of accretion, followed by erosion of the beach at the down-drift end, a well-known phenomenon observed at many a harbor entrance (see Chapter 8 on Coastline Models). Short-wave non-linearity is important in a number of aspects. One aspect is the onshore transport of mass, associated with the Stokes drift, which is compensated by an offshore-directed return flow (see Sections 4.4.2 and 4.5). In shallow water, but prior to breaking, both bound higher and lower harmonics are generated through triad interactions [Hasselmann (1962)]. The presence of higher harmonics, which are initially in phase with the primary waves, results in more peaked wave crests and flatter troughs [Stokes (1847)]. This generally results in a short-wave averaged onshore sediment transport (see Section 4.4.1) [Roelvink and Stive (1989)]. Note that additional complexity is present due to the vertical structure of both the sediment concentration and wave and flow-dynamics which can lead to offshore directed sediment transport (see Chapter 4 on sediment transport). Near breaking, the waves assume a see-saw shape, as the phase shift between the primary waves and its super harmonics increases. At this point intrawave pressure gradients can become important in the sediment response ([Madsen (1975)], [Drake and Calantoni (2001)], [Nielsen and Callaghan (2003)], [Foster et al. (2006)]). At breaking, turbulence gets injected into the water column, providing additional stirring of the sediment ([Roelvink and Stive (1989)], [Steetzel (1993)], [van Thiel de Vries et al. (2008)]). In summary, detailed knowledge of the wave dynamics is required in predicting the morphodynamic behavior of any beach.

Below we briefly outline the key aspects of wave generation and propagation in view of their relevance for coastal modeling. For more detailed descriptions on the subject, including expressions derived from linear wave theory, refer to other textbooks such as [Phillips (1977)], [Mei (1989)], [Dean and Dalrymple (1991)], [Dingemans (1997)], [Svendsen (2006)], [Holthuijsen (2007)] and others.

2.2 Wave generation, propagation and dissipation

Waves are generated by the exchange of momentum between wind and underlying water surface. The initial transfer of wind energy to wave energy assumes the presence of wind pressure variations with length scales and propagation speeds which match the dispersion relation of the wave like disturbances ([Phillips (1957)]) resulting in a resonant transfer of energy form the wind to the waves. The initial wave conditions are characterized by very short waves propagating in all directions. According to linear theory, the rate at which short waves can draw energy from the wind is much higher than for the longer waves because momentum transfer is proportional to the wave steepness (the square of the surface slope). As the waves grow in amplitude the water surface begins to interact with the wind resulting in an increased growth rate of the waves and a preferential direction aligned with the wind direction ([Miles (1957)]). Experimental observations have shown that the combined mechanisms of [Phillips (1957)] and [Miles (1957)] are not able to explain the observed growth rate, which is an order of magnitude larger. This effect is commonly attributed to the occurrence of weak non-linear four-wave (quadruplet) interactions between individual wave components, which transfers energy from the saturated high frequency waves toward the lower frequency waves ([Hasselmann (1962)]). Even though the interactions are relatively weak, i.e. the Gaussian description still holds, the large distances traveled by waves makes their cumulative effect very prominent, resulting in a significant increases in wave height and wave period. The continuous input of wind energy and the concurrent non-linear interactions result in a wave field with both frequency and directional spreading.

Dissipation in deeper water occurs through wave breaking, known as white capping, which transfers wave energy into small-scale turbulence (e.g. [Melville and Matusov (2002)]). Given the fact that wave breaking at deeper water is steepness dependent ([Miche (1944)]) this process is most relevant for the saturated high frequency wave components. As the water depth becomes less, the waves start interacting with the bottom and energy is lost within the bottom boundary layer through bottom friction. At these depths, Bragg scattering also becomes important ([Ardhuin *et al.* (2003)]) increasing the directional spreading of the short waves in the presence of bathymetric variability. In contrast, the alongcrest differences in wave celerity refract the waves toward the shore normal, decreases the directional spreading.

The most dramatic changes occur in the nearshore. The presence of near-resonant three-wave (triad) interactions quickly transfer energy to waves of both higher frequency (super-harmonics), thereby changing the wave shape, and lower frequencies (sub-harmonics) generating infragravity wave energy. As the water depth becomes even less, depth-limited wave breaking becomes important, transferring wave energy into surface rollers. The energy associated with the rollers is then transferred to small-scale turbulence. Depth-limited wave breaking is probably the most important (and least understood) process in the nearshore zone as it transfers energy and momentum from the breaking wind waves to wave-driven currents, infragravity waves, turbulence and concurrent sediment transport.

2.3 Spectral description

Instead of describing the irregular wave surface elevation in a deterministic sense the wave field is often described in a spectral fashion. The transformation of the time domain description to the frequency domain is generally performed with a Fast Fourier Transform (FFT, [Cooley and Tukey (1965)]). The FFT yields an amplitude, a, and phase, ϕ, for each Fourier component with frequency f. Given the random character of the wave field the phase information changes with each realization and is typically ignored. Instead the wave energy density is collected as function of frequency:

$$E(f) = \frac{1}{2}\frac{a^2(f)}{\Delta f} \tag{2.1}$$

where Δf represents the frequency band width for which the FFT is calculated resulting in the frequency spectrum $E(f)$. The important details of the spectral analyses used to translate time series into spectra, and specifically for wind waves, can be found in many text books such as [Holthuijsen (2007)].

The frequency spectrum captures the temporal surface elevation variability only. To capture the spatial variability, associated with the directional spreading of the wind waves, additional analyses are required. In the case of a spatially homogeneous wave field a spatial FFT can be applied which yields the amplitude and phase as function of the cross-shore and alongshore wave numbers k_x and k_y. Using the linear dispersion relation relation (neglecting surface tension and amplitude effects):

$$\sigma^2 = gk \tanh kh \tag{2.2}$$

with the intrinsic angular frequency $\sigma = 2\pi f, k = \sqrt{k_x^2 + k_y^2}$, and h the water depth, the frequency-directional spectrum, $E(f,\theta)$, can be constructed. To perform the spatial FFT the surface elevation has to be known over a large area at regularly spaced intervals. This information is generally not available in the field, where surface elevation measurements are typically obtained with sparse in-situ instruments. Notable exceptions are observations of the surface elevation obtained with radar and video (e.g. [van Dongeren et al. (2008)] and references therein). However, by using a combination of in-situ instruments $E(f,\theta)$ can also be constructed by examining the cross-spectral variance between signals obtained at the individual intsruments. Generally the Maximum Entropy Method (MEM) is used to construct $E(f,\theta)$ of wind waves measured with a directional wave rider or combined pressure sensor and current velocity meter ([Lygre and Krogstad (1986)]). Alternatively, a spatial array of individual pressure sensors or flow meters can be used to estimated $E(f,\theta)$ (e.g. [Pawka (1983)], [Reniers et al. (2010a)]). An overview of the different methods to estimate $E(f,\theta)$ for wind waves is given by [Benoit et al. (1997)].

Spectral analyses of measurements of sea states have shown the existence of preferred shapes, which depend on the evolution of the sea state. Well known examples are the Pierson-Moskovitz (PM) spectrum for fully developed seas and the more peaked Jonswap spectrum for young seas ([Hasselmann et al. (1973)]):

$$E(\sigma) = \alpha g^2 (2\pi)^{-4} \sigma^{-5} \exp\left[-\frac{5}{4}\left(\frac{\sigma}{\sigma_p}\right)^{-4}\right] \gamma_0^{\exp\left[-\frac{1}{2}\left\{\frac{\sigma - \sigma_p}{\epsilon \sigma_p}\right\}\right]} \tag{2.3}$$

where α is an energy scaling coefficent, σ_p is the peak radial frequency, γ_0 is the peak enhancement factor and ϵ determines the spectral width around the peak. For

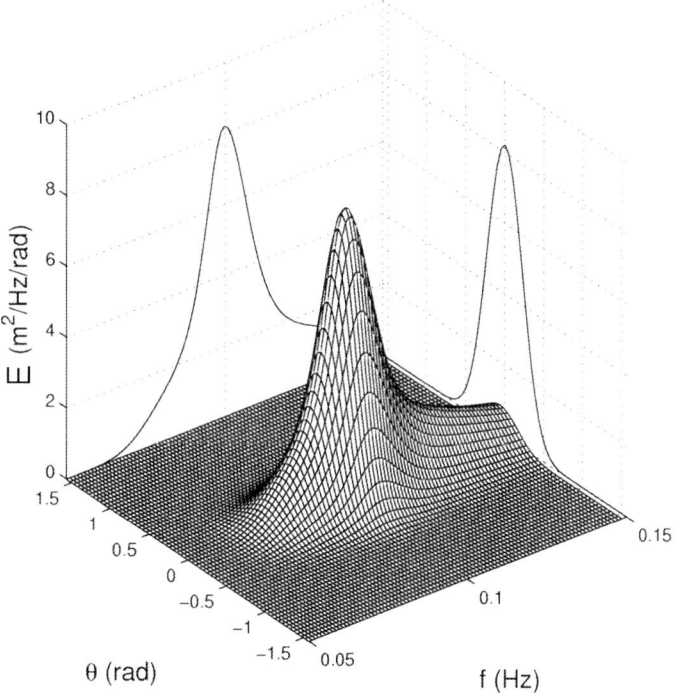

Fig. 2.1 Example of an energy density distribution of a random sea state according to a Jonswap frequency spectrum and $\cos^m$ directional distribution with m = 20 with a peak frequency of 0.1 Hz and a peak direction of 0 degrees.

the PM spectrum $\alpha = 0.0081$ and γ_0 is equal to 1. For the Jonswap spectrum γ_0 is increased to 3.3 and $\epsilon = 0.07$ for $\sigma < \sigma_p$ and $\epsilon = 0.09$ for $\sigma > \sigma_p$.

Examples for directional distributions are also available such as the $\cos^m$ distribution:

$$D(\theta) = A_m \cos(\theta - \theta_p)^m \qquad (2.4)$$

where A_m is a proportionality factor to ensure unity of the integrated directional distribution function and θ_p is the peak direction. Small (large) values of m correspond to broad (narrow) directional distribution. The frequency-directional spectrum can now be constructed according to:

$$E(f, \theta) = E(f) D(\theta) \qquad (2.5)$$

An example of a synthetic frequency-directional distribution for wind waves is shown in Figure 2.1. If detailed information on the spectra is not available these standard spectral shapes can be used as boundary conditions for model calculations. In the case of deterministic wave modeling, a random phase model can be used to generate representative time series of the surface elevation by summing a large number of sinusoidal wave components with amplitude $\hat{\eta}_j$ calculated from the frequency spectrum, eq. (2.3), and a random phase, ϕ_j, assigned to each frequency component ([Miles and Funke (1989)]):

$$\eta(x,y,t) = \text{Re}\left[\sum_{j=1}^{N} \hat{\eta}_j e^{i(\sigma_j t - k_{x,j} x - k_{y,j} y + \phi_j)}\right] \quad (2.6)$$

The linear dispersion relation, eq. (2.2), is utilized to relate the wave number k_j to the angular frequency σ_j and the corresponding individual incidence angle is obtained from a probability function based on the directional distribution function, eq. (2.4). Note that each individual realization is deterministically different but statistically equal due to the random phase and direction assignment.

2.4 Wave conditions

In the following we consider the wave conditions associated with sea and swell. Sea waves are considered to be locally generated by the wind and depend on fetch, local water depth and wind speed ([Sverdrup and Munk (1947)]). Swell waves are not correlated to the local wind conditions, but are waves which arrive from distant storms and have travelled long distances before breaking on a shore ([Barber and Ursell (1948)], [Snodgrass et al. (1966)]). Due to the dispersion that occurs over a large distance swell waves are generally narrow in both frequency and directional spreading. In contrast, sea waves have both broad directional and frequency spreading. Depending on the location and orientation of the beach both sea and swell waves can be present. At deeper water sea waves are characterized by relatively steep surface slopes, defined by the ratio of wave height over wave length, whereas the longer period swell display much milder slopes. Their morphological impact is generally expected to be quite different due to differences in the non-linear evolution of sea and swell waves as they enter shallow water. Swell waves become significantly more skewed than the shorter sea waves prior to breaking, resulting in significant asymmetry-driven sediment transport (see Section 4.4.1). In contrast, the shorter sea waves become sea-saw-shaped over a short distance offshore of the breakpoint, hence their morphological contribution is mostly associated with the offshore return flow transport induce transport due to mass-flux compensation (see Section 4.5).

Local wave conditions change continuously due to changes in the environmental conditions such as wind speed and direction, tidal elevation, tidal currents, etc. As a result the wave conditions for a specific site are generally represented by probability density distributions where wave information has been collected over many years. Based on these long term data sets representative wave conditions can be obtained with statistical tools (see Section 6.3.4). The observations are generally restricted to deeper water obtained from ship observations, measuring platforms, wave buoys or satellite. To predict changes in the coastal morphology the offshore wave conditions thus have to be translated to the nearshore area of interest. This translation is performed with wave models (discussed below).

2.5 Wave modeling

The need for wave information in modeling the coastal morphology is clear. A number of approaches have been considered in the past. The first wave models were developed during the second world war and consider the waves to be locally generated by the wind and depend on fetch, local wind speed, wind duration ([Sverdrup and Munk (1947)]). The first generation of wave energy balance models considered the evolution of a frequency spectrum subject to wind input and white capping ([Gelci et al. (1956)]). Second generation spectral models also explicitly include the effects induced by the non-linear wave interactions described by [Hasselmann (1962)] but also additional processes such as directional spreading, refraction and shoaling, bottom friction, wave current interaction and depth limited wave breaking over arbitrary bathymetry (e.g. [Holthuijsen et al. (1989)]). Present-day third generation wave models fully resolve the frequency-directional evolution of the energy density (e.g. WAM, WAVEWATCH, SWAN) thereby implicitly including non-linear wave interactions. Given the fact that the phase information of the individual waves is not retained, this modeling approach allows for relatively large spatial computational steps and as such is suitable to compute the wave conditions in a large domain. The down side of this approach is that bound sub-and super harmonics (which are phase coupled) are not accounted for and as such the intra-wave non-linearity cannot be reconstructed from these models. A notable exception is the fourth-generation spectral wave model by ([Janssen et al. (2006)]), which includes a phase evolution equation. An alternative approach to resolve the non-linear intra-wave kinematics of wind waves in intermediate and shallow water depths is given by Boussinesq models ([Dingemans (1997)], and references therein). These models typically do not take into account the generation or growth of waves due to wind energy, but do include shoaling, refraction, diffraction, non-linear interactions, depth-limited wave breaking and wave-current interaction, often all the way up to the water line thereby including the swash. However, the vertical distribution of the flow is not resolved and the bottom slopes are expected to be mild. Both time-domain and spectral boussinesq models are available. Another example are the full three dimensional quasi-hydrostatic models ([Stelling and Zijlema (2003)]) which can resolve the veritical flow-dynamics on the intra-wave scale on an arbitrary slope. Note that the computational effort associated with these intra-wave models is large, which can make their application to either large scale domains and/or long durations impractible.

In principle, given the wind conditions out on the ocean, the wave conditions close to shore can be predicted, through the application of a sequence of wave prediction models (Figure 2.2) going from ocean scale (e.g. WAM or WAVEWATCH

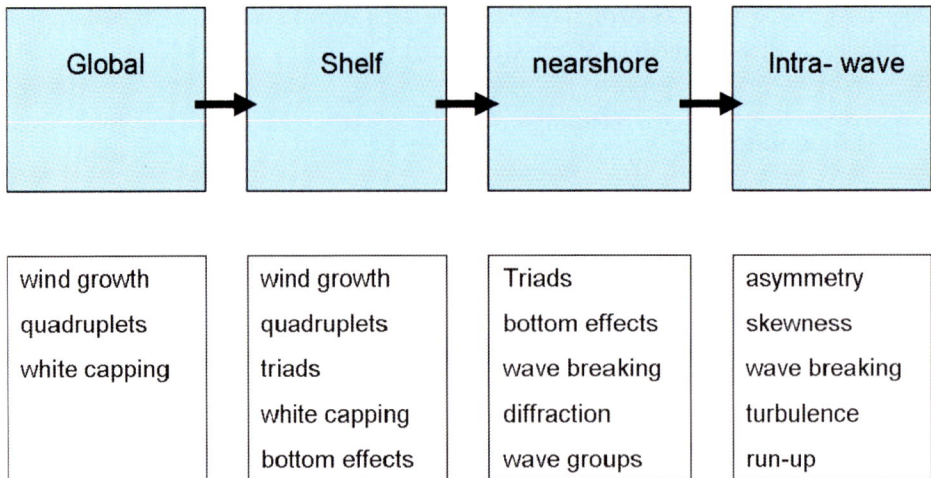

Fig. 2.2 Flow chart of wave modeling from global scale to intra-wave with corresponding key model elements.

models), which give the required boundary conditions for the coastal shelf modeling (e.g. SWAN or STWAVE) which feed into nearshore models (e.g. nested SWAN or XBEACH) which in turn can be coupled to phase resolving models to predict surfzone and swash motions (e.g. Boussinesq and non-hydrostatic models such as FUNWAVE and SWASH3D). Alternatively, if boundary conditions are known closer to shore one could initiate the nearshore model with these measurements, which is preferable from a computational point of view.

The present morphodynamic modeling is focussed on the nearshore with a minimum domain length scale in the order of a kilometer and typical time-scale in the order of a week. This clearly invokes restrictions on the wave modeling. Complex models that are available to compute the intra-wave kinematics within the nearshore require significant computational time and are at this point less suitable for the morphodynamic computations that are considered here. The focus in the wave modeling is therefore on the phase-averaged and short-wave-averaged modeling discussed next. Note that the intra-wave scale is no longer resolved, which may require parameterisation of intra-wave effects in the sediment transport (see Section 4.4.1).

2.6 Governing equations

2.6.1 *Wave action balance*

A classical approach to wave modeling is by considering individual wave rays traveling from deep water to shore taking into account the physical processes such as wind input, propagation and dissipation. Collecting the individual wave rays then yields information on nearshore wave conditions. However, in the presence of irregular bathymetry this approach is know to generate caustics, i.e. locations where individual wave rays cross, resulting in unrealistic wave heights. To avoid the effect of caustics either the bathymetry or the resulting wave field (or both) have to be spatially smoothed ([Bouws and Battjes (1982)]), which is non-trivial given the fact that the smoothing length scale has important consequences for the interpretation of the model outcome (e.g. [Plant et al. (2009)]). The Eulerian alternative to the Lagrangian approach described above is by considering the changes in wave action on a fixed grid ([Hasselmann et al. (1973)]):

$$\frac{\partial N}{\partial t} + \frac{\partial}{\partial x}c_x N + \frac{\partial}{\partial y}c_y N + \frac{\partial}{\partial \sigma}c_\sigma N + \frac{\partial}{\partial \theta}c_\theta N = \frac{S}{\sigma} \qquad (2.7)$$

where N represents the action density as function of angular frequency, σ, and direction, θ:

$$N(\sigma, \theta) = \frac{E(\sigma, \theta)}{\sigma} \qquad (2.8)$$

and c_i represents the transport velocities in x,y, σ and θ-space and E represents the energy density. In the following we use expressions derived from linear wave theory to describe the relevant wave action balance components. The wave energy in a single free surface gravity wave component (dropping σ and θ) is given by the sum of its potential and kinetic energy:

$$E_w = \frac{1}{2}\rho g a^2 \qquad (2.9)$$

in which ρ presents the density of water and g the gravitational acceleration. Each wave component travels with its own velocity c:

$$c = \frac{\sigma}{k} \qquad (2.10)$$

in which k is dictated by the linear dispersion relation eq. (2.2). The group velocity, at which the wave energy travels, is given by:

$$c_g = \frac{\partial \sigma}{\partial k} = \frac{1}{2}\left[1 + \frac{2kh}{\sinh 2kh}\right]c \qquad (2.11)$$

Taking into account the direction of the individual wave component, the propagation speed of the wave energy in geographical space (in the presence of a current) becomes:

$$c_x = c_g \cos(\theta) + u \qquad (2.12)$$

$$c_y = c_g \sin(\theta) + v \qquad (2.13)$$

with u and v the cross-shore and alongshore Stokes-depth-averaged Eulerian velocities respectively. Given the depth dependence of the frequency, eq. (2.2), temporal changes in water depth result in a frequency shift, which can be expressed with a propagation in σ-space :

$$c_\sigma = \frac{\partial \sigma}{\partial h}\left(\frac{\partial h}{\partial t} + \vec{u}\cdot\nabla h\right) - c_g \vec{k}\cdot\frac{\partial \vec{u}}{\partial s} \qquad (2.14)$$

where s is in the wave propagation direction and m is nomal to the wave propagation direction. In the case of an obliquely incident wave, the phase and group velocity will vary along the wave-crest given the differences in depth and/or currents, which leads to refraction. This effect is expressed as a propagation speed in θ-space:

$$c_\theta = -\frac{1}{k}\left(\frac{\partial \sigma}{\partial h}\frac{\partial h}{\partial m} + \vec{k}\cdot\frac{\partial \vec{u}}{\partial m}\right) \qquad (2.15)$$

The wave number k is obtained from the eikonal equations:

$$\frac{\partial k_x}{\partial t} + \frac{\partial \omega}{\partial x} = 0 \qquad (2.16)$$

$$\frac{\partial k_y}{\partial t} + \frac{\partial \omega}{\partial y} = 0 \qquad (2.17)$$

and ω represents the absolute angular frequency given by

$$\omega = \sigma + k_x u + k_y v \qquad (2.18)$$

which completes the set of equations to describe the advection of wave action (i.e. left hand side of eq. (2.7)). The S-term on the right hand side of eq. (2.7) represents source and sink terms associated with wind growth, non-linear interactions and wave dissipation. As the waves propagate, these processes affect the action density distribution, thus changing the spectral shape of the energy density (see [Holthuijsen (2007)]).

2.6.2 Wave energy balance

The spectral wave action balance in eq. (2.7) describes the slow variation in time and space of the full wave spectrum. This equation can be simplified by integrating over all frequencies, which yields the time-varying HISWA equations ([Holthuijsen et al. (1989)]), which can be applied to model the propagation of (grouped) directionally spread waves ([Roelvink et al. (2009)]). A further reduction yielding the energy balance can be made if the spectrum is directionally narrow-banded and the peak frequency is constant in space:

$$\frac{\partial E_w}{\partial t} + \frac{\partial}{\partial x}\left(E_w c_g \cos(\theta_m)\right) + \frac{\partial}{\partial y}\left(E_w c_g \sin(\theta_m)\right) = -D_w - D_f \qquad (2.19)$$

where the wave energy is equivalent to:

$$E_w = \tfrac{1}{8}\rho g H_{rms}^2 \qquad (2.20)$$

and D_w and D_f represent wave energy dissipation due to wave breaking and bottom friction respectively. This simple equation is useful to describe the propagation and dissipation of the total wave energy for given mean wave direction θ_m. For a stationary wave field on a planar beach, i.e. alongshore uniform, in the absence of wave dissipation the wave energy balance reduces to:

$$\frac{\partial}{\partial x}(E_w c_g \cos\theta_m) = 0 \qquad (2.21)$$

which shows that changes in the group velocity, as a result of cross-shore depth variations, results in shoaling and refraction:

$$\frac{E_{w,i}}{E_{w,0}} = \frac{c_{g,0}}{c_{g,i}} \frac{\cos(\theta_{m,0})}{\cos(\theta_{m,i})} = K_s K_r \qquad (2.22)$$

and the subscript 0 refers to a position offshore where the wave direction is known and the subcript i refers to an arbitrary cross-shore position. On a planar beach the incidence angle can be obtained with Snel's law:

$$\frac{\sin\theta_{m,i}}{c_i} = \frac{\sin\theta_{m,0}}{c_0} \qquad (2.23)$$

hence the local wave energy and direction can be obtained given the conditions at some reference point (Figure 2.3). Simple expressions like these can be used to verify the correct implementation of more complex models. Note that the shoaling factor becomes infinite in the limit of $h \to 0$ and subsequently $H_{rms} \to \infty$. In reality, as the wave height over water depth ratio increases, the incident waves become unstable and start to break. Hence, these expressions are only valid outside the surf zone.

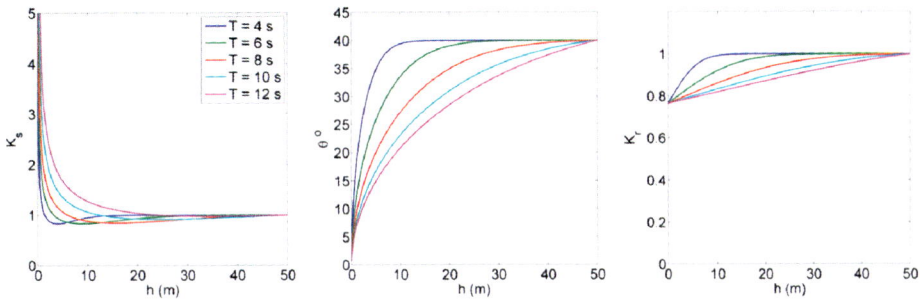

Fig. 2.3 Shoaling factor for a planar beach as function of wave period (left panel). Incidence angle calculated with Snel's law (middle panel) and corresponding refraction factor (right panel) as function of wave period.

Though the process of wave breaking is very complex and cannot be modelled exactly, there are several formulations available that describe D_w as a function of wave energy, period and water depth with reasonable accuracy (see [Apotsos et al. (2008)] for an overview).

[Battjes and Janssen (1978)] were the first to propose a model for the breaking dissipation in random waves, where they used a bore analogy to model the dissipation in a single wave, and a simple model of the wave height distribution (a so-called clipped Rayleigh distribution) to provide the probability that waves are breaking. As all breaking waves were assumed to have the same maximum wave height (a function of depth and period) and the total wave dissipation was simply the product of the dissipation in a single wave and the probability that waves are breaking or fraction of breaking waves:

$$D_w = Q_b D_b = Q_b \frac{1}{4} \rho g \alpha f_p H_{\max}^2 \tag{2.24}$$

Here the fraction of breaking waves is given by the implicit relation:

$$Q_b = \exp\left(-\left(\frac{H_{max}^2}{H_{rms}^2}(1-Q_b)\right)\right) \tag{2.25}$$

and the maximum wave height is given by the Miche criterion:

$$H_{\max} = \gamma h \tag{2.26}$$

With suitable calibration (generally keeping $\alpha = 1$ and making γ dependent on the wave steepness) this model accurately describes the dissipation of energy due to breaking [Battjes and Stive (1985)], though sub-parts of the model such as the wave height distribution or the fraction of breaking waves may be wrong. Other models have been proposed with improved breaking wave height distributions (e.g. [Thornton and Guza (1986)], [Baldock et al. (1998)], [Janssen and Battjes (2007)]) but the BJ78 model is still widely used, both in profile models and in spectral wave models where the dissipation is distributed in proportion to the spectral density. A recent review of wave dissipation models and corresponding optimal parameter settings show that the expected error in the wave height transformation is in the order of 10 % ([Apotsos et al. (2008)]). Note that these models are valid for the average dissipation over a train of random waves and are less suited to represent the wave dissipation in wave groups for which alternative breaking wave dissipation models are available ([Roelvink (1993)]).

Although the wave energy dissipation due to bottom friction is weak compared with wave breaking it becomes important over large propagation distances on coastal shelves (e.g. [Ardhuin et al. (2003)]). A general expression is given by:

$$D_f = \rho C_f \overline{|u_{rms}|^3} \tag{2.27}$$

where C_f is a friction coefficient of $O(0.01)$ and the near-bed root mean square orbital motion is obtained from linear wave theory:

$$u_{rms} = \frac{\omega a}{\sqrt{2} \sinh kh} \tag{2.28}$$

and a is the wave amplitude.

2.6.3 Roller energy balance

While the wave energy balance adequately describes the propagation and decay of organized wave energy, it has often been found that there is a delay between the point where the waves start to break and the point where the wave set-up and longshore current start to build. This 'transition zone' effect is generally attributed to the temporary storage of shoreward momentum in the surface rollers. Several authors have analyzed the typical dimensions of such rollers and their associated momentum (e.g. [Svendsen (1984)], [Roelvink and Stive (1989)], [Nairn et al. (1990)], [Deigaard (1993)] and [Stive and de Vriend (1990)]). The rollers can be represented as a blob of water with cross-sectional area R that slides down the front slope of a breaking wave. The roller exerts a shear stress on the water beneath it equal to:

$$\tau_{roller} = \frac{\rho g R}{L} \beta_s \tag{2.29}$$

where β_s is the slope of the breaking wave front, and L is the wave length. The roller has a kinetic energy equal to:

$$E_r = \frac{1}{2} \frac{\rho R \overline{(U_{roller}^2 + W_{roller}^2)}}{L} \tag{2.30}$$

and negligible potential energy. We can now formulate an energy balance for the roller as follows:

$$\frac{dE_r}{dt} = \frac{\partial E_r}{\partial t} + \frac{\partial E_r c \cos\theta_m}{\partial x} + \frac{\partial E_r c \sin\theta_m}{\partial y} = D_w - D_r \tag{2.31}$$

where D_w is the loss of organised wave motion due to breaking (e.g. eq.(2.24)) and D_r is the roller energy dissipation. The latter is equal to the work done by the shear stress between the roller and the wave [Deigaard (1993)]:

$$D_r = \tau_{roller} c \tag{2.32}$$

Given the complex motion in the breaking waves, we can only give very approximate estimates of the order of magnitude of the parameters in eqs. (2.29) thru (2.32). To close the roller energy balance we need to express the roller area R as a function of E_r. This can be done by introducing:

$$\overline{(U_{roller}^2 + W_{roller}^2)} = \beta_2 C^2 \tag{2.33}$$

Combining this with eq.'s (2.29) and (2.32) we then find:

$$D_r = 2\frac{\beta_s}{\beta_2}\frac{g}{C} E_r \tag{2.34}$$

The coefficients β_s and β_2 are usually lumped together into one coefficient of order 0.1 ([Reniers and Battjes (1997)]), which may vary with distance through the surf zone.

2.7 Propagation of wave groups

As discussed earlier the presence of frequency and directional spreading in the incident waves results in an irregular wave field. As a result waves tend to impinge on a beach in groups, a well known fact by many surfers waiting at the outer edge of the surf zone for set of high waves to arrive. The groupiness is a function of both the frequency distribution and the directional distribution, where (broad-) narrow-banded spectra display (weak) strong groupiness. Although relevant from a recreational point of view, the grouped structure of the incident waves also has important consequences for the morphological response of the beach, as it forces infragravity waves and unsteady surfzone circulations (see Section 3.6).

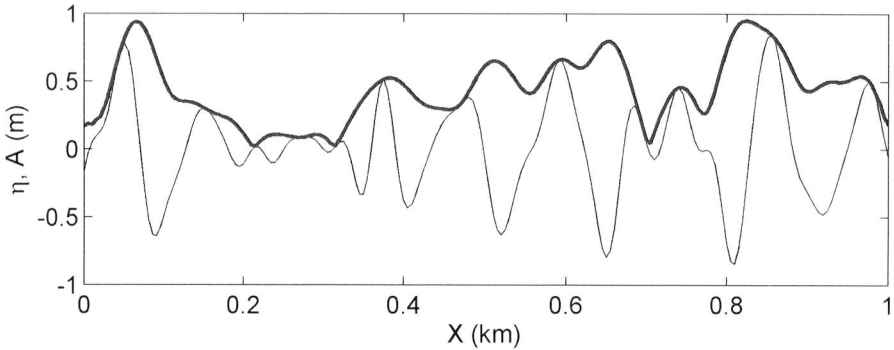

Fig. 2.4 Transect of surface elevation (thin line) and corresponding wave envelope (thick line) calculated with eq.'s 2.6 and 2.35.

Starting with the surface elevation (thin line in Figure 2.4), the corresponding wave envelope can be calculated with the Hilbert transform:

$$A(\vec{x}, t) = |H(\eta(\vec{x}, t))| \qquad (2.35)$$

thus revealing the group-structure of the wave field (Figure 2.4). The corresponding wave energy is given by:

$$E_w(\vec{x}, t) = \frac{1}{2}\rho g A^2(\vec{x}, t) \qquad (2.36)$$

which propagates with the group velocity. Provided the spectrum is narrow banded, eq. (2.19) can be used to describe the slow variation in time and space of random wave groups. To account for the groupiness in wave breaking [Roelvink (1993)] presented a formulation for the slowly-varying dissipation as a function of the local, slowly varying wave energy, which reads:

$$D_w = 2\alpha f_m E_w \left(1 - \exp\left(-\left(\frac{\sqrt{8E_w/(\rho g)}}{\gamma h}\right)^n\right)\right) \qquad (2.37)$$

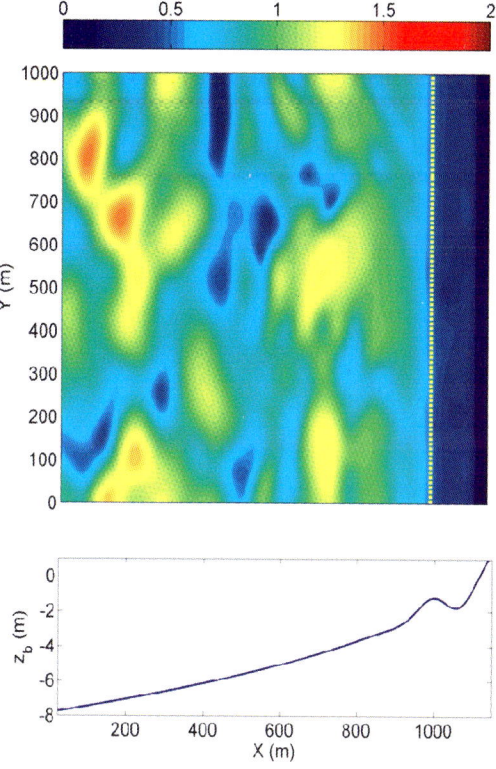

Fig. 2.5 Snapshot of the wave height at the wave-group scale (upper panel) calculated with the wave energy balance, eq. 2.19, for normally incident waves corresponding to a Jonswap frequency spectrum and $\cos^m$ directional distribution with m = 20 with a peak frequency of 0.1 Hz over a single barred planar beach (lower panel). Wave breaking occurs predominantly over the bar (bar crest indicated by dashed yellow line) and near the shore line.

where f_m is the mean frequency. Given $E(f,\theta)$, the wave-group varying energy at the offshore boundary can be calculated with eqs. (2.6), (2.35) and (2.36). The wave-group energy transformation is then calculated with the energy balance, eq. (2.19), taking into account shoaling, refraction and dissipation due to wave breaking.

A snapshot of the spatial distribution of the wave height at the group scale shows significant variation in both the cross-shore and alongshore direction (upper panel of Figure 2.5). Outside the surfzone the cross-shore modulation is a result of the frequency spreading, whereas the alongshore length scale of the wave groups is related to the directional spreading, eq.(2.4), with (short) long length scales for (large) small spreading. Inside the surfzone the modulation decreases as the higher waves are breaking first with near saturation on top of the bar crest. In most cases this is a good representation of what happens in the surfzone. However, in the presence of a (wide) trough the modulation can increase again as the different harmonic wave components propagate with different velocity (e.g. [Dingemans (1997)] and references therein).

2.8 Wave propagation over complex bathymetry

The combined effects of refraction, shoaling and wave breaking over a complex bathymetry can be modelled with the wave action balance eq. (2.7). An example of this is shown in Figure 2.6 using SWAN initialized by offshore measurements of the frequency-directional spectrum obtained by a CDIP bouy (http://cdip.ucsd.edu) approximately 10 km offshore of the Scripps pier (located at $(X,Y) = (0,0)$ m in panel A of Figure 2.6). Wave directions are defined with respect to North (indicated by the black arrow in panel a).

The offshore spectrum is characterized by a strong swell peak around $\theta = 270^o$ and a much weaker peak incident from the South-West around $\theta = 200^o$ (panel B in Figure 2.6). The mean offshore wave direction at deeper water ($h > 100$ m) is closely aligned with the X-axis. However, wave refraction over the canyon walls results in (de-) focusing of the incident waves, resulting in a strong variability of the wave height along the shore line (panel A of Figure 2.6). The refractive (de-)focusing is also observed in $E(f,\theta)$ at the Scripps canyon head, (indicated by the black dot in panel A), where energy is transferred away from the main direction resulting in a bi-modal directional spectrum (panel C of Figure 2.6). Given the fact that refraction is a function of wave period with stronger refraction for lower frequencies (see Figure 2.4 and eq. (2.23)), $E(f,\theta)$ is no longer symmetric around $\theta = 270^o$.

In the absence of strong bathymetric variability the wave height up to the break point can be well predicted with linear shoaling (panel D), where a mean wave period of 10 s was used. The wave height at the canyon head is generally well predicted for a wide range in wave and tidal conditions during a two week period (panel E of Figure 2.6), suggesting that the refractive and shoaling processes are well captured by SWAN.

2.9 Wave blocking

During severe conditions strong nearshore currents interact with the incoming waves. This results in current refraction, wave current interaction with energy transfer from the currents to the incident waves and potentially wave blocking. In the following unidirectional harmonic waves propagating on strong opposing currents are considered. The wave action balance, eq. (2.7), then reduces to:

$$\frac{\partial N}{\partial t} + \frac{\partial c_x N}{\partial x} = \frac{-D}{\sigma} \qquad (2.38)$$

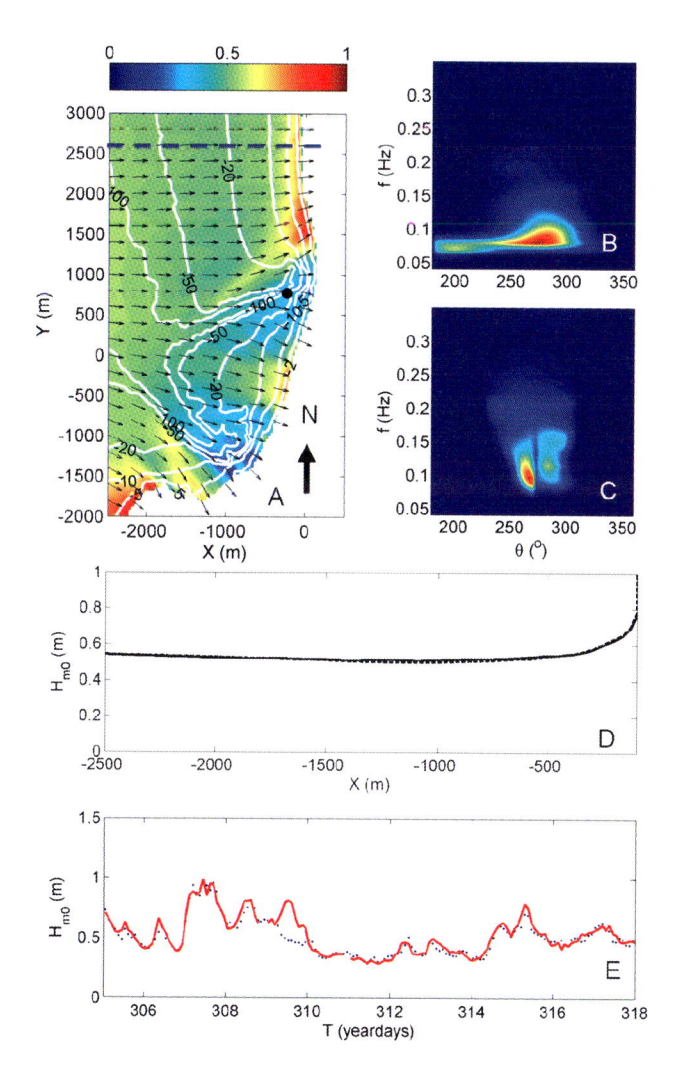

Fig. 2.6 Example of SWAN results. Panel A: Significant wave height (color scale in m) and mean wave direction (arrows) over complex canyon bathymetry (white bottom contour lines in m). Panel B: Offshore f-θ spectrum (θ with respect to North). Panel C: f-θ spectrum near Scripps canyon (at black dot). Panel D: SWAN wave height (solid line) compared with linear shoaling (dashed line) at transect y = 2600 m (blue dashed line in panel A). Panel E: Comparison of measured (dots) and computed significant wave height (solid red line) over a two week period (at black dot location).

where the wave action is given by:

$$N(x,t) = \frac{E_w(x,t)}{\sigma(x,t)} \quad (2.39)$$

and E_w represents the wave energy and σ the intrinsic wave frequency. The propagation velocity is given by:

$$c_x = c_g + u \quad (2.40)$$

where c_g is given by eq. (2.11). Wave dissipation due to wave breaking is represented by eq. (2.37), where the maximum wave height is now also limited by steepness:

$$H_{\max} = \frac{\gamma}{k} \tanh kh \tag{2.41}$$

The wave height is obtained from the wave energy:

$$H = \sqrt{\frac{8E_w}{\rho g}} \tag{2.42}$$

In the following deep water waves propagating over an increasing counter current are considered. In that case there is an analytical solution of the wave height transformation given by [Brevik and Aas (1980)]:

$$\frac{H}{H_0} = \left[\frac{1+G_0}{\left(\frac{k_0}{k} - \frac{u}{c_0}\right)(1+G) + \frac{2u}{c_0}} \left(1 - \frac{ku}{k_0 c_0}\right) \right]^{0.5} \tag{2.43}$$

where the subscript 0 refers to the conditions in absence of a current and G is a transfer function.

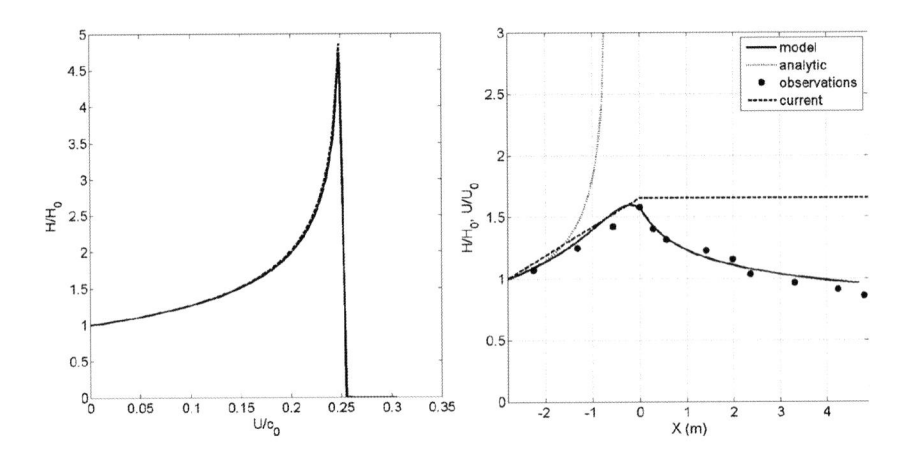

Fig. 2.7 Left panel: Analytical (dashed line) and numerical model (solid line) solution of wave transformation over a linearly increasing opposing current velocity. Right panel: Effect of non-linear dispersion on wave blocking. Observations by [Chawla and Kirby (2002)] (dots) of normalized wave transformation compared with analytical solution of [Brevik and Aas (1980)] (dotted line) and numerical model solution of eq. 2.38 using the non-linear dispersion eq. 2.44 (solid line). Opposing normalized current velocity indicated by the dashed line.

Considering an incident wave, with a height of 0.063 m and a wave period of 2.06 s, propagating on a linearly increasing current and numerically solving for the wave action equation, eq. (2.38), shows good correspondence with the analytical results of [Brevik and Aas (1980)] (left panel of Figure 2.7). As the waves get closer to the blocking point both the wave height and wave number increase (not shown), and consequently the wave steepness increases. Once the waves become too steep, wave breaking sets in and the model prediction deviates from the (unbounded) analytical solution.

In addition to wave breaking wave steepness also affects the dispersion and propagation of the waves:

$$\sigma = \sqrt{gk\tanh(kh)}(1+(ka)^2) \qquad (2.44)$$

The group velocity is now obtained from this third-order Stokes dispersion relation, eq. (2.44):

$$c_g = \frac{\partial \sigma}{\partial k} = c\left(\frac{1}{2} + \frac{kh}{\sinh 2kh}\right)\sqrt{1+(ka)^2} + ka^2 \frac{\sqrt{gk\tanh kh}}{\sqrt{(1+(ka)^2}} \qquad (2.45)$$

This non-linear effect, which is generally negligible, becomes important close to the blocking as demonstrated by [Chawla and Kirby (2002)] in their laboratory experiments. They obtained an increasing velocity from 0.32 m/s to approximately 0.53 m/s by gradually decreasing the width of the flume in the direction of wave propagation for a constant discharge. As a result the incident waves not only experience a counter current but also a funnelling effect which has to be accounted for in the wave action balance by including the flume width B:

$$\frac{\partial NB}{\partial t} + \frac{\partial c_x NB}{\partial x} = \frac{-DB}{\sigma} \qquad (2.46)$$

Considering waves with a mean wave period of 1.357 s, a wave height of 7 cm, wave blocking did not occur in the wave tank, which is well represented by the numerical model, showing a good match with the observations, provided that the non-linear dispersion relation is used (right panel of Figure 2.7). Note that using the linear dispersion relation leads to blocking under these conditions (corresponding to the dotted line in the right panel of Figure 2.7) showing the need to include the effect of wave steepness on the wave propagation for strong currents.

Chapter 3

Currents

3.1 Introduction

In this chapter we will discuss a number of flow situations that are typical of coastal environments. We need to understand these situations, the dominant forces governing them and some solutions to schematized problems in order to be able set up more complex models, recognize their behaviour and judge whether they behave well or not. We will start with a discussion of the shallow water equations and their main assumptions, since these equations are extremely useful in coastal environments where horizontal scales are typically orders of magnitude larger than vertical scales. We will then first apply the depth-averaged version of these equations to describe time-varying and horizontally varying processes of tidal currents, wind-driven currents and wave-driven currents, and then end up by studying vertical structure of these flows.

3.2 Governing equations

Before we go into particular situations, we will discuss the 3D and depth-averaged shallow water equations, with which we will confront most of the morphological problems in the rest of this guide.

3.2.1 *3D Shallow water equations*

The shallow water equations can be derived from the more general Navier-Stokes equations, consisting of the momentum balance:

$$
\begin{aligned}
\frac{\partial \rho u}{\partial t} + u\frac{\partial \rho u}{\partial x} + v\frac{\partial \rho u}{\partial y} + w\frac{\partial \rho u}{\partial z} - f_{cor}\rho v &= \left(\frac{\partial \sigma_{xx}}{\partial x} + \frac{\partial \tau_{xy}}{\partial y} + \frac{\partial \tau_{xz}}{\partial z} - \frac{\partial p}{\partial x}\right) \\
\frac{\partial \rho v}{\partial t} + u\frac{\partial \rho v}{\partial x} + v\frac{\partial \rho v}{\partial y} + w\frac{\partial \rho v}{\partial z} + f_{cor}\rho u &= \left(\frac{\partial \tau_{yx}}{\partial x} + \frac{\partial \sigma_{yy}}{\partial y} + \frac{\partial \tau_{yz}}{\partial z} - \frac{\partial p}{\partial y}\right) \\
\frac{\partial \rho w}{\partial t} + u\frac{\partial \rho w}{\partial x} + v\frac{\partial \rho w}{\partial y} + w\frac{\partial \rho w}{\partial z} \phantom{+ f_{cor}\rho u} &= \left(\frac{\partial \tau_{zx}}{\partial x} + \frac{\partial \tau_{zy}}{\partial y} + \frac{\partial \sigma_{zz}}{\partial z} - \frac{\partial p}{\partial z}\right) - \rho g
\end{aligned} \quad (3.1)
$$

and the mass balance:

$$\frac{\partial \rho}{\partial t} + \frac{\partial \rho u}{\partial x} + \frac{\partial \rho v}{\partial y} + \frac{\partial \rho w}{\partial z} = 0 \tag{3.2}$$

These equations are valid for any flow type, where the velocity in x, y and z-direction is given by u, v and w respectively, ρ is the water density, σ and τ are the normal and shear stress tensors related to molecular viscosity and p the pressure. The term $f_{cor} = 2\Omega \sin \varphi$, with Ω the Earth rotation angular frequency at $2\pi/\text{day}$ and φ the latitude, represents the Coriolis force resulting from the fact that our frame of reference is fixed to the rotating Earth. There are 8 unknowns in these 4 equations, so we will need some simplifications to solve the problem. First we assume that the flow is incompressible and density is uniform, so that the mass balance reduces to a volume balance:

$$\frac{\partial u}{\partial x} + \frac{\partial v}{\partial y} + \frac{\partial w}{\partial z} = 0 \tag{3.3}$$

Next the flow field is decomposed into mean, oscillatory and turbulent motions:

$$u = \bar{u} + \tilde{u} + u' \tag{3.4}$$

with similar expressions for the other quantities. The following step is to average the equations over the short wave motions introducing both turbulent and wave-related Reynolds-stresses:

$$\begin{aligned}
&\overline{\frac{\partial u}{\partial t}} + \overline{u\frac{\partial u}{\partial x}} + \overline{v\frac{\partial u}{\partial y}} + \overline{w\frac{\partial u}{\partial z}} = \\
&\frac{\partial \bar{u}}{\partial t} + \bar{u}\frac{\partial \bar{u}}{\partial x} + \bar{v}\frac{\partial \bar{u}}{\partial y} + \bar{w}\frac{\partial \bar{u}}{\partial z} + \\
&\overline{\tilde{u}\frac{\partial \tilde{u}}{\partial x}} + \overline{\tilde{v}\frac{\partial \tilde{u}}{\partial y}} + \overline{\tilde{w}\frac{\partial \tilde{u}}{\partial z}} + \\
&\overline{u'\frac{\partial u'}{\partial x}} + \overline{v'\frac{\partial u'}{\partial y}} + \overline{w'\frac{\partial u'}{\partial z}}
\end{aligned} \tag{3.5}$$

where we have used the fact that the different velocity components are independent. Using continuity we can rewrite this for both the turbulent and wave related motions to:

$$\begin{aligned}
\overline{u'\frac{\partial u'}{\partial x}} + \overline{v'\frac{\partial u'}{\partial y}} + \overline{w'\frac{\partial u'}{\partial z}} &= \overline{\frac{\partial u'u'}{\partial x}} + \overline{\frac{\partial v'u'}{\partial y}} + \overline{\frac{\partial w'u'}{\partial y}} \\
\overline{\tilde{u}\frac{\partial \tilde{u}}{\partial x}} + \overline{\tilde{v}\frac{\partial \tilde{u}}{\partial y}} + \overline{\tilde{w}\frac{\partial \tilde{u}}{\partial z}} &= \overline{\frac{\partial \tilde{u}\tilde{u}}{\partial x}} + \overline{\frac{\partial \tilde{v}\tilde{u}}{\partial y}} + \overline{\frac{\partial \tilde{w}\tilde{u}}{\partial z}}
\end{aligned} \tag{3.6}$$

For the turbulent motions we assume that the turbulent stress along a plane is proportional to the gradient of the velocity across that plane (now including the y-and z-dimensions):

$$\overline{\rho u'u'} = \sigma_{xx} = \nu_h \frac{\partial \bar{u}}{\partial x}, \overline{\rho v'u'} = \tau_{xy} = \nu_h \frac{\partial \bar{u}}{\partial y}, \overline{\rho w'u'} = \tau_{xz} = \nu_v \frac{\partial \bar{u}}{\partial z},$$

$$\overline{\rho u'v'} = \tau_{yx} = \nu_h \frac{\partial \bar{v}}{\partial x}, \overline{\rho v'v'} = \sigma_{yy} = \nu_h \frac{\partial \bar{v}}{\partial y}, \overline{\rho w'v'} = \tau_{yz} = \nu_v \frac{\partial \bar{v}}{\partial z} \quad (3.7)$$

$$\overline{\rho u'w'} = \tau_{zx} = \nu_v \frac{\partial \bar{w}}{\partial x}, \overline{\rho v'w'} = \tau_{zy} = \nu_v \frac{\partial \bar{w}}{\partial y}, \overline{\rho w'w'} = \sigma_{zz} = \nu_v \frac{\partial \bar{w}}{\partial z}$$

In most coastal areas horizontal scales are very much larger than vertical scales, which is why a different turbulence viscosity ν_h is used for the horizontal shear stresses than the viscosity ν_v used for the vertical shear stresses.

The effect of averaging over the wave motions also leads to extra terms in the momentum balance but also in a non-zero time-averaged wave-related pressure ([Longuet-Higgins and Stewart (1962)]) which is included in the wave forcing expressed as:

$$-W_x = \frac{\partial \left[\rho \overline{\tilde{u}\tilde{u}} + \bar{\tilde{p}}\right]}{\partial x} + \frac{\partial \rho \overline{\tilde{v}\tilde{u}}}{\partial y} + \frac{\partial \rho \overline{\tilde{w}\tilde{u}}}{\partial z},$$

$$-W_y = \frac{\partial \rho \overline{\tilde{u}\tilde{v}}}{\partial x} + \frac{\partial \left[\rho \overline{\tilde{v}\tilde{v}} + \bar{\tilde{p}}\right]}{\partial y} + \frac{\partial \rho \overline{\tilde{w}\tilde{v}}}{\partial z}, \quad (3.8)$$

$$-W_z = \frac{\partial \rho \overline{\tilde{u}\tilde{w}}}{\partial x} + \frac{\partial \rho \overline{\tilde{v}\tilde{w}}}{\partial y} + \frac{\partial \left[\rho \overline{\tilde{w}\tilde{w}} + \bar{\tilde{p}}\right]}{\partial z}$$

The two horizontal momentum equations can now be written as:

$$\frac{\partial \bar{u}}{\partial t} + \bar{u}\frac{\partial \bar{u}}{\partial x} + \bar{v}\frac{\partial \bar{u}}{\partial y} + \bar{w}\frac{\partial \bar{u}}{\partial z} - f_{cor}\bar{v} = \frac{\partial}{\partial x}\left(\nu_h \frac{\partial \bar{u}}{\partial x}\right) + \frac{\partial}{\partial y}\left(\nu_h \frac{\partial \bar{u}}{\partial y}\right) + \frac{\partial}{\partial z}\left(\nu_v \frac{\partial \bar{u}}{\partial z}\right) - \frac{1}{\rho}\frac{\partial \bar{p}}{\partial x} + \frac{W_x}{\rho}$$

$$\frac{\partial \bar{v}}{\partial t} + \bar{u}\frac{\partial \bar{v}}{\partial x} + \bar{v}\frac{\partial \bar{v}}{\partial y} + \bar{w}\frac{\partial \bar{v}}{\partial z} + f_{cor}\bar{u} = \frac{\partial}{\partial x}\left(\nu_h \frac{\partial \bar{v}}{\partial x}\right) + \frac{\partial}{\partial y}\left(\nu_h \frac{\partial \bar{v}}{\partial y}\right) + \frac{\partial}{\partial z}\left(\nu_v \frac{\partial \bar{v}}{\partial z}\right) - \frac{1}{\rho}\frac{\partial \bar{p}}{\partial y} + \frac{W_y}{\rho}$$

$$\{1\} \quad \{\quad\quad 2 \quad\quad\} \quad \{3\} \quad \{\quad\quad\quad 4 \quad\quad\quad\} \quad \{5\} \quad \{6\} \quad \{7\}$$
$$(3.9)$$

The terms in this equation represent {1} inertia, {2} advection, {3} Coriolis effect, {4} horizontal viscosity, {5} vertical viscosity, {6} pressure gradient and {7} wave forcing.

Consistent with the eddy viscosity scaling the wave-averaged vertical velocities are much smaller than the horizontal velocities, and therefore the turbulent shear stresses and the vertical acceleration are very small compared with the acceleration of gravity g. In addition, for a horizontal bottom and non-dissipative waves the $\tilde{u}, \tilde{v}$ orbital velocities are 90 degrees out of phase with $\tilde{w}$ (except within the wave boundary layers, see Section 4.4.3) and the mean wave related pressure is in balance with the vertical Reynolds wave stress. Using these assumptions, the vertical momentum balance reduces to the so-called hydrostatic balance:

$$\frac{\partial \bar{p}}{\partial z} = -\rho g \quad (3.10)$$

This can be integrated easily to obtain the pressure at elevation z, for a water level $\bar{\eta}$ and an atmospheric pressure p_a:

$$\bar{p} = p_a + g \int_z^{\bar{\eta}} \rho \, dz \tag{3.11}$$

In case of density differences (e.g. due to salinity or temperature differences) this has to be integrated numerically; for a homogeneous fluid, we obtain the simple expression:

$$\bar{p} = p_a + \rho g(\bar{\eta} - z) \tag{3.12}$$

Together with the mass balance equation (3.3) we now have 4 unknowns $\bar{u}, \bar{v}, \bar{w}$ and $\bar{\eta}$ and 4 equations, which can be solved, given closure relations for the turbulent eddy viscosity and appropriate boundary conditions at the bottom, where $z = -d$:

$$\begin{aligned} w &= -\bar{u}\frac{\partial d}{\partial x} - \bar{v}\frac{\partial d}{\partial y} \\ \rho \nu_v \frac{\partial \bar{u}}{\partial z} + w_x &= \tau_{bx} \\ \rho \nu_v \frac{\partial \bar{v}}{\partial z} + w_y &= \tau_{by} \end{aligned} \tag{3.13}$$

and at the surface, where $z = \bar{\eta}$:

$$\begin{aligned} \bar{w} &= \frac{\partial \bar{\eta}}{\partial t} + \bar{u}\frac{\partial \bar{\eta}}{\partial x} + \bar{v}\frac{\partial \bar{\eta}}{\partial y} \\ \rho \nu_v \frac{\partial \bar{u}}{\partial z} + w_x &= \tau_{sx} \\ \rho \nu_v \frac{\partial \bar{v}}{\partial z} + w_y &= \tau_{sy} \end{aligned} \tag{3.14}$$

The rate of change of the short-wave averaged surface elevation $\partial \bar{\eta}/\partial t$ can be found by integrating the mass balance eq. (3.3) over the depth (using the Leibnitz rule):

$$\begin{aligned} &\int_{-d}^{\eta} \frac{\partial u}{\partial x} dz + \int_{-d}^{\eta} \frac{\partial v}{\partial y} dz + \int_{-d}^{\eta} \frac{\partial w}{\partial z} dz = 0 \\ &\Leftrightarrow \frac{\partial}{\partial x}\int_{-d}^{\eta} u \, dz - u_\eta \frac{\partial \eta}{\partial x} - u_{-d}\frac{\partial d}{\partial x} + \frac{\partial}{\partial y}\int_{-d}^{\eta} v \, dz - v_\eta \frac{\partial \eta}{\partial y} - v_{-d}\frac{\partial d}{\partial y} + w_\eta - w_{-d} = 0 \\ &\Leftrightarrow \frac{\partial}{\partial x}\int_{-d}^{\eta} u \, dz + \frac{\partial}{\partial y}\int_{-d}^{\eta} v \, dz + \frac{\partial \eta}{\partial t} = 0 \wedge w_{-d} = -u_{-d}\frac{\partial d}{\partial x} - v_{-d}\frac{\partial d}{\partial y}, \\ &w_\eta = \frac{\partial \eta}{\partial t} + u_\eta \frac{\partial \eta}{\partial x} + v_\eta \frac{\partial \eta}{\partial y} \end{aligned} \tag{3.15}$$

Next we introduce the temporal scales again, eq. (3.4) and retain the non-zero terms:

$$\frac{\partial}{\partial x}\overline{\int_{-d}^{\eta}\bar{u}+\tilde{u}+u'dz}+\frac{\partial}{\partial y}\overline{\int_{-d}^{\eta}\bar{v}+\tilde{v}+v'dz}+\overline{\frac{\partial \bar{\eta}+\tilde{\eta}+\eta'}{\partial t}}=0$$

$$\Leftrightarrow \frac{\partial \bar{\eta}}{\partial t}+\frac{\partial}{\partial x}U(d+\bar{\eta})+\frac{\partial}{\partial y}V(d+\bar{\eta})+\frac{\partial}{\partial x}\overline{\int_{-d}^{\eta}\tilde{u}dz}+\frac{\partial}{\partial y}\overline{\int_{-d}^{\eta}\tilde{v}dz}=0 \qquad (3.16)$$

$$\Leftrightarrow \frac{\partial \bar{\eta}}{\partial t}+\frac{\partial}{\partial x}Uh+\frac{\partial}{\partial y}Vh=0$$

where (U,V) represent the depth- and wave-averaged flow including the wave related volumetric transport due to Stokes drift in the (x,y) direction, similar to [Mei (1989)]. In the Eulerian reference frame this wave related mass flux is located between the wave trough and crest and in the Lagrangian frame it is distributed over the vertical (see Section 3.2.3). In summary, the following set of equations represents the so-called 3D shallow water equations or 3D hydrostatic model (dropping the over bar):

$$\frac{\partial u}{\partial t}+u\frac{\partial u}{\partial x}+v\frac{\partial u}{\partial y}+w\frac{\partial u}{\partial z}-f_{cor}v=\frac{\partial}{\partial x}\left(\nu_h\frac{\partial u}{\partial x}\right)+\frac{\partial}{\partial y}\left(\nu_h\frac{\partial u}{\partial y}\right)+\frac{\partial}{\partial z}\left(\nu_v\frac{\partial u}{\partial z}\right)-\frac{1}{\rho}\frac{\partial p}{\partial x}+\frac{W_x}{\rho}$$

$$\frac{\partial v}{\partial t}+u\frac{\partial v}{\partial x}+v\frac{\partial v}{\partial y}+w\frac{\partial v}{\partial z}+f_{cor}u=\frac{\partial}{\partial x}\left(\nu_h\frac{\partial v}{\partial x}\right)+\frac{\partial}{\partial y}\left(\nu_h\frac{\partial v}{\partial y}\right)+\frac{\partial}{\partial z}\left(\nu_v\frac{\partial v}{\partial z}\right)-\frac{1}{\rho}\frac{\partial p}{\partial y}+\frac{W_y}{\rho}$$

$$\frac{\partial Uh}{\partial x}+\frac{\partial Vh}{\partial x}+\frac{\partial \eta}{\partial t}=0$$

$$p=p_a+\int_{z}^{\eta}\rho g dz$$

$$\frac{\partial u}{\partial x}+\frac{\partial v}{\partial y}+\frac{\partial w}{\partial z}=0$$

$$(3.17)$$

3.2.2 Depth-averaged shallow water equations

The use of the full 3D equations is necessary where there are strong variations in flow characteristics through the vertical. This is the case when there are strong density gradients, i.e. near a river mouth, when there is a strong curvature of the flow, as in river bends, or when the focus is on cross-shore wave-induced currents, where the top part of the vertical may be onshore directed and the bottom part offshore directed. In many cases, however, a good first approximation of the flow is given by the depth-averaged version of the shallow water equations.

By averaging the set of equations (3.17) over the water depth we get the following momentum balance:

$$\frac{\partial U}{\partial t}+U\frac{\partial U}{\partial x}+V\frac{\partial U}{\partial y}-f_{cor}V=\frac{\partial}{\partial x}D_h\frac{\partial U}{\partial x}+\frac{\partial}{\partial y}D_h\frac{\partial U}{\partial y}+\frac{\tau_{sx}}{\rho h}-\frac{\tau_{bx}}{\rho h}-\frac{1}{\rho}\frac{\partial p_a}{\partial x}-g\frac{\partial \bar{\eta}}{\partial x}+\frac{F_x}{\rho h}$$

$$\frac{\partial V}{\partial t} + U\frac{\partial V}{\partial x} + V\frac{\partial V}{\partial y} + f_{cor}U = \frac{\partial}{\partial x}D_h\frac{\partial V}{\partial x} + \frac{\partial}{\partial y}D_h\frac{\partial V}{\partial y} + \frac{\tau_{sy}}{\rho h} - \frac{\tau_{by}}{\rho h} - \frac{1}{\rho}\frac{\partial p_a}{\partial y} - g\frac{\partial \bar{\eta}}{\partial y} + \frac{F_y}{\rho h} \quad (3.18)$$

and the volume balance is given by eq. 3.16. D_h is the depth-averaged horizontal turbulence viscosity. The integration of the vertical shear stress gradient over the depth results in the surface shear stress τ_s minus the bed shear stress τ_b. The water level gradient terms follow from the pressure gradients through the hydrostatic pressure assumption (eq. 3.12).

3.2.3 Stokes drift

The presence of waves leads to a mass flux in the direction of wave propagation. This can be seen by evaluating the cross-shore positon of a water parcel taking into account the velocity changes as the parcel changes position (i.e. following the parcel in a Lagrangian fashion):

$$\xi(t; x_0, z_0) \approx \int_{t_0}^{t} u(x_0, z_0, t')dt' + \int_{t_0}^{t}\left(\int_{t_0}^{t} \vec{u}(x_0, z_0, t')dt'\right)\cdot\nabla\vec{u}(x_0, z_0, t')dt' \quad (3.19)$$

For a progressive normally incident harmonic wave the velocity field according to linear wave theory is described by:

$$u(x,t) = \omega a \frac{\cosh k(h+z)}{\sinh kh}\sin(kx - \omega t) \quad (3.20)$$

where a is the wave amplitude, k the wave number and ω the angular frequency. Substituting this expression in to eq. 3.19, integrating over a wave cycle and subsequently dividing by the wave period yields the Lagrangian drift velocity or Stokes drift ([Stokes (1847)], [Phillips (1977)]):

$$\bar{u}_l = \omega k a^2 \frac{\cosh 2k(h+z)}{2\sinh^2 kh} \quad (3.21)$$

The Stokes drift thus represents a wave-averaged velocity in the direction of wave propagation with maximum velocity near the surface and is important for the transport of floating and suspended matter (see also Section 4.4.2).

3.2.4 Wave forcing

3.2.4.1 Interior flow

The evaluation of the wave forcing in the presence of a current is non-trivial. Most of the complications arise from the fact that in the Eulerian reference-frame the interface between water and air moves up and down with the waves. As a result the velocity field is ill defined leading to errors in the estimation of the wave forces for finite amplitude waves. One way to overcome this problem is by considering the fluid motion in the Lagrangian reference frame. More specifically, [Andrews and McIntyre (1978)] use the Generalized Lagrangian Mean (GLM) to describe the combined wave and flow field where the mean lagrangian particle trajectories are given by the combined flow and Stokes drift [Stokes (1847)]. The use of this approach in the surf zone, where waves are breaking and mass transport by surface rollers is important, has not been resolved. Additional methods to describe the combined wave and flow motion include [Mellor (2003)], [Ardhuin et al. (2008)], [McWilliams et al. (2004)] and [Newberger and Allen (2007)] and is an area of ongoing research.

Here we use a relatively simple approach where the wave forces are applied as a shear stress at the surface associated with wave breaking combined with a depth-invariant body force. To that end the wave forces W_i are integrated over the vertical to yield the depth-integrated wave forces F_i.

$$-F_x = \frac{\partial}{\partial x}\left(\overline{\int_{-d}^{\eta} \rho\tilde{u}\tilde{u} + \tilde{p}dz}\right) + \frac{\partial}{\partial y}\left(\overline{\int_{-d}^{\eta} \rho\tilde{v}\tilde{u}dz}\right) + \frac{\partial}{\partial z}\left(\overline{\int_{-d}^{\eta} \rho\tilde{w}\tilde{u}dz}\right)$$

$$-F_y = \frac{\partial}{\partial y}\left(\overline{\int_{-d}^{\eta} \rho\tilde{v}\tilde{v} + \tilde{p}dz}\right) + \frac{\partial}{\partial x}\left(\overline{\int_{-d}^{\eta} \rho\tilde{u}\tilde{v}dz}\right) + \frac{\partial}{\partial z}\left(\overline{\int_{-d}^{\eta} \rho\tilde{w}\tilde{v}dz}\right)$$

(3.22)

The third term on the right hand side in F_x and F_y can become important at locations where the vertical and horizontal velocities are not in quadrature. This occurs within the wave boundary layers (both at the bed [Longuet-Higgins (1953)], Section 3.2.5, and to a lesser extent near the surface [Longuet-Higgins (1960)]), in the case of dissipative waves [Deigaard and Fredsoe (1989)], shoaling waves [De Vriend and Kitou (1990)] and breaking waves [Zou et al. (2003)], but also in the presence of a vertically varying mean flow [Peregrine (1976)], [Rivero and Arcilla (1995)] as well as Coriolis forcing [Hasselmann (1970)], [Xu and Bowen (1993)]. On a horizontal bed and depth-invariant mean flow the vertical and horizontal wave velocities of non-dissipative waves are in quadrature and yield a zero time-averaged contribution. In that case the wave forces are often described as radiation stresses:

$$-F_x = \frac{\partial}{\partial x}\left(\overline{\int_{-d}^{\eta} \rho\tilde{u}\tilde{u} + \tilde{p}dz}\right) + \frac{\partial}{\partial y}\left(\overline{\int_{-d}^{\eta} \rho\tilde{v}\tilde{u}dz}\right) = \frac{\partial S_{xx}}{\partial x} + \frac{\partial S_{yx}}{\partial y}$$

$$-F_y = \frac{\partial}{\partial y}\left(\overline{\int_{-d}^{\eta} \rho\tilde{v}\tilde{v} + \tilde{p}dz}\right) + \frac{\partial}{\partial x}\left(\overline{\int_{-d}^{\eta} \rho\tilde{u}\tilde{v}dz}\right) = \frac{\partial S_{yy}}{\partial y} + \frac{\partial S_{xy}}{\partial x}$$

(3.23)

where the corresponding values, in the Eulerian frame work, can be approximated using linear wave theory ([Longuet-Higgins and Stewart (1962)], [Longuet-Higgins and Stewart (1964)], [Phillips (1977)], [Mei (1989)]) as function of the wave energy:

$$S_{xx} = \left[\tfrac{c_g}{c}\cos^2\theta + \left(\tfrac{c_g}{c} - \tfrac{1}{2}\right)\right] E_w$$
$$S_{yy} = \left[\tfrac{c_g}{c}\sin^2\theta + \left(\tfrac{c_g}{c} - \tfrac{1}{2}\right)\right] E_w \qquad (3.24)$$
$$S_{xy} = S_{yx} = \left[\tfrac{c_g}{c}\sin\theta\cos\theta\right] E_w$$

In the presence of breaking waves the radiation stresses should be extended with additional momentum associated with the wave roller ([Longuet-Higgins and Turner (1974)], [Svendsen (1984)]) which can be expressed as a function of the roller energy:

$$S_{xx,r} = \cos^2\theta E_r$$
$$S_{yy,r} = \sin^2\theta E_r \qquad (3.25)$$
$$S_{xy,r} = S_{yx,r} = \cos\theta\sin\theta E_r$$

The part of the wave forcing that is responsible for generating currents is applied as a shear stress at the surface ([Dingemans *et al.* (1987)], [Deigaard (1993)]):

$$\tau_{sx} = \tfrac{D_r}{c}\cos\theta$$
$$\tau_{sy} = \tfrac{D_r}{c}\sin\theta \qquad (3.26)$$

and the depth-invariant part of the wave forcing is then obtained by subtracting the surface stress from the total wave forcing:

$$-F_{w,x} = \left(\tfrac{\partial S_{xx}}{\partial x} + \tfrac{\partial S_{yx}}{\partial y}\right) + \tau_{s,x}$$
$$-F_{w,y} = \left(\tfrac{\partial S_{yy}}{\partial y} + \tfrac{\partial S_{xy}}{\partial x}\right) + \tau_{s,y} \qquad (3.27)$$

In the case of depth-averaged flow modeling the total depth integrated forcing, eq. (3.23), is often used.

3.2.4.2 Wave boundary layer

The wave energy dissipation in the boundary layer due to bottom friction results in an additional net forcing in the direction of wave propagation [Longuet-Higgins (1953)]. This forcing is associated with the non-zero wave–averaged coupling, denoted here with $<>$, between the horizontal and vertical orbital velocities $<uw>$ within the viscous wave boundary layer, resulting from the fact that the flow motion within the wave boundary layer is no longer irrotational. This can be understood by considering unidirectional waves propagating in the x-direction, where the corresponding linearized momentum balance equation is given by:

$$\frac{\partial \vec{u}}{\partial t} = -\nabla \left(\frac{p}{\rho}\right) + v\nabla^2 \vec{u} \qquad (3.28)$$

The velocity field is given by:

$$\vec{u} = \begin{bmatrix} u \\ w \end{bmatrix} = \begin{bmatrix} \tilde{u} + u' \\ \tilde{w} + w' \end{bmatrix} \qquad (3.29)$$

separating the irrotational wave motion, $(\tilde{u}, \tilde{w})$, with expressions given by linear wave theory, and the rotational orbital flow velocities, (u', w'). Taking the curl of the total wave-related velocity field, thus eliminating the irrotational part of the flow, yields:

$$\frac{\partial \Omega_y}{\partial t} = v\nabla^2 \Omega_y \Leftrightarrow \frac{\partial}{\partial t}\left(\frac{\partial u'}{\partial z} - \frac{\partial w'}{\partial x}\right) = v\nabla^2\left(\frac{\partial u'}{\partial z} - \frac{\partial w'}{\partial x}\right) \Leftrightarrow \frac{\partial u'}{\partial t} = v\frac{\partial^2 u'}{\partial z^2} \qquad (3.30)$$

where we used the fact that gradients in the vertical are much larger, $O(\delta^{-1})$ with δ being the thickness of the boundary layer, than in the horizontal direction, $O(L^{-1})$ with L the wave length. The boundary conditions are given by the zero velocity at the bed, corresponding to $u' = -\tilde{u}_{z=0}$, and the free stream velocity away from the boundary layer, $u_\infty = \tilde{u}|_{z>>\delta}$ hence $u' = 0|_{z>>\delta}$. Solving for u then gives ([Longuet-Higgins (1953)]):

$$u(z,t) = \hat{u}_\infty \left[\cos \omega t - e^{-\beta z} \cos(\omega t - \beta z)\right] \qquad (3.31)$$

and

$$\beta = \sqrt{\frac{\omega}{2v}} \qquad (3.32)$$

where the corresponding vertical velocity can be obtained from continuity [Longuet-Higgins (1953)].

An example of the near-bed intra-wave velocity profiles for an incident wave height of 0.8 m with a wave period T of 6 seconds at 5 m water depth shows the strong near-bed velocity gradients of the leading velocity within boundary layer (left panel of Figure 3.1 with increasing phase going from left to right).

With u and w resolved the corresponding wave-related Reynolds stress can be evaluated:

$$<uw> = \frac{\omega^2 a^2 k}{4\beta \sinh^2 kh} \left(2\beta z e^{-\beta z} \sin(\beta z) - 1 + 2e^{-\beta z} \cos(\beta z) - e^{-\beta 2z}\right) \qquad (3.33)$$

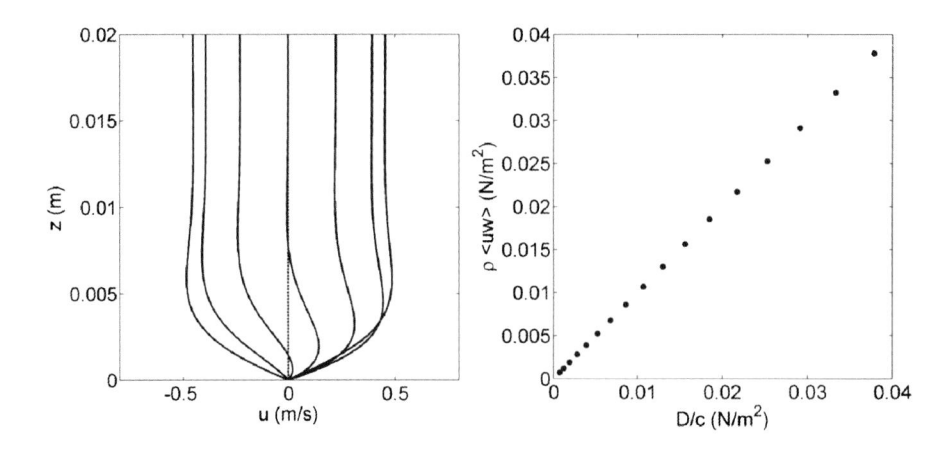

Fig. 3.1 Left panel: Instantaneous velocity profiles at $30°$ phase intervals calculated with eq. 3.31 for a regular wave with amplitude $a = 0.4$ m and $T = 6$ s at 5 m water depth. Right panel: wave-related Reynolds stress at the top of the boundary layer for increasing wave amplitudes $a = [0.1:0.025:0.5]$ m utilizing eqs. 3.33 and 3.34.

Alternatively the wave-related Reynolds stresses can simply be obtained from the wave energy dissipation within the wave boundary layer due to bottom friction ([Fredsoe and Deigaard (1992)], [Reniers et al. (2004b)]) which is readily available from the wave energy balance:

$$<uw> = f_D \frac{D_f}{c} = f_D \frac{\rho f_w}{2\sqrt{\pi}} \frac{u_{orb}^3}{c} \frac{z}{\delta} \qquad (3.34)$$

where $u_{orb} = \tilde{u}_\infty \sqrt{2}$ and the friction coefficient is calculated from the bed roughness ([Soulsby (1997)]):

$$f_w = 1.39 \left(\frac{u_{orb}}{\omega z_0}\right)^{-0.52} \qquad (3.35)$$

with:

$$z_0 = \frac{k_s}{33} \qquad (3.36)$$

and k_s is the Nikuradse roughness and the boundary layer thickness is given by [Nielsen (1992)]:

$$\delta = \frac{1}{2} f_w \frac{u_{orb}}{\omega h} \qquad (3.37)$$

This approach provides a good match with eq. (3.33) for f_D equals 1.25 and the turbulent eddy viscosity within the wave boundary layer is given by:

$$v = \frac{f_w^2 u_{orb}^2}{4\omega} \tag{3.38}$$

to calculate β in eq. (3.33) (right panel of Figure 3.1). The advantage of using eq. (3.34) over eq. (3.33) is that the wave-related Reynolds stresses are consistent with the frictional losses within larger scale wave and flow modeling and can easily be added to the other wave-forces represented by eq. (3.23).

3.2.5 *Bottom shear stress*

The bottom shear stress is a function of the mean current velocity and the orbital velocity. For current-only situations, the bottom shear stress is described by:

$$\tau_b = \rho C_f \, |\vec{u}| \, \vec{u} \tag{3.39}$$

The friction coefficient C_f will generally depend on the local grain material and bed forms. As no generally accepted models exist that describe this dependency, very simple models are applied in practice:

- A prescribed value of C_f or Chézy coefficient C, where $C_f = g/C^2$
- A prescribed Manning value n, where $C_f = gn^2/h^{1/3}$
- A prescribed Nikuradse roughness height k_s, where $C_f = 0.03 \left(\log \frac{12h}{k_s} \right)^{-2}$

The choice of keeping one of these parameters constant over an area has important consequences for the flow distribution between shallow and deep areas, as can be seen from Figure 3.2. For a constant Chézy value, the friction coefficient is also constant with depth, whereas for constant Nikuradse roughness or Manning value, the friction coefficient increases rapidly with decreasing depth. This will tend to shift the flow to deeper water and to reduce velocities in the nearshore, with important consequences for the longshore (sediment) transport. For a constant friction coefficient C_f, the Nikuradse roughness increases linearly with depth. This simple 'model' at least represents the situation often found where tidal flats and shallow coastal areas are relatively smooth and channels and deeper areas exhibit dunes and/or wave ripples. [Ruessink *et al.* (2001)] found good agreement between measured and modelled longshore currents using a constant friction coefficient (see Section 3.5).

Various models have been developed that describe the wave-current interaction near the bed. Depending on the choice of model, the relative effect of waves becomes more or less important. This is the main reason to pay some attention to this choice. [Bijker (1988)] estimated the mean current velocity and the orbital velocity at a certain reference level, where they could be added up; this way a time series of the total friction velocity could be generated and converted to a time series of the bed shear stress; after averaging over a wave cycle this leads to the mean shear stress due to current and waves. This method overestimates the effect of waves on shear stress

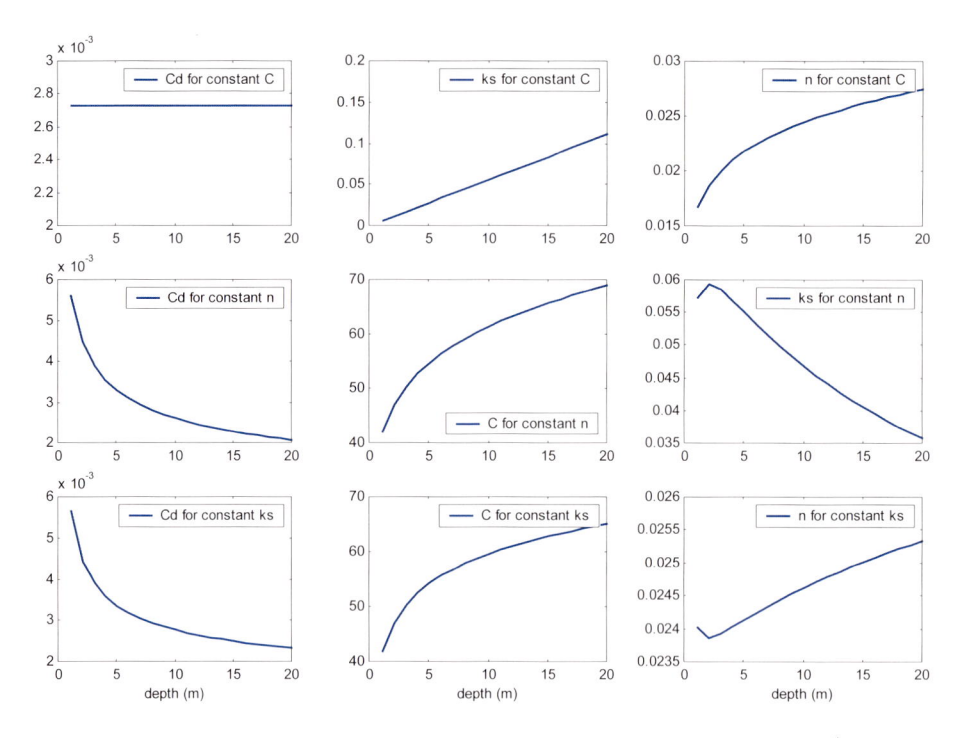

Fig. 3.2 Dependency of various friction parameters on depth; constant $C = 60 \text{ m}^{\frac{1}{2}}/\text{s}$, constant $n=0.024$, constant $k_s=0.06$ m

as it does not take into account that the current near the bed is reduced due to the increased bed shear. Various models have been developed since then, solving the momentum balance inside the boundary layer using different approximations of the near-bed viscosity or solving it using a mixing-length model or more sophisticated turbulence models.

The larger-scale effect on the current outside this boundary layer is that the roughness appears to be enhanced by the waves; the velocity profile outside the wave boundary layer follows:

$$u = \frac{u_{*,cw}}{\kappa} \ln \frac{z}{z_{0a}} \qquad (3.40)$$

[Van Rijn (1993)] gives a direct expression for the apparent roughness, z_{0a}, based on experimental data of flumes with rippled beds. He finds a rather strong but not fully understood effect of the angle between currents and waves. In general, he predicts much smaller effect of the waves on bottom shear stress.

[Soulsby et al. (1993)] carried out a very useful inter-comparison of the most common wave-current interaction models and made a parameterisation that is easy to implement for all these models. The clever thing about the parameterisation is that it expresses the mean shear stress and maximum shear stress over a wave cycle

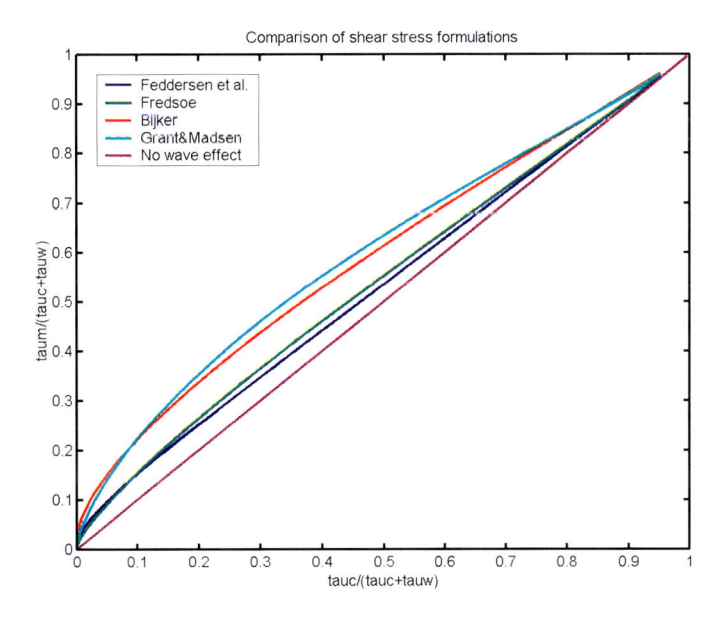

Fig. 3.3 Comparison of various formulations of mean shear stress due to current and waves

in a dimensionless form, where both parameters are made dimensionless with the sum of the current shear stress and the wave shear stress. They are expressed as function of the ratio of current shear stress to the sum of current and wave shear stresses:

$$\frac{\tau_{wci}}{\tau_c + \tau_w} = func\left(\frac{\tau_c}{\tau_c + \tau_w}\right) \quad (3.41)$$

For the mean shear stress, $func$ goes to zero when this ratio goes to zero, and approaches 1 as the ratio approaches 1. In between, $func$ is always greater than 1 as the effect of adding waves is to enhance the shear stress due to current only. It turns out that when plotted this way, most models have very similar curves; the earlier ones of [Bijker (1988)] and [Grant and Madsen (1978)] are significantly different; the others all converge to the same shape. From a 'consumer-test' comparison against measured data the model of [Fredsoe (1984)] appeared to give the most robust results. [Soulsby et al. (1993)] also gives an even simpler expression, based on a fit to data, which yields:

$$\tau_m = \tau_c \left[1 + 1.2\left(\frac{\tau_w}{\tau_c + \tau_w}\right)^{3.2}\right] \quad (3.42)$$

These models were all basically derived for monochromatic waves, and consider in increasing detail what goes on in the wave boundary layer (which is in the order of mm's to cm's thick). [Soulsby (1997)] gives advice on how to apply the models for random waves, which is to use the orbital velocity amplitude according to the root mean square wave height as equivalent amplitude. Some of the models consider the effect of the direction of the waves relative to the current direction. We will discuss some aspects of this further on.

[Feddersen et al. (2000)] focus on the effects of randomness and directional spreading of realistic wave fields, and derive a practical formulation which is approximately valid for all angles between waves and currents. However, they do not consider the wave boundary layer in any detail, but simply add up the instantaneous depth-averaged current and orbital velocity vectors at some distance above the boundary layer and obtain the instantaneous shear stress from:

$$\vec{\tau}_{cw} = \rho C_f \left| \vec{u} + \vec{\tilde{u}} \right| \left(\vec{u} + \vec{\tilde{u}} \right) \quad (3.43)$$

They find that the time-averaged shear stress is only weakly dependent on the wave angle relative to the current, and the wave directional spreading, and present an approximate formula for the mean bed shear stress due to current and waves.

$$\bar{\tau}_{cw} = \rho C_f \vec{u} \sqrt{(1.16s)^2 + |\vec{u}|^2} \quad (3.44)$$

where s is the standard deviation of the velocity, a good statistical measure of the orbital velocity in random waves.

3.2.6 *Turbulent eddy viscosity*

The small scale turbulent mixing is modeled with a turbulent eddy viscosity concept. In the case of 3D flow modeling the eddy viscosity can be obtained from algebraic expression or additional turbulence models. For a full description of turbulence models frequently used in coastal modeling we refer to [Rodi (1984)]. A frequently used turbulence model is the $k - \varepsilon$-model which takes into account the generation, transport and dissipation, ε, of turbulent kinetic energy, k. Turbulence is generated in the bottom boundary layer, at the surface in the presence of wind and at times when waves are breaking. The latter can result in pulses of energetic turbulence that reach the bottom and subsequently stir up large amounts of sediment ([Steetzel (1993)], [Roelvink and Stive (1989)], [van Thiel de Vries et al. (2008)]). The turbulent eddy viscosity is related to the turbulent kinetic energy by:

$$\nu_v = c_1 \ell \sqrt{k} \quad (3.45)$$

where the mixing length is obtained from:

$$\ell = c_2 \frac{k^{\frac{3}{2}}}{\varepsilon} \quad (3.46)$$

and c_1 and c_2 are calibration coefficients. To evaluate the vertical current velocity distribution under quasi-stationary conditions along the open coast we will frequently use a much simpler parabolic distribution:

$$\nu_v = -\kappa v_* z \frac{(h+z)}{h} \quad (3.47)$$

where κ is von Karman's constant and the shear velocity is given by:

$$v_* = \sqrt{\frac{\tau}{\rho}} \quad (3.48)$$

In the depth-averaged model equations the turbulent eddy viscosity is often described as a combination of background eddy viscosity and a wave breaking induced eddy viscosity [Battjes (1975)]:

$$D_h = v_\infty + \alpha h \left(\frac{D_r}{\rho}\right)^{\frac{1}{3}} \quad (3.49)$$

where α is a calibration coefficient in the order of 1 and v_∞ is the back ground viscosity of $O(0.1)$ m^2/s.

This concludes the description of the governing equations which will be applied in the next sections to examine a variety in coastal flows.

3.3 Tidal currents

Tidal currents obviously play an important role in and around tidal inlets and estuaries, but also along many open coasts significant tidal currents occur. Especially where the tidal motion is obstructed by headlands, dams, harbour moles or other topographic features, the tidal currents may be strong enough to cause significant bottom changes. In this section we will discuss some typical features of the tidal propagation and tidal currents and provide some easy simplified models to enable you to make first-order estimates of the tidal motion.

3.3.1 *Open coasts*

The propagation of the tides along the ocean coasts and into shallow seas and estuaries can be described by the shallow water equations, as tidal wave lengths are very long compared to the water depth. We will discuss some typical schematised cases in order to explain the most important phenomena that govern the generation of tidal currents along our coasts.

3.3.1.1 *Open ocean, Kelvin wave*

The tidal motion on the open ocean is dominated by inertia and the Coriolis force; advection terms, horizontal diffusion and wave effects may be ignored:

$$\frac{\partial u}{\partial t} + u\frac{\partial u}{\partial x} + v\frac{\partial u}{\partial y} - f_{cor}v = \frac{\partial}{\partial x}D_h\frac{\partial u}{\partial x} + \frac{\partial}{\partial y}D_h\frac{\partial u}{\partial y} + \frac{\tau_{sx}}{\rho h} - \frac{\tau_{bx}}{\rho h} - \frac{1}{\rho}\frac{\partial p_a}{\partial x} - g\frac{\partial \eta}{\partial x} + \frac{F_x}{\rho h}$$

$$\frac{\partial v}{\partial t} + u\frac{\partial v}{\partial x} + v\frac{\partial v}{\partial y} + f_{cor}u = \frac{\partial}{\partial x}D_h\frac{\partial v}{\partial x} + \frac{\partial}{\partial y}D_h\frac{\partial v}{\partial y} + \frac{\tau_{sy}}{\rho h} - \frac{\tau_{by}}{\rho h} - \frac{1}{\rho}\frac{\partial p_a}{\partial y} - g\frac{\partial \eta}{\partial y} + \frac{F_y}{\rho h}$$
$$(3.50)$$

If we now consider that the depth is large compared with the tidal amplitude, and the friction is relatively unimportant and may be linearized, we get the following reduced set of equations:

$$\frac{\partial u}{\partial t} - f_{cor} v = -\frac{\lambda}{h} u - g\frac{\partial \eta}{\partial x}$$

$$\frac{\partial v}{\partial t} + f_{cor} u = -\frac{\lambda}{h} v - g\frac{\partial \eta}{\partial y} \quad (3.51)$$

$$\frac{\partial \eta}{\partial t} + h\left(\frac{\partial u}{\partial x} + \frac{\partial v}{\partial y}\right) = 0$$

Here λ is a friction parameter corresponding to the linearized bed shear stress:

$$\frac{\tau_{by}}{\rho h} = \frac{\rho C_f |\vec{u}| v}{\rho h} \approx \frac{\lambda}{h} v \quad (3.52)$$

with a similar expression for the cross-shore bottom shear stress. Since our objective is to model coastal hydrodynamics, we are most interested in tidal waves propagating along a closed boundary. Throughout this chapter we will have the coast at the eastern side at x=0, where we may take $u = 0$. On the open sea, in relatively deep water, friction effects may be ignored. The equations then reduce to:

$$f_{cor} v = g\frac{\partial \eta}{\partial x}$$

$$\frac{\partial v}{\partial t} = -g\frac{\partial \eta}{\partial y} \quad (3.53)$$

$$\frac{\partial \eta}{\partial t} + h\frac{\partial v}{\partial y} = 0$$

These equations have a solution known as the Kelvin wave; we can find this solution by assuming that v and η are alongshore propagating waves described by:

$$\eta = \widehat{\eta} \cos(\omega t - ky), \quad v = \widehat{v} \cos(\omega t - ky) \quad (3.54)$$

which upon substitution in (3.53) gives:

$$\widehat{v}\omega = gk\widehat{\eta}$$

$$\widehat{\eta}\omega = hk\widehat{v} \quad (3.55)$$

From which it follows that:

$$\widehat{v} = \widehat{\eta}\sqrt{\frac{g}{h}} \quad (3.56)$$

and:

$$c = \omega/k = \sqrt{gh} \quad (3.57)$$

By substituting (3.56) into (3.57) we find:

$$f_{cor}\widehat{\eta}\sqrt{\frac{g}{h}} = g\frac{\partial \widehat{\eta}}{\partial x} \Rightarrow \frac{\partial \widehat{\eta}}{\partial x} = \frac{f_{cor}}{\sqrt{gh}}\widehat{\eta} - \frac{f_{cor}}{c}\widehat{\eta} \qquad (3.58)$$

This differential equation describes the cross-shore variation of the water level amplitude, with the solution:

$$\widehat{\eta} = \widehat{\eta}_0 \exp\left(\frac{f_{cor}}{c}x\right) \qquad (3.59)$$

and the complete expression for the water level variation is now:

$$\eta = \eta_0 \exp\left(\frac{f_{cor}x}{c}\right)\cos\left(\omega t - ky\right) \qquad (3.60)$$

The water level amplitude exponentially decays away from the coast where x<0. Such waves propagate along the coast at the usual shallow water celerity; in the northern hemisphere they have the coast at their right-hand side, in the southern hemisphere they travel in opposite direction. The length scale of the amplitude decay is $c/|f_{cor}|$. This is in the order of 2000 km in deep oceans (depth 4 km, f_{cor} approx. 1.10^{-4}) to 200 km in a shallow sea with a typical depth of 40 m (see Figure 3.4).

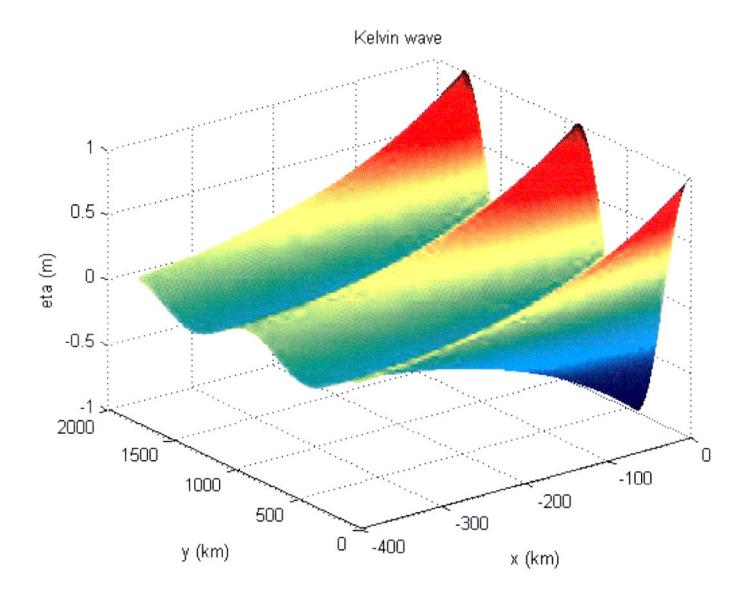

Fig. 3.4 Kelvin wave for a water depth of 40 m, at a latitude of 51 deg. and a tidal period of 12.5 hrs.

3.3.1.2 Tidal flow structure in the nearshore area

If we now consider the flow in a relatively narrow strip (say, 10 km wide) along the coast, we may approximate (3.60) by a linear decay of the water level amplitude:

$$\eta = \eta_0 \left(1 + \frac{f_{cor}x}{c}\right) \cos\left(\omega t - ky\right) \qquad (3.61)$$

This is consistent with assuming that the alongshore pressure gradient $g\,\partial\eta/\partial y$ does not vary with distance from shore. Having removed this complexity we can now add friction, in order to look at the tidal current behaviour in the nearshore area.

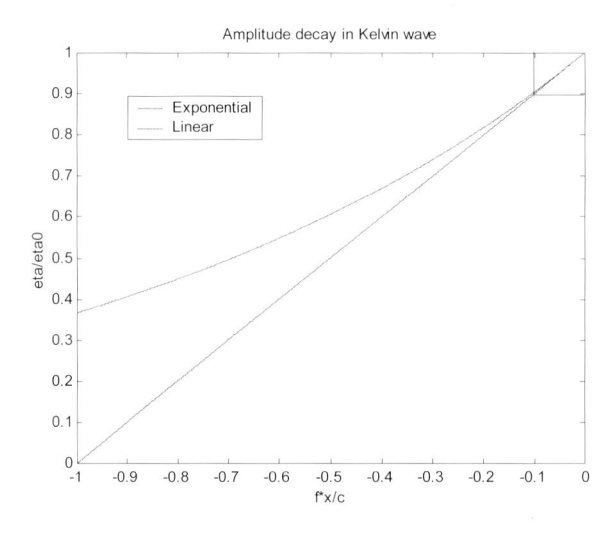

Fig. 3.5 Amplitude decay in a Kelvin wave compared with a linear decay in a narrow coastal strip (indicated by the boxed area

Again we neglect the u-component of the velocity, so that the most important terms in the momentum equations become:

$$g\frac{\partial \eta}{\partial x} = f_{cor}v \qquad (3.62)$$

$$\frac{\partial v}{\partial t} = -g\frac{\partial \eta}{\partial y} - \frac{\tau_{by}}{\rho h} \qquad (3.63)$$

The first equation relates the cross-shore slope of the water level to the Coriolis force generated by the alongshore current; the second describes the dynamics of the alongshore current, which is driven by the alongshore pressure gradient and restrained by the bottom shear stress. If we now consider a typical coastal model with a domain that covers an area of five kilometres in cross-shore direction, the cross-shore water level difference due to the Coriolis force, given a alongshore velocity of 1 m/s, is in the order of $10^{-4}*1/9.81*5000 = 0.05$ m. This is small compared with the tidal amplitude of typically 1 m, which means that for models with such a limited cross-shore extent we may assume that the alongshore pressure gradient is constant in the cross-shore direction (see Figure 3.5).

We can now solve this set of equations for a given cross-shore profile, if we know the water level and the alongshore waterlevel gradient at the offshore boundary of the profile. This is relatively easy if we know the tidal components of the water level in some points nearby; we may get them from tide gauges, from existing model databases or by performing harmonic analyses on some points in a regional model.

Let us now assume that we know how a tidal component propagates along the coast:

$$\eta = \widehat{\eta} \cos(\omega t - ky) \qquad (3.64)$$

with $\widehat{\eta}$ the amplitude, ω the angular frequency and k the alongshore wavenumber. If we again linearise the bed shear stress, eq. (3.52), we can solve equation (3.63) analytically, yielding:

$$v = \frac{gk}{\omega} \frac{1}{1+\left(\frac{\lambda}{\omega h}\right)^2} \widehat{\eta} \cos(\omega t - ky) - \frac{gk}{\omega} \frac{\frac{\omega h}{\lambda}}{1+\left(\frac{\omega h}{\lambda}\right)^2} \widehat{\eta} \sin(\omega t - ky) \Rightarrow$$

$$v = \frac{gk\widehat{\eta}}{\omega} \left(\frac{1}{1+\varphi^2} \cos(\omega t - ky) - \frac{\varphi}{1+\varphi^2} \sin(\omega t - ky) \right), \quad \varphi = \frac{\lambda}{\omega h} \Rightarrow \qquad (3.65)$$

$$v = \frac{1}{\sqrt{(1+\varphi^2)}} \frac{gk\widehat{\eta}}{\omega} \left(\cos(\omega t - ky + \arctan(\varphi)) \right)$$

From this solution it becomes clear that in deeper water, where the friction effect is small, the alongshore velocity is in phase with the water level, whereas in very shallow water, where the inertia effect is small, the velocity is governed by the balance between the alongshore water level gradient and the bottom friction, and is therefore in phase with the negative water level gradient.

Another obvious result of this analysis is that the alongshore velocity is proportional to k, or inversely proportional to the alongshore tidal wavelength. This explains why on open ocean coasts, where the tidal wave races along the coast at hundreds of kilometres per hour, and the alongshore wavelength is thousands of kilometres, longshore tidal velocities are very small. In shallow seas such as the southern North Sea the tidal wave length is in the order of hundreds of kilometres and therefore longshore velocities are in the range of 0.5-2 m/s, depending on the local tidal range and alongshore wavelength. Note that we cannot estimate the alongshore tidal wave propagation speed based on the local water depth in a nearshore model: it follows from the tidal propagation in deeper water.

To illustrate the behaviour of the tide-induced longshore velocity we plot time series of the water level, the water level gradient and the longshore velocity at three different depths (Figure 3.6). We see that indeed in deeper water the velocity is almost in phase with the water level, whereas in shallow water the velocity follows the negative water level gradient.

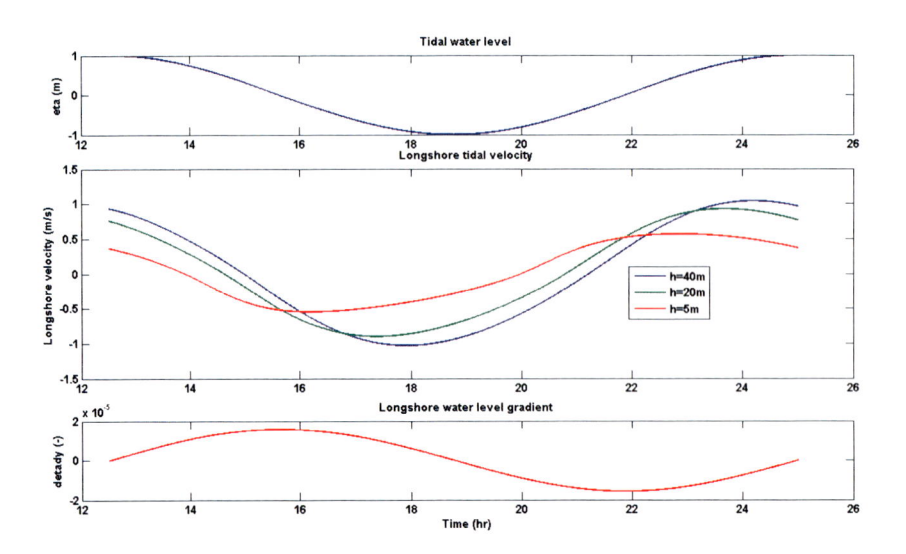

Fig. 3.6 Time series of longshore tidal velocity (middle panel) at different depths, compared with time series of water level (top panel) and longshore water level gradient (bottom panel).

3.3.2 Propagation into estuaries

Let us start with the simplest case, that of a tidal wave propagating in x-direction over a uniform sea bed. We use the linearised version of the shallow water equations:

$$\frac{\partial u}{\partial t} = -\frac{\lambda}{h}u - g\frac{\partial \eta}{\partial x} \qquad (3.66)$$

$$\frac{\partial \eta}{\partial t} + h\frac{\partial u}{\partial x} = 0 \qquad (3.67)$$

By introducing solutions of the form:

$$\eta = \widehat{\eta}\cos(\omega t - kx)$$
$$u = \widehat{u}_1 \cos(\omega t - kx) + \widehat{u}_2 \sin(\omega t - kx) \qquad (3.68)$$

we get:

$$\omega/k = c = \sqrt{gh}\frac{1}{\sqrt{1+\varphi^2}} \qquad (3.69)$$

and:

$$u = \frac{c}{h}\widehat{\eta} = \sqrt{\frac{g}{h}}\frac{1}{\sqrt{1+\varphi^2}}\widehat{\eta}\cos(\omega t - kx) - \sqrt{\frac{g}{h}}\frac{\varphi}{\sqrt{1+\varphi^2}}\widehat{\eta}\sin(\omega t - kx) \qquad (3.70)$$

where:

$$\varphi = \frac{\lambda}{\omega h} \approx \frac{\pi/4\ C_f \widehat{u}}{2\pi h/T} = \frac{C_f \widehat{u} T}{8h} \qquad (3.71)$$

This result of course is similar to that of (3.65), except that now the wave celerity is determined by the local water depth. The dimensionless friction coefficient φ is approximately 1 for a water depth of 10 m, which means that friction is an important factor in the propagation of a tidal wave into an estuary.

3.3.3 Resonance

Tides propagating into an approximately rectangular basin (Figure 3.7) are often reflected at the end, so that a (partially) standing wave pattern develops.

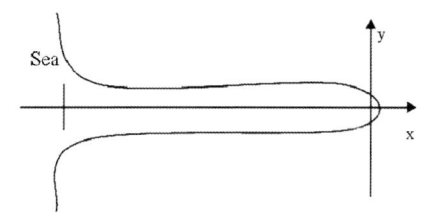

Fig. 3.7 Depiction of narrow tidal inlet

This can be described by considering two waves propagating in opposite directions, where the wave travelling in seaward direction has an amplitude r times the incoming wave. If we take the landward boundary of the estuary at the right-hand side and at $x = 0$ (Figure 3.7), we get for the total elevation η:

$$\eta = \widehat{\eta}_{in} \cos(\omega t - kx) + r\widehat{\eta}_{in} \cos(\omega t + kx) \qquad (3.72)$$
$$= 2r\widehat{\eta}_{in} \cos(\omega t) \cos(kx) + (1-r)\widehat{\eta}_{in} \cos(\omega t - kx)$$

We can see that the reflected wave causes a standing wave pattern, whereas the part that is not reflected propagates through the estuary.

An example of this pattern is shown in the left panel of Figure 3.8. The total amplitude varies as a function of the distance from the landward boundary in the following manner:

$$\widehat{\eta} = \widehat{\eta}_{in} \sqrt{(1-r)^2 + 4r \cos^2(kx)} \qquad (3.73)$$

From this equation we can derive the amplification factor, the ratio between maximum amplitude (at the landward side) and the amplitude at the seaward end:

$$\frac{\widehat{\eta}_{land}}{\widehat{\eta}_{sea}} = \frac{\sqrt{(1-r)^2 + 4r}}{\sqrt{(1-r)^2 + 4r \cos^2(2\pi X/L)}} \qquad (3.74)$$

where X is the length of the estuary and L the tidal wavelength. This relation is shown in the right panel of Figure 3.8. Clearly, resonance occurs at ratios of estuary length over tidal wavelength of $(2n-1)/4$ with n any whole number.

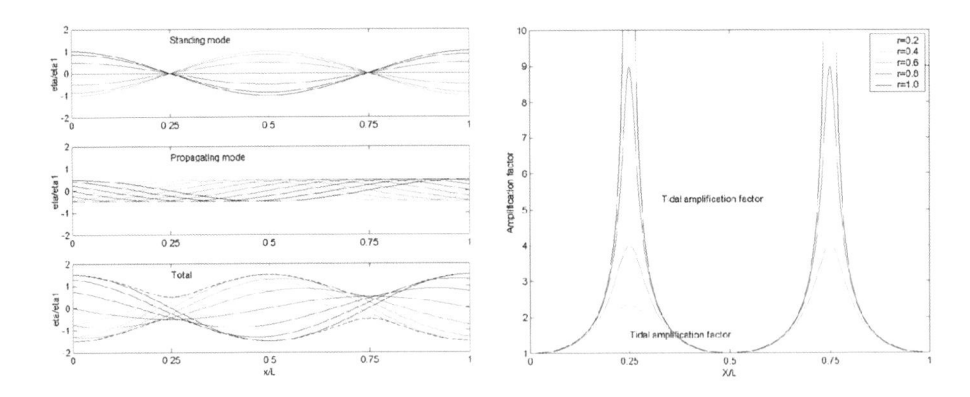

Fig. 3.8 left panel: Standing and propagating modes and total surface elevation pattern for reflection coefficient of 0.5; drawn lines at 1/12 period intervals. Dashed line: envelope of the total tidal elevation. Right panel: Amplification of tidal amplitude as a function of estuary length and reflection coefficient.

3.3.4 Funnelling effect

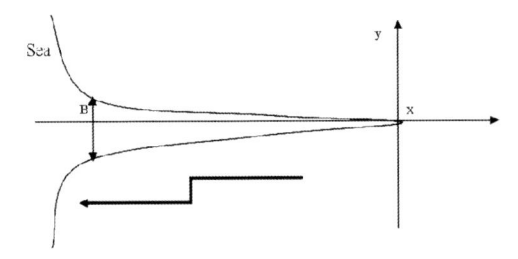

Fig. 3.9 Funnel shaped tidal inlet

Another mechanism for an increasing amplitude in a tidal channel occurs when the channel width or depth gradually decreases (Figure 3.9). If we ignore reflection because the changes in depth or width are so gradual, we may assume that the tidal energy flux is conserved along the channel:

$$\frac{\partial}{\partial x}\left(E_{tide}Bc\right) = \frac{\partial}{\partial x}\left(\frac{1}{2}\rho g \widehat{\eta}^2 B\sqrt{gh}\right) = 0 \qquad (3.75)$$

where B is the channel width and E_{tide} the tidal wave energy. From this relation it follows that the amplitude increases as:

$$\widehat{\eta} \propto B^{-1/2} h^{-1/4} \qquad (3.76)$$

Apparently, reducing the width leads to a stronger increase in tidal amplitude than reducing the depth.

3.3.5 Short, wide basins

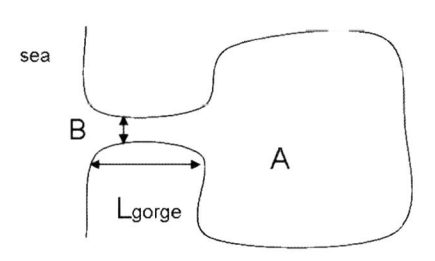

Fig. 3.10 Lagoon shaped inlet.

For tidal inlets that are relatively short compared to the tidal wavelength (Figure 3.10), we may assume for a first estimate that the water level within the tidal basin with area A goes up and down in a uniform way. The discharge required to fill and empty the basin is then:

$$Q = Bhu = A\frac{\partial \eta_{inside}}{\partial t} \quad (3.77)$$

Inside the narrow gorge, a water level gradient is created by the bottom friction:

$$\frac{\lambda}{h}u = -g\frac{\partial \eta}{\partial x} \approx g\frac{\eta_{outside} - \eta_{inside}}{L_{gorge}} \quad (3.78)$$

where for simplicity we have again linearised the bottom friction. We can now combine equations (3.77) and (3.78) to yield:

$$\frac{\partial \eta_{inside}}{\partial t} = \frac{Bh^2}{A}\frac{g}{\lambda L_{gorge}}\left(\eta_{outside} - \eta_{inside}\right) \quad (3.79)$$

If we prescribe:

$$\eta_{outside} = \widehat{\eta}_{outside}\cos(\omega t) \quad (3.80)$$

we find the following solution:

$$\eta_{inside} = \frac{1}{\sqrt{1+(\omega/\mu)^2}}\widehat{\eta}_{out}\cos\left(\omega t - \arctan(\omega/\mu)\right),$$
$$\mu = \frac{Bh^2}{A}\frac{g}{\lambda L_{gorge}} \quad (3.81)$$

With this simple relation we can estimate how much the tide inside is damped and how big the phase lag is between the tide inside and outside the inlet.

3.3.6 Tidal currents around structures

Around structures such as harbour moles, the tidal flow is forced into more complicated patterns. We will discuss typical features by looking at the flow around IJmuiden harbour in the Netherlands. This harbour sticks out approx. 2.5 km in an otherwise long stretch of uninterrupted coast and displays many typical features of such harbours. The tide range is approx. 2 m and the tide propagates in northward direction at a leisurely pace of some 15-20 m/s. Typical peak velocities are 0.7 m/s and peak ebb velocities of 0.5 m/s, the difference being due to the M4 component that is partly in phase with the M2 tide.

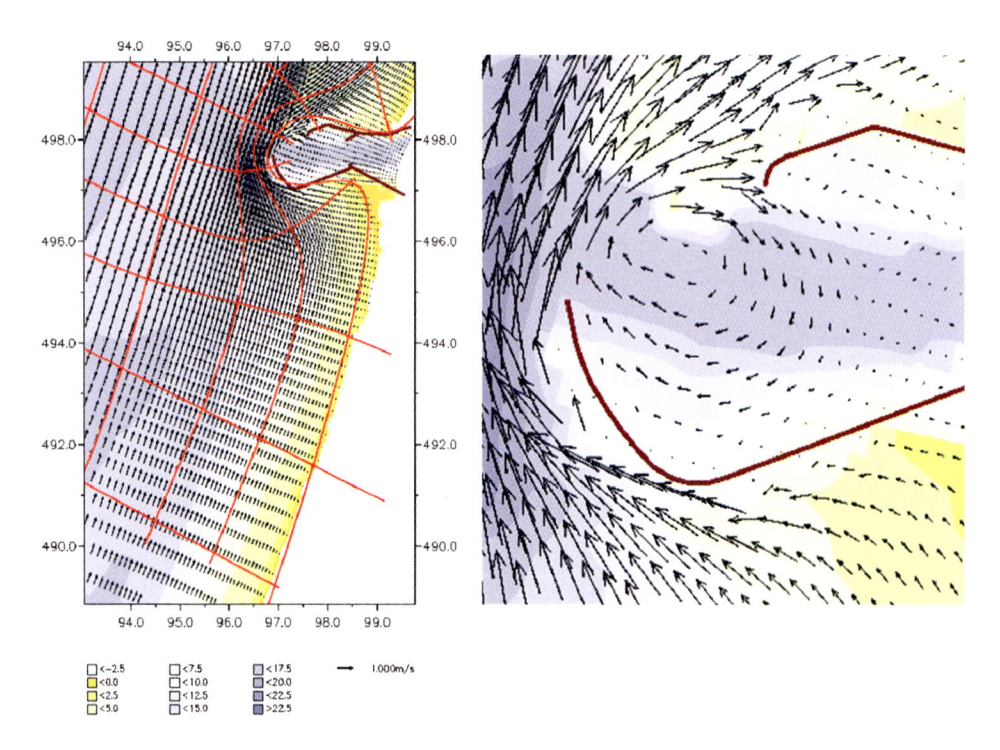

Fig. 3.11 Left panel: Predicted flow pattern at Ijmuiden harbour during maximum flood with estimated stream tubes overlain in red. Right panel: Detail showing the harbour entrance gyre during maximum flood. (depth in m indicated by the colors, cross-shore and alongshore distance in km.)

In Figure 3.11 we have plotted flow patterns due to the tide during maximum flood only. We will now discuss the characteristics of different regions:

3.3.6.1 Convergence area

This area is dominated by the horizontal advection terms and relatively little horizontal turbulence. The flow behaves much like potential flow, and a first estimate of the flow pattern can be made by sketching a curved orthogonal grid and following the fluxes through stream tubes, starting from the undisturbed upstream area where we know how to estimate the velocity from the previous paragraphs.

The velocity in a stream tube of local width B and local depth h can be estimated using the fact that no water passes the streamlines and hence the flux Q through it is constant. It then follows that,

$$uhB = Q = u_0 h_0 B_0 \qquad (3.82)$$

The strong contraction of the flow in front of the entrance (Figure 3.11), in combination with locally enhanced turbulence, leads to a scour pit in front of the entrance, which may be in the order of the undisturbed water depth. As we see from equation (3.82) this will help reduce the velocities until a new equilibrium is found.

3.3.6.2 Harbour entrance gyre

With a geometry such as this harbour, a gyre is often driven by the tidal current (right panel of Figure 3.11). This gyre leads to an exchange of water and the sediment in it. The dominant terms that determine its strength are the horizontal advection and the horizontal turbulent diffusion. The latter is dominated by turbulent eddies that are generated in the area of strong shear just after the separation point. Bottom friction plays only a small role as the harbour basin is usually relatively deep and the current velocities are small.

In these areas where the flow is strongly curved, the vertical flow structure is not logarithmic but shows a helical pattern, where the velocity field at the top layers has an outward directed component and the near-bed velocity has an inward component. In the centre of gyres the velocity tends to be upward to compensate for this imbalance. This may be an important mechanism for moving gyres to transport sediment over some distance, as the vertical velocities can be of the order of the sediment fall velocity.

3.3.6.3 Downstream wake area

Downstream of the detachment point there is a turbulent mixing layer that may extend for some 7 times the length of the breakwaters downstream, for a sharp dam in uniform depth; the presence of a sloping profile and a curved breakwater layout may reduce this distance considerably. In the case of IJmuiden, a long shoal north of the northern mole at present reduces the gradients in velocity and directs the flow shoreward by current refraction, which leads to a reattachment of the tidal flow within a few times the breakwater length.

3.3.7 *Flow patterns around a realistic inlet*

As an example of the kind of flow patterns one can expect around a tidal inlet, we take a natural inlet in the Dutch Wadden Sea, the Amelander Zeegat (Figure 3.12). The inlet's behaviour has been the subject of many phenomenological studies and modeling studies. The inlet has two main channels, the Westgat on the western side and the Borndiep on the eastern side (North is up). As the tide propagates from west to east along the sea side, the strongest water level gradients and hence currents occur on the western side of the inlet, which may well explain why the preferred orientation of both channels is towards the northwest.

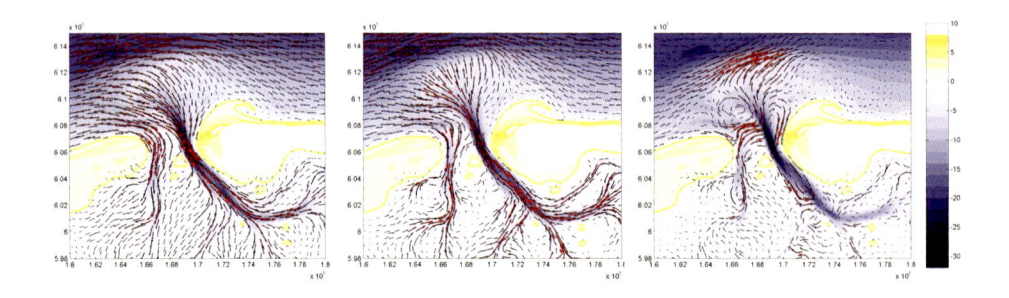

Fig. 3.12 Amelander zeegat flow patterns. Left panel: Maximum flood flow. Arrows denote trajectories following flow for 0.5 hour. Middle panel: Siilar for maximum ebb flow. Right panel: Residual flow where arrows denote trajectory following residual current for 3 hours. Depth in m indicated by the colorbar on the right.

The left panel in Figure 3.12 shows the flow pattern during maximum flood tide; the curved arrows follow the flow and denote the distance travelled with the flow in half an hour, starting from a given location. The overall flow direction at sea is towards the east; on top of this we see a flow towards the inlet from all sides. The flow velocities are concentrated in the two channels, but there are significant velocities along the coast at each side of the inlet gorge. Inside the inlet we see the flood current flowing onto the tidal flats.

In the middle panel of Figure 3.12 the flow pattern during maximum ebb tide is visualized. We see the water flowing off the tidal flats into the channels and outward; the concentrated flow in the channels fans out over the shallow ebb delta. At sea, the flow direction has reversed and is towards the west now.

The average flow velocity over the tidal cycle is shown in the right panel of Figure 3.12. We now see the typical pattern of flood-dominance near the sides of the inlet and ebb dominance in the main channels. Also, we see a number of tide-averaged circulation cells: two cells on either side of the outflow of the Borndiep, with opposite directions. The eastward residual flow on the ebb shield can be explained by the fact that the (eastward) flood current is drawn into the inlet whereas the westward ebb current is pushed out by the ebb jet. On the flanks of the Borndiep and Westgat there is a counter-clockwise circulation that may be explained by the Coriolis force pushing the flow towards its right-hand side in the northern hemisphere. We will get back to this inlet as we discuss typical effects of wind and waves and resulting sediment transport patterns and morphological changes.

3.3.8 *Current pattern across a trench*

In this section we will look at the current magnitude and direction changes as it crosses a trench or navigation channel at an angle. This is important both for navigation and for sedimentation and migration of the channel.

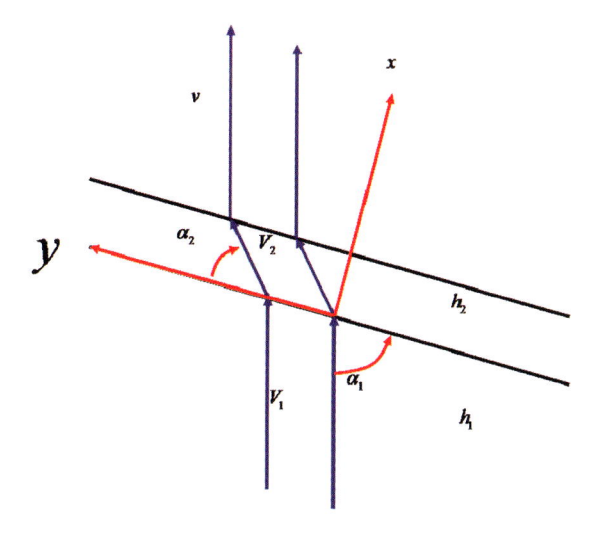

Fig. 3.13 Principle sketch flow across a channel

In Figure 3.13 we see the principle sketch of the situation. We can analyse it by assuming that the channel is very long, depth outside the channel is uniform at h_1 and inside it is uniformly h_2. Furthermore we assume that the flow is stationary and uniform in the channel direction. We take the x-axis perpendicular to the channel direction and the y-axis in the channel direction. The velocity components outside the channel are:

$$u_1 = V_1 \sin \alpha, \quad v_1 = V_1 \cos \alpha \tag{3.83}$$

For the flow component across the trench we can now solve the depth-averaged current component u_2 from the continuity equation:

$$\frac{\partial \eta}{\partial t} + \frac{\partial hu}{\partial x} + \frac{\partial hv}{\partial y} = 0 \Rightarrow hu = const \Rightarrow h_2 u_2 = h_1 u_1 \Rightarrow u_2 = \frac{h_1}{h_2} u_1 \tag{3.84}$$

Here the gradient in time is zero because we consider the stationary solution and the gradient with y is zero as we assume constant conditions along the channel.

For the flow component along the channel we use the fact that the water level gradient in the direction of the channel must be the same outside and inside the channel, because otherwise the *cross-channel* water level gradient would not be uniform. We can then use the along-channel momentum balance to solve the along-channel velocity inside the channel:

$$\frac{\partial v}{\partial t} + u\frac{\partial v}{\partial x} + v\frac{\partial v}{\partial y} + f_{cor} u = \frac{\partial}{\partial x} D_h \frac{\partial v}{\partial x} + \frac{\tau_{sy}}{\rho h} - \frac{\tau_{by}}{\rho h} - \frac{1}{\rho}\frac{\partial p_a}{\partial y} - g\frac{\partial \eta}{\partial y} + \frac{F_y}{\rho h} \quad (3.85)$$

The horizontal advection and dispersion terms are non-zero, as there are gradients in v across the channel; they lead to a gradual adaptation of the along-channel flow to the depth in the channel. For the moment we focus on the velocity some distance into the channel, where these effects are attenuated, and we are left with the balance between the along-channel water level gradient and the bed shear stress:

$$\frac{\tau_{by}}{\rho h} = -g\frac{\partial \eta}{\partial y} = -g i_y \quad (3.86)$$

The along-channel water level gradient is computed from:

$$|V_1|^2 = C^2 h_1 i_1 \Rightarrow i_1 = \frac{|V_1|^2}{C^2 h_1} \quad (3.87)$$

$$i_{y,2} = i_{y,1} = i_1 \cos(\alpha_1)$$

We can solve eq. (3.87) using Chezy's friction law:

$$\frac{g}{C^2 h_2}|V_2| v_2 = \frac{g}{C^2 h_2} v_2 \sqrt{u_2^2 + v_2^2} = g i_{y,2} \Rightarrow$$

$$\Rightarrow v_2^2 (u_2^2 + v_2^2) = C^4 h_2^2 i_{y,2}^2 \Rightarrow \quad (3.88)$$

$$\Rightarrow (v_2^2)^2 + u_2^2 v_2^2 - C^4 h_2^2 i_{y,2}^2 = 0$$

This is a quadratic equation in v_2^2; all other terms are known by now. The solution is therefore simple:

$$v_2^2 = \frac{-u_2^2 +/- \sqrt{u_2^4 + 4C^4 h_2^2 i_{y,2}^2}}{2} \Rightarrow$$

$$v_2 = \sqrt{\frac{-u_2^2 + \sqrt{u_2^4 + 4C^4 h_2^2 i_{y,2}^2}}{2}} \quad (3.89)$$

Finally the angle of the flow with respect to the channel axis is given by:

$$\alpha_2 = \arctan\left(\frac{u_2}{v_2}\right) \quad (3.90)$$

In Figure 3.14 the current angle in the channel and the ratio of the velocity magnitudes in the channel and outside it are plotted as a function of the incident angle, for the case of $h_1 = 10m$, $h_2 = 15m$, $C = 65 m^{1/2}/s$, $V_1 = 1\,m/s$. We see that the current refracts in the direction of the channel axis, in this case by approx. 20 degrees. This is explained by the fact that the cross-channel flow velocity decreases because of the increased depth, while the along-channel flow increases due to the reduced friction (again as a result of the increased depth). As for the magnitude of the current inside the channel, it is reduced for flow perpendicular to the channel and enhanced for flow in the direction of the channel. Of course this has important consequences for the sedimentation of the channel. For further clarification we included a Matlab script **trench.m** used to generate Figure 3.14.

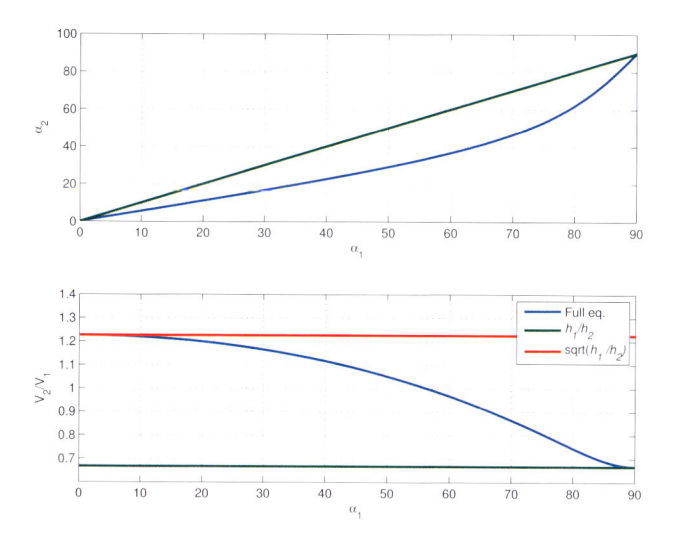

Fig. 3.14 Current angle and current magnitude ratio in channel as function of undisturbed angle of current with respect to the channel axis.

3.4 Wind-driven longshore current and set-up on an alongshore uniform coast

3.4.1 *Wind-driven longshore current*

The wind-driven longshore current along an approximately uniform coast is governed by the balance between the alongshore wind shear stress, the bed friction and the inertia:

$$\frac{\partial v}{\partial t} + u\frac{\partial v}{\partial x} + v\frac{\partial u}{\partial y} + fu = \frac{\partial}{\partial x}\left(D_h \frac{\partial v}{\partial x}\right) + \frac{\partial}{\partial y}\left(D_h \frac{\partial v}{\partial y}\right) + \frac{\tau_{w,y}}{\rho h} - \frac{\tau_{by}}{\rho h} - g\frac{\partial \eta}{\partial y} \quad (3.91)$$

As the wind starts blowing the balance is entirely between inertia and wind stress; in the equilibrium situation the balance is between wind stress and bed shear stress. We can then easily estimate the equilibrium longshore velocity as:

$$v_\infty = \sqrt{\frac{\tau_{w,y}}{\rho_w C_f}} \qquad (3.92)$$

Interestingly, the longshore wind-driven current speed at equilibrium is not dependent on water depth. We can also estimate a time-scale for the adaptation of the velocity to equilibrium:

$$T_{wind} = \frac{v_\infty}{\partial v/\partial t|_{t=0}} = h\sqrt{\frac{\rho_w}{\tau_{w,y} C_f}} \qquad (3.93)$$

The adaptation time-scale is proportional to the water depth and inversely proportional to the wind velocity. For the typical case shown in Figure 3.15 with a wind velocity of 10 m/s in water depth of up to 16 m it takes almost a day for the wind-driven current to spin up completely. With weaker winds and/or greater depth this takes even longer.

The wind stress magnitude is given by:

$$\rho_a |\tau_w| = \rho_a C_d |W_{10}|^2 \qquad (3.94)$$

Here ρ_a is the density of air (approx. 1.25 kg/m3), C_d the wind drag coefficient (approx. 0.002) and W_{10} the wind velocity at 10 m above the surface. Considerable uncertainty exists over appropriate values for C_d, especially for relatively low wind speeds and relatively high wind speeds, though its general tendency appears to be to increase from approx. 0.001 to 0.002 for wind speeds from 0 to 20 m/s, and to further increase to a maximum in the order of 0.003 for wind speeds in excess of 30 m/s ([Powell *et al.* (2003)] and [Donelan *et al.* (2004)]).

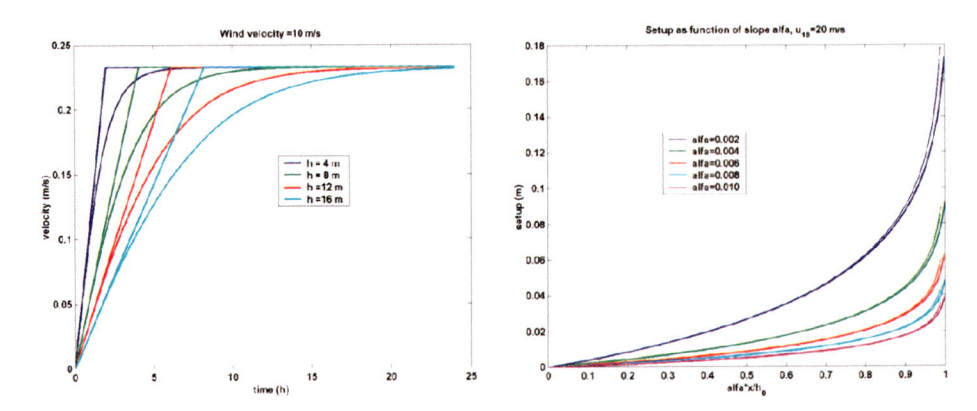

Fig. 3.15 left panel: Adaptation of longshore wind-driven current for different depths. Straight lines indicate initial gradient and time scale of adaptation. Right panel: Wind-induced set-up as function of dimensionless cross-shore distance, for various slopes and wind velocity of 20 m/s Thick lines: numerical solution, thin lines: analytical approximation

3.4.2 Wind-driven set-up

The cross-shore set-up is governed by the slope term and the cross-shore wind stress:

$$\frac{\partial u}{\partial t} + u\frac{\partial u}{\partial x} + v\frac{\partial u}{\partial y} - fv = \frac{\partial}{\partial x}D_h\frac{\partial u}{\partial x} + \frac{\partial}{\partial y}D_h\frac{\partial u}{\partial y} + \frac{\tau_{w,x}}{\rho h} - \frac{\tau_{b,x}}{\rho h} - g\frac{\partial \eta}{\partial x} \quad (3.95)$$

On a horizontal bed, this means that the slope is linearly proportional to the wind stress and inversely proportional to the water depth. Therefore wind set-up and set-down can be significant in wide shallow basins, more so than in deep seas. For a horizontal bottom, the difference in water level between two points at a cross-shore distance L is given by:

$$\Delta \eta = \frac{\tau_{w,x}}{\rho g h}L \quad (3.96)$$

On a profile with variable depth, the total setup can be estimated by adding the setup contributions over sections with approximately constant water depth.

For the case of a plane sloping beach we can estimate the total set-up by a simple analytical expression, if we neglect the effect of the set-up on the total water depth. The water depth is then given by:

$$h = h_0 - \alpha(x - x_0) \quad (3.97)$$

Combining (3.96) and (3.97) we find:

$$\eta = -\frac{\tau_{w,x}}{\rho g \alpha} \ln\left(\alpha \frac{x - x_0}{h_0}\right) \quad (3.98)$$

which compares well with the numerical solution taking into account the effect of the set-up on the water depth (right panel of Figure 3.15).

3.5 Wave-driven longshore current and set-up on an uniform coast

3.5.1 Wave-driven longshore current

Waves approaching the coast at an angle carry a flux of momentum or 'radiation stress' that has an alongshore directed component (eq. 3.23). As the waves break, they lose this radiation stress and the momentum is transferred to the longshore current. For a longshore uniform beach at $x = 0$, we have $u = 0$, $\partial/\partial y = 0$, so the depth-averaged current velocity is described by:

$$\frac{\partial v}{\partial t} + u\frac{\partial v}{\partial x} + v\frac{\partial u}{\partial y} + fu = \frac{\partial}{\partial x}D_h\frac{\partial v}{\partial x} + \frac{\partial}{\partial y}D_h\frac{\partial v}{\partial y} - g\frac{\partial \eta}{\partial y} + \frac{F_y}{\rho h} - \frac{\tau_{b,y}}{\rho h} \quad (3.99)$$

The forcing by the radiation stress gradient, eq. 3.23, is given by the combined wave-and roller related contributions (eq.'s 3.24 and 3.25):

$$\begin{aligned}F_y &= -\frac{\partial S_{xy}}{\partial x} = -\frac{\partial}{\partial x}\left[\frac{C_g}{C}\left(E\cos(\alpha)\sin(\alpha)\right) + E_r\cos(\alpha)\sin(\alpha)\right] \\ &= -\frac{\partial}{\partial x}\left[\frac{\sin(\alpha)}{C}\left(EC_g\cos(\alpha) + E_rC\cos(\alpha)\right)\right] \\ &= -\left(EC_g\cos(\alpha) + E_rC\cos(\alpha)\right)\frac{\partial}{\partial x}\left[\frac{\sin(\alpha)}{C}\right] \\ &\quad - \frac{\sin(\alpha)}{C}\frac{\partial}{\partial x}\left(EC_g\cos(\alpha) + E_rC\cos(\alpha)\right)\end{aligned} \qquad (3.100)$$

In a longshore uniform situation, according to Snel's law, the first term on the right-hand side equals zero; the second term exactly equals the sum of the wave energy dissipation and the roller energy input and dissipation terms, so the forcing term reduces to:

$$F_y = \frac{D_w + (-D_w + D_r)}{C}\sin(\alpha) = \frac{D_r}{C}\sin(\alpha) \qquad (3.101)$$

consistent with eq. 3.26. For simplified turbulent mixing analytical expressions for the longshore current on an alongshore beach of constant slope can be derived [Longuet-Higgins (1970)]. However, for more realistic profiles a numerical evaluation is required as outlined in Section 3.5.3.

3.5.2 *Wave-driven set-up*

For this alongshore uniform situation, the depth-averaged cross-shore momentum balance reads:

$$\frac{\partial u}{\partial t} + u\frac{\partial u}{\partial x} + v\frac{\partial u}{\partial y} - fv = \frac{\partial}{\partial x}D_h\frac{\partial u}{\partial x} + \frac{\partial}{\partial y}D_h\frac{\partial u}{\partial y} - g\frac{\partial \eta}{\partial x} + \frac{F_x}{\rho h} - \frac{\tau_{bx}}{\rho h} \qquad (3.102)$$

with:

$$F_x = -\frac{\partial S_{xx}}{\partial x} = -\frac{\partial}{\partial x}\left[\left(\frac{C_g}{C}\left(1 + \cos^2(\theta)\right) - \frac{1}{2}\right)E_w + \cos^2(\theta)E_r\right] \qquad (3.103)$$

In a stationary case, this balance is dominated by the forcing and the surface slope term, which leads to wave set-down outside of the breaker zone, where the radiation stress increases due to shoaling, and set-up inside the breaker zone, where it reduces due to breaking of the waves. Compared to the wave forcing in the surf zone, the cross-shore bed shear stress and Coriolis force are small and has been neglected. A quick estimate of the maximum set-up on a planar beach can be obtained by using the shallow water approximation for the radiation stress (ignoring the roller contribution):

$$F_x = -\frac{\partial S_{xx}}{\partial x} \approx -\frac{\partial}{\partial x}\frac{3}{2}E_w = -\frac{3}{8}\rho g\gamma^2 h\frac{dh}{dx} \qquad (3.104)$$

which yields:
$$\eta_{\max} = -\frac{3}{8}\gamma H_{br} \tag{3.105}$$
where the wave height at the onset of wave breaking is given by $H_{br} = \gamma h_b$ with h_b the depth at breaking.

3.5.3 Numerical evaluation

To fully evaluate the forcing, eqs. (3.101) and (3.103), the cross-shore distribution of the wave and roller energy has to be known. This can be obtained from the combined wave and roller energy balances, eq.'s (2.19) and (2.31) which have to be solved numerically.

As in many coastal profile models, we will assume that the wave conditions and water level vary slowly in time, compared to the time it takes for waves to travel through the surf zone, and that inertia effects in the current may be neglected. We will also neglect wave height variations on the time-scale of wave groups and consider the time-averaged distribution of waves only. The wave energy and roller energy balances for a uniform coast then reduce to:

$$\frac{\partial}{\partial x}\left(E_w c_g \cos(\theta_m)\right) = -D_w \tag{3.106}$$

This equation can be solved numerically by writing it as the following difference equation:

$$\frac{E_{w,i} C_{g,i} \cos(\theta_{m,i}) - E_{w,i-1} C_{g,i-1} \cos(\theta_{m,i-1})}{\Delta x} = -\frac{D_{w,i} + D_{w,i-1}}{2} \Rightarrow$$

$$\Rightarrow E_{w,i} = \frac{E_{w,i-1} C_{g,i-1} \cos(\theta_{m,i-1}) - \frac{1}{2}(D_{w,i} + D_{w,i-1})\Delta x}{C_{g,i} \cos(\theta_{m,i})} \tag{3.107}$$

We see that the wave energy in a point depends on the wave energy in the previous point and on group velocity and wave direction in both points. For second-order accuracy, we have taken the dissipation as the average dissipation between the new and the old point, which means that to evaluate the dissipation we need to know the energy in the new point i. We can overcome this by first estimating the wave dissipation in the new point i based on the wave energy in point $i-1$ and the depth h in point i and then repeating the calculation of eq. (3.107), a so-called predictor-corrector method. Similarly, we can solve the simplified roller energy balance:

$$\frac{\partial E_r c \cos\theta_m}{\partial x} = D_w - D_r \tag{3.108}$$

by writing it in discretised form:

$$\frac{E_{r,i} C_i \cos(\theta_{m,i}) - E_{r,i-1} C_{i-1} \cos(\theta_{m,i-1})}{\Delta x} = \frac{D_{w,i-1} + D_{w,i}}{2} - \frac{D_{r,i-1} + D_{r,i}}{2} \Rightarrow$$

$$\Rightarrow E_{r,i} = \frac{E_{r,i-1} C_{i-1} \cos(\theta_{m,i-1}) + \frac{1}{2}(D_{w,i-1} + D_{w,i} - D_{r,i-1} - D_{r,i})\Delta x}{C_i \cos(\theta_{m,i})} \tag{3.109}$$

The boundary conditions are the known wave energy at the offshore boundary (from a wave buoy or a large-scale wave model) and a roller energy that is equal to zero outside the surf zone. The wave celerity and group velocity only depend on the constant wave period (usually the peak period) and the water depth. From the local wave celerity the wave direction can be calculated based on the offshore wave direction and celerity, using Snel's law (eq. 2.23). The water depth is influenced by the wave-induced setup; therefore it is necessary to solve this equation at the same time, using the simplified cross-shore momentum balance eq. (3.102). This equation can be discretised as follows, using eq. (3.103):

$$\frac{\eta_i - \eta_{i-1}}{\Delta x} = -\frac{2}{\rho g (h_{i-1} + h_i)} \left[\frac{S_{xx,i} - S_{xx,i-1}}{\Delta x} \right] \Rightarrow$$
$$\eta_i = \eta_{i-1} - \frac{2}{\rho g (h_{i-1} + h_i)} (S_{xx,i} - S_{xx,i-1}) \quad (3.110)$$

Again, the solution in point i partly depends on values in point i that have to be solved iteratively, using the same predictor-corrector approach.

In the enclosed Matlab function **balance_1d.m** you can see how such a scheme can be programmed. The function computes the solution for a given set of increasing x-values, for given bottom level z_b and boundary conditions for wave height, direction, wave period and water level, and given calibration coefficients *gamma* and *beta* and cut-off water depth h_{min}. Once we know the cross-shore distribution of the longshore component of the wave force, we can solve the longshore current velocity, using eq. (3.99).

For the stationary (equilibrium) case, and neglecting the horizontal viscosity, we simply get:

$$\tau_{by} = F_y \quad (3.111)$$

We have to realise that the bed shear stress here is a function of the (unknown) longshore velocity v and the wave-induced shear stress, according to one of the formulations discussed under 'bed shear stress' earlier. The procedure to follow is:

- to compute the wave-induced shear stress;
- to solve the current-related shear stress *tauc* from *tauw* and *taum* iteratively;
- to compute v from *tauc*.

This procedure is illustrated by the enclosed Matlab functions **longshore_current.m** and **soulsby.m**. In case we do not want to neglect the horizontal viscosity, we can take another approach, where we solve the non-stationary problem according to eq. (3.99). A simple scheme that would do the job is the following:

$$\frac{v_i^{n+1} - v_i^n}{\Delta t} = \frac{(D_{h,i} + D_{h,i+1})(v_{i+1} - v_i) - (D_{h,i-1} + D_{h,i})(v_i - v_{i-1})}{2\Delta x^2} + \frac{F_{y,i}}{\rho h_i} - \frac{\tau_{by,i}}{\rho h_i} \Rightarrow$$
$$v_i^{n+1} = v_i^n + \Delta t \left(\frac{(D_{h,i} + D_{h,i+1})(v_{i+1} - v_i) - (D_{h,i-1} + D_{h,i})(v_i - v_{i-1})}{2\Delta x^2} + \frac{F_{y,i}}{\rho h_i} - \frac{\tau_{by,i}}{\rho h_i} \right)$$
$$(3.112)$$

Note that this explicit scheme is only conditionally stable, for sufficiently small time steps that satisfy the courant criterion (see Section 9.4). We can keep updating the velocity for a given time period or while a convergence criterion has not yet been met. A Matlab function that solves this scheme, **longshore_current_t.m**, has been enclosed.

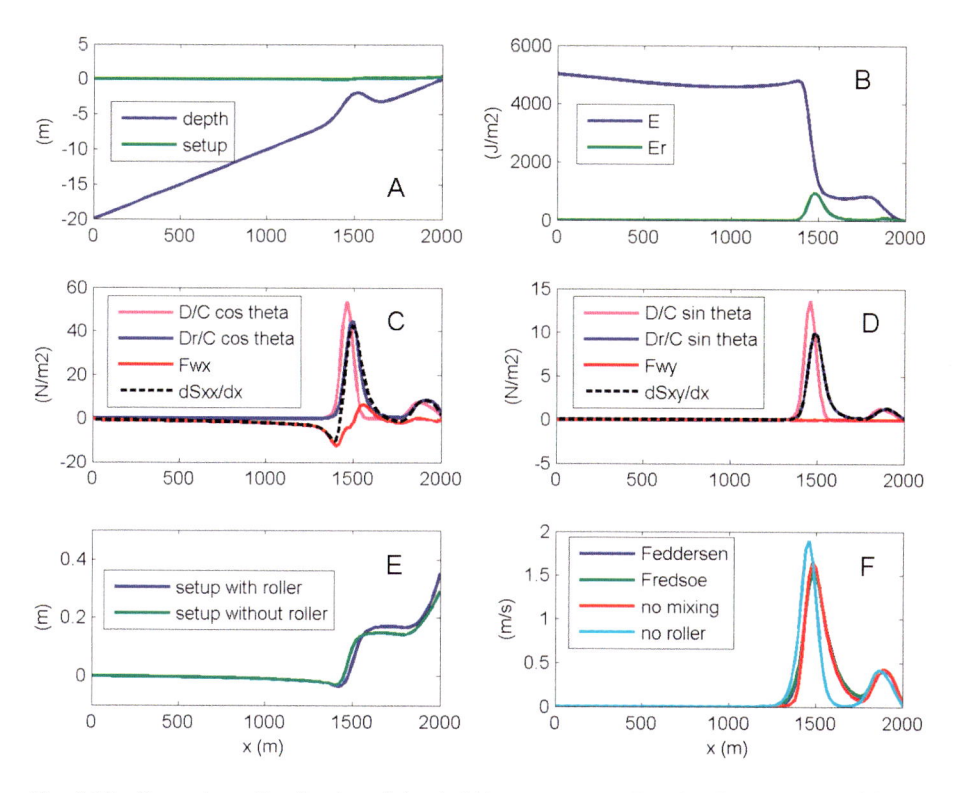

Fig. 3.16 Cross-shore distribution of depth (A), wave energy E and roller energy Er (B), contributions to the forcing in cross-shore direction (C) and longshore direction (D), wave-induced setup (E) and longshore velocity (F).

To illustrate the cross-shore behaviour of the forcing and the longshore current we solved eq. (3.99) for a longshore uniform profile at a slope of 1:100 starting at 20 m depth at $x=0$ m with a Gaussian bar superimposed with a height of 3 m and a length scale of approximately 200 m, at $x=1500$ m. The incident wave height is 2 m, peak period is 7 s and the angle of incidence is 30 degrees from the normal (240 degrees in nautical convention). The equilibrium situation is shown in Figure 3.16. The roller energy acts as a buffer: when waves break, their momentum is first transferred to the roller which then through the shear stress it acts on the water surface transfers it on to the current. In the distribution of the longshore forcing this is evident: to illustrate this we have also plotted the forcing without the roller effect (panels C and D in Figure 3.16). The effect of the rollers is to reduce the peak and shift it in landward direction, delaying both the set-up [Nairn et al. (1990)] and the longshore current which increases significantly in the trough [Reniers and Battjes (1997)] (see panels E and F in Figure 3.16).

The contribution of the depth-invariant body force, $\vec{F}_w$ given by eq. (3.27), is compared with the total forcing, $\vec{F}$ given by eq. (3.23) and the surface shear stress, $\vec{\tau}_s$ eq. (3.26). This example shows that the body force mainly acts outside of the surfzone creating a set-down of the mean water level (see panels C and E). Inside the surfzone the body force is small (panel C) for the cross-shore forcing and absent in the alongshore forcing (panel D in Figure 3.16) compared with the surface shear stress forcing the set-up and longshore current respectively.

Using the bottom shear stress formulation by [Feddersen et al. (2000)], [Ruessink et al. (2001)] show good agreement for wave-driven currents along approximately uniform beaches at Egmond in the Netherlands and Duck, NC. They apply the depth-averaged velocity in eq. (3.99). In panel F of Figure 3.16 we give a comparison between the [Feddersen et al. (2000)] formulations and those by some of the formulations parameterised by [Soulsby et al. (1993)]. Apparently the simple formulation by [Feddersen et al. (2000)] gives very comparable results to those by the parameterised [Fredsoe (1984)] model, applied to random waves. The inclusion of a constant turbulent mixing, with $v_t = 0.5$ m2/s, has only a moderate effect on the longshore current distribution reducing the peak velocity and increasing the flows at the offshore side of the bar (panel F in Figure 3.16).

3.5.4 Shear instabilities

In the section we examine the time-mean properties of the wave-driven longshore current. Depending on the cross-shore distribution of the wave-driven longshore current velocity, low-frequency modulations in the horizontal velocity field may be present. These oscillations result from the shear instability of the longshore current. Small perturbations in the velocity field grow at the expense of the potential vorticity associated with the longshore current. The theoretical background for this was introduced by [Bowen and Holman (1989)]. They obtained a linear stability equation for longshore currents, showing with a simplified test case (horizontal bottom) that the backshear of the longshore current is very important in the generation of shear instabilities. A strong back shear makes the longshore current unstable to a wide range of small perturbations. This was confirmed by [Dodd and Thornton (1993)] and [Putrevu and Svendsen (1992)] using more realistic longshore current velocity and bottom profiles. Later studies [Falques et al. (1994)] also showed that the distance of the longshore current velocity maximum to the shoreline, X_b, is important in the generation of the longshore instabilities. An increase in X_b results in a wider range of unstable wave numbers.

The generation of shear instabilities is strongly affected by the presence of dissipative effects. [Putrevu and Svendsen (1992)] and [Dodd and Thornton (1993)] showed that the number of possible unstable modes and their corresponding growth rate is reduced if dissipation due to bottom friction is increased. The combined effect of bottom friction and eddy viscosity on the stability of wave-driven longshore currents was investigated by [Falques et al. (1994)], who show that for eddy viscosity and bottom friction giving rise to similar damping, eddy viscosity gives a stronger reduction of the span of unstable wave numbers compared to bottom friction.

Here we consider the potential generation of longshore current shear instabilities over a barred profile depicted in Figure 3.16. To allow for the growth of the instabilities we consider the time-varying momentum equations without turbulent mixing:

$$\frac{\partial u}{\partial t} + u\frac{\partial u}{\partial x} + v\frac{\partial u}{\partial y} - fv = \frac{\partial}{\partial x}D_h\frac{\partial u}{\partial x} + \frac{\partial}{\partial y}D_h\frac{\partial u}{\partial y} - g\frac{\partial \eta}{\partial x} + \frac{F_x}{\rho h} - \frac{\tau_{bx}}{\rho h} \qquad (3.113)$$

$$\frac{\partial v}{\partial t} + u\frac{\partial v}{\partial x} + v\frac{\partial u}{\partial y} + fu = \frac{\partial}{\partial x}D_h\frac{\partial v}{\partial x} + \frac{\partial}{\partial y}D_h\frac{\partial v}{\partial y} - g\frac{\partial \eta}{\partial y} + \frac{F_y}{\rho h} - \frac{\tau_{by}}{\rho h} \qquad (3.114)$$

The basic state, i.e. in absence of shear instabilites, is given by the longshore current velocity profile, $V(x)$, and set-up $\eta(x)$ shown in Figure 3.16. Next the shear instability perturbations are introduced as:

$$e^{ik(y-ct)}\left(u'(x), v'(x), \eta'(x)\right) \qquad (3.115)$$

Upon substitution we obtain the momentum balance for the mean motion, eq. (3.113) and (3.114), corresponding to the basic state, and the momentum balance for the shear instabilities:

$$ik(V-c)u' + g\frac{\partial \eta'}{\partial x} = \frac{\tau'_{x,b}}{\rho h} \qquad (3.116)$$

$$u'\frac{dV}{dx} + ik(V-c)v' + ikg\eta' = -\frac{\tau'_{y,b}}{\rho h} \qquad (3.117)$$

The corresponding continuity equation for the shear instabilities is given by:

$$\frac{\partial hu'}{\partial x} + ikhv' + ik(V-c)\eta' = 0 \qquad (3.118)$$

After introduction of a stream function:

$$u' = -\frac{1}{h}\frac{\partial \psi}{\partial y}$$
$$v' = \frac{1}{h}\frac{\partial \psi}{\partial x} \qquad (3.119)$$

these equations can be combined into a single equation [Dodd et al. (2000)]:

$$\left(v - c - \frac{i\mu}{kh}\right)\left(\frac{\partial^2 \psi}{\partial x^2} - \frac{1}{h}\frac{dh}{dx}\frac{\partial \psi}{\partial x} - k^2\psi\right) = h\frac{d}{dx}\left(\frac{1}{h}\frac{dV}{dx}\right)\psi \qquad (3.120)$$

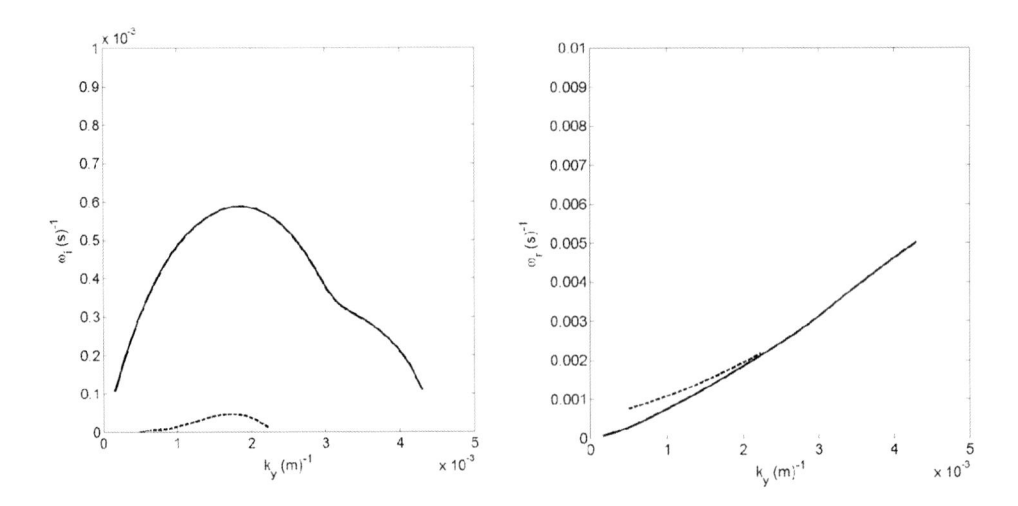

Fig. 3.17 Left panel: Predicted growth rate of shear instabilities for longshore current velocity profile shown in panel F in Figure without (solid line) and with bottom friction (dashed line). Right panel: Corresponding dispersion relation without (solid line) and with bottom friction (dashed line).

where the bottom friction has been linearized to:

$$\vec{\tau}_b = \rho \mu V \vec{u}' \tag{3.121}$$

For a given longshore current velocity and bottom profile this equation can be solved for a range in alongshore wave numbers (corresponding Matlab-code to calculate growth rates and dispersion relation is included as **shearinstab.m**). This will yield a solution for the stream function, ψ, and corresponding (complex) phase speed, c, for each of these wave numbers. If the imaginary part of the phase speed is positive (negative) the shear instabilities will grow (dampen):

$$u' e^{ik(y-ct)} = -\frac{1}{h} \frac{\partial \psi}{\partial y} e^{iky} e^{-ikct} = -\frac{1}{h} \frac{\partial \psi}{\partial y} e^{iky} e^{-i\omega_r t} e^{\omega_i t} \tag{3.122}$$

$$v' e^{ik(y-ct)} = \frac{1}{h} \frac{\partial \psi}{\partial x} e^{iky} e^{-ikct} = \frac{1}{h} \frac{\partial \psi}{\partial x} e^{iky} e^{-i\omega_r t} e^{\omega_i t} \tag{3.123}$$

where the imaginary part of the angular frequency, ω_i, corresponds to the growth rate. Plotting the growth rate as function of the alongshore wave number, the fastest growing mode (FGM) can be retrieved at $k \approx 0.0018 \text{m}^{-1}$ (see left panel of Figure 3.17) corresponding to a wave length of approximately 550 m. The inclusion of bottom friction, with $\mu = 0.01$, clearly reduces the growth rate and limits the range of unstable modes (compare solid and dashed lines in the left panel of Figure 3.17). The shear instability dispersion relation is obtained from the real part

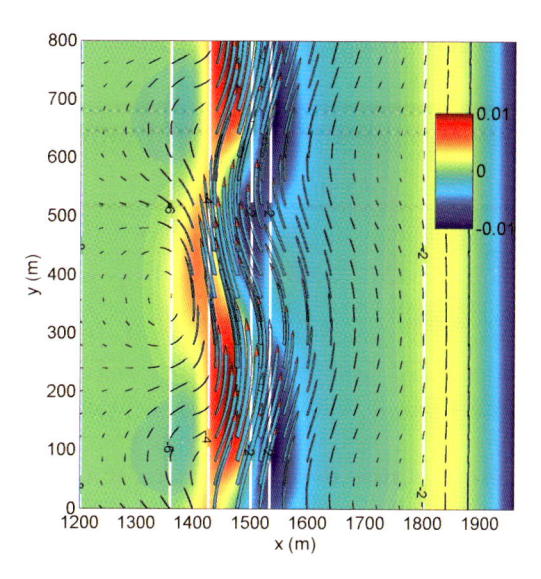

Fig. 3.18 Superposition of FGM shear instability velocity pattern on the mean longshore current velocity profile resulting in the meandering of the longshore current. Arrows correspond to particle trajectories integrated over a period of 100 seconds. Corresponding vorticity in s-1 is given by the color scale.

of the angular frequency, ω_r, showing a more or less constant alongshore propagation speed, $\text{Re}\{c\} = \text{Re}\{\frac{\omega}{k}\}$, of approximately 1 m/s corresponding to the mean longshore current velocity over the bar (panel F in Figure 3.16).

Superimposing the velocity field of the FGM, obtained from the stream function solution at maximum growth rate with an (arbitrary) amplitude of 200 m^3/s, on the steady alongshore current shows the meandering of the current at the 550 m length scale associated with the presence of shear instabilities (see Figure 3.18).

Keep in mind that the linear stability analysis is only valid for shear instabilities of infinitesimal amplitude, i.e. initial conditions. More detailed analytical and numerical analyses were made using weakly non-linear ([Dodd and Thornton (1992)]) and fully non-linear modeling (e.g. [Nadaoka and Yagi (1993)], [Deigaard et al. (1994)], [Falques et al. (1994)], [Allen et al. (1996)], [Özkan-Haller and Kirby (1999)] and [Slinn et al. (1998)]) to describe the development of shear instabilities in wave-driven longshore currents to finite amplitude values. A key finding from these studies is that the shear instabilities induce a cross-shore mixing of the mean momentum resulting in smoother mean longshore current velocity profiles, which affects the sediment transport [Deigaard et al. (1994)]. Also, for low damping, the instabilities can exhibit chaotic behaviour with eddies ejecting from the longshore current causing strong offshore directed flows (e.g. [Slinn et al. (1998)]), generating potentially dangerous conditions for swimmers ([Dalrymple et al. (2011)]). A comprehensive review of shear instabilities in longshore current instabilities, including field [Oltman-Shay et al. (1989)] and laboratory [Reniers et al. (1997)] observations, is given by [Dodd et al. (2000)].

3.6 Wave-group driven motions

3.6.1 *Introduction*

In the previous descriptions of wave-driven currents the wave forcing was assumed to be constant in time. As shown in Chapter 2 the frequency and directional spreading in the incident wave field leads to the presence of wave-groups. This in turn results in a modulation of the wave forcing at wave-group time-and spatial scales. This modulation forces both long waves, also known as infragravity motions or Low Frequency motions (LFs), as well as surfzone vortices, also known as wave-group-breaking induced Very Low Frequency motions (VLFs).

Infragravity waves have periods between 25 s and 250 s and can have significant effect on coastal morphodynamics. [Roelvink and Stive (1989)] have shown the importance with respect to morphology where the coupling between the wave groups and the underlying infragravity waves results in preferential sediment transport directions (see Section 4.4.4). Other important areas are wave runup [van Gent (2001)], dune erosion ([van Thiel de Vries et al. (2008)] and [Roelvink et al. (2009)]) and overwash ([McCall et al. (2010)]).

Wave-group-breaking induced vortical motions have significantly longer periods in the range of 4 to 30 minutes and their potential impact on coastal hydro- and morpho-dynamics is still being explored. [Reniers et al. (2004a)] numerically found VLFs within the surf zone that initiated a quasi-rhithmic evolution of rip-channel morphology (see Section 10.5). It is evident that wave-group-breaking can generate transient rip currents on both planar and rip-channelled beaches ([Fowler and Dalrymple (1990)], [Reniers et al. (2004a)], [Johnson and Pattiaratchi (2006)], [Reniers et al. (2007)], [Long and Özkan-Haller (2009)]) which strongly affect mixing of surfzone matter ([Spydell and Feddersen (2009)], [Brown et al. (2009)]) and are capable of transporting this matter out of the surfzone ([Reniers et al. (2009)] and [Reniers et al. (2010b)]).

The first observations where infragravity waves were linked to surface elevations on a wave group scale were done by [Munk (1949)] and [Tucker (1952)]. The latter observed a significant positive correlation at a negative time lag and observed a smaller negative correlation at zero time lag. [Biesel (1952)] showed that bound infragravity waves propagate with the group velocity of the short wave groups with a phase lag of 180° thus explaining the negative correlation at zero time lag. [Longuet-Higgins and Stewart (1962)] and [Longuet-Higgins and Stewart (1964)] found a possible explanation for the larger positive correlation at negative time lag suggesting that bound infragravity waves, non-linearly forced by the spatial changes of short wave momentum flux, increase strongly in amplitude while traveling with the shoaling wave groups to the shore. These bound infragravity waves then get released at breaking and subsequently reflect at the shore line towards deeper water as free infragravity waves experiencing weaker inverse shoaling. Due to the stronger refraction of the free infragravity waves, not all reflected infragravity waves propagate to the deeper water, but some refract back to the shore where reflection takes place again [Herbers et al. (1995)]. The infragravity waves that make it out to deeper water are called leaky waves, whereas the trapped waves are called edge waves [Ursell (1952)].

Field observations and understanding of VLFs has lagged with respect to infragravity waves starting with [Tang and Dalrymple (1989)]. This can partly be explained by the fact that the surface elevation signature of VLFs is minimal, i.e. they cannot be detected with pressure sensors or wave gauges. Furthermore, the frequent co-existence and potential interaction with shear instabilities (see Section 3.5.4) and wave breaking induced vortices masks the VLF-response to the grouped incident waves [Haller et al. (1999)]. A recent study combining field observations and numerical modeling indicate that VLFs, like infragravity waves, are ubiquitous in the surf zone [MacMahan et al. (2010b)].

3.6.2 *Wave group induced bound long waves*

The randomness in the wave field is characterized by the presence of wave groups (see Section 2.7) and the corresponding variation in wave momentum that forces bound long waves can be illustrated by considering wave groups made up of a bichromatic wave composed of two normally incident waves of equal amplitude but different frequency, propagating over a horizontal bed:

$$\eta_s(x,t) = a\sin(\omega_1 t - k_1 x) + a\sin(\omega_2 t - k_2 x) \tag{3.124}$$

which can be written as the rapid fluctuation of a slowly varying wave group amplitude, A:

$$\eta_s(x,t) = 2a\cos\left(\frac{\Delta\omega}{2}t - \frac{\Delta k}{2}x\right)\sin\left(\frac{\Sigma\omega}{2}t - \frac{\Sigma k}{2}x\right) = A(x,t)\sin\left(\frac{\Sigma\omega}{2}t - \frac{\Sigma k}{2}x\right) \tag{3.125}$$

where the mean angular frequeny and wave number are given by:

$$\omega_m = \frac{\Sigma\omega}{2} = \frac{\omega_1 + \omega_2}{2}, k_m = \frac{\Sigma k}{2} = \frac{k_1 + k_2}{2} \tag{3.126}$$

and the difference angular frequency and wave number by:

$$\Delta\omega = \omega_2 - \omega_1, \Delta k = k_2 - k_1 \tag{3.127}$$

Next we examine the long wave flow response due to the presence of the wave groups. Starting with the linarized short-wave averaged cross-shore momentum equation for an alongshore uniform beach without bottom friction:

$$\frac{\partial u}{\partial t} = -g\frac{\partial \eta}{\partial x} - \frac{1}{\rho h}\frac{\partial S_{xx}}{\partial x} \tag{3.128}$$

where the radiation stress is given by the wave groups:

$$S_{xx} = \frac{1}{2}\rho g A^2(x,t)(2n - 0.5) = \rho g a^2(1 + \cos(\Delta\omega t - \Delta k x))(2n - 0.5) \quad (3.129)$$

where eq. (3.24) has been used and n is the ratio of the group velocity over phase speed. The radiation stress thus consists of a mean part, and a modulation which propagates with the group velocity given by:

$$c_g = \frac{\Delta\omega}{\Delta k}\bigg|_{\lim \Delta k, \Delta\omega \to 0} = \frac{d\omega}{dk}$$

Using continuity to relate the velocity to the surface elevation on a horizontal bed:

$$\frac{\partial^2 u}{\partial x \partial t} = -\frac{1}{h}\frac{\partial^2 \eta}{\partial t^2} \quad (3.130)$$

taking the x-derivative of eq. (3.128) and substituting eq. (3.130) yields the long wave equation on a horizontal bed:

$$\frac{-\partial^2 \eta}{\partial t^2} + gh\frac{\partial^2 \eta}{\partial x^2} = \left(\frac{-1}{\rho}\frac{\partial^2 S_{xx}}{\partial x^2}\right) \quad (3.131)$$

Introducing solutions with $\Delta\omega$ and Δk periodicity for the bound long wave:

$$\eta = \hat{\eta}_b \cos(\Delta\omega t - \Delta k x) \quad (3.132)$$

which upon substitution in eq. (3.131) yields:

$$(\Delta\omega^2 - gh\Delta k^2)\hat{\eta}_b \cos(\Delta\omega t - \Delta k x) = \Delta k^2 g a^2 (2n - 0.5)\cos(\Delta\omega t - \Delta k x) \quad (3.133)$$

The bound long wave surface elevation amplitude is then given by ([Longuet-Higgins and Stewart (1962)]):

$$\hat{\eta}_b = \frac{\Delta k^2 g a^2 (2n - 0.5)}{(\Delta\omega^2 - gh\Delta k^2)} = \frac{-g a^2 (2n - 0.5)}{(gh - \frac{\Delta\omega^2}{\Delta k^2})} = \frac{-g a^2 (2n - 0.5)}{(gh - c_g^2)} \quad (3.134)$$

The corresponding bound long wave velocity amplitude can be obtained from continuity yielding:

$$\hat{u}_b = \frac{\Delta\omega}{\Delta k}\frac{\hat{\eta}_b}{h} \quad (3.135)$$

The long wave travels with the wave groups, at the group velocity, and the bound long wave trough coincides with a group of high short waves, i.e. 180° out of phase with the wave group signal. It is clear from eq. (3.134) that the bound long wave amplitude increases with decreasing depth as the group velocity, c_g, approaches the free long wave velocity $\sqrt{gh}$. This depth dependence in shallow water can be explored by expressing the group velocity as [Battjes et al. (2004)]:

$$c_g^2 = gh\left[1 - (k_m h)^2 + O(k_m h)^4\right] \quad \text{and} \quad k_m h << 1 \quad (3.136)$$

which upon substitution in eq (3.134) gives

$$\hat{\eta}_{l,b} = \frac{-\left(2n - \frac{1}{2}\right)ga^2}{gh(k_m h)^2} \tag{3.137}$$

For shallow water the dispersion relation becomes:

$$\omega_m^2 \approx gk_m \tan k_m h \approx gk_m^2 h \tag{3.138}$$

which upon substitution gives:

$$\hat{\eta}_{l,b} = \frac{-\left(\frac{3}{2}\right)ga^2}{\omega_m^2 h^2} \tag{3.139}$$

The short wave amplitude depth-dependence is governed by the (shallow water) wave energy balance (in absence of dissipation):

$$\frac{dc_g a^2}{dx} \approx \frac{d\sqrt{gh}a^2}{dx} = 0 \Rightarrow a^2(x) = a_0^2 h^{-\frac{1}{2}} \tag{3.140}$$

where a_0 represents the short wave amplitude at some reference point in shallow water. Hence we have the following expression for the bound long wave amplitude solely as function of depth [Longuet-Higgins and Stewart (1962)]:

$$\hat{\eta}_{l,b} \approx -\frac{3}{2}ga_0^2 \omega_m^{-2} h^{-\frac{5}{2}} \tag{3.141}$$

As pointed out by [Longuet-Higgins and Stewart (1962)] this formulation only holds for mildly sloping beds outside of the surf zone (i.e. wave breaking has not been considered in the wave energy balance). In the case of an upward sloping bottom the phase lag between the shoaling wave groups and bound long waves decreases, allowing the transfer of energy from the short waves to the bound long wave ([van Dongeren (1997)], [Janssen et al. (2003)]). For increasing slope the energy transfer is limited ([Longuet-Higgins and Stewart (1964)]) and the equilibrium solution given by eq. (3.134) is no longer valid. [Battjes et al. (2004)] and [van Dongeren et al. (2007)] established that for a relative slope:

$$\beta_H = \frac{\frac{dh}{dx}}{\Delta k h_b} < 0.1 \tag{3.142}$$

the equilibrium solution, eq. (3.134) still holds (with h_b the depth at short wave breaking). For increasing slope additonal terms in the momentum equation should be included leading to lower bound infragravity wave height estimates. Note that bound long waves typically only contribute 30 percent to the total infragravity variance in the nearshore with the rest related to free leaky and edge waves discussed next ([Herbers et al. (1995)]). This can in part be explained by the fact that once the bound long wave is released within the surfzone it adheres to the un-forced wave energy balance, eq. (3.140), with a weaker depth-dependence, $h^{-\frac{1}{2}}$. As a result the free infragravity energy generally dominates the bound infragravity energy at intermediate water depths (e.g. [Herbers et al. (1994)]), except in the cases where the beach slope is mild enough for long wave energy dissipation due to non-linear interactions ([Henderson et al. (2004)], [Thomson et al. (2006)]) and breaking to occur [van Dongeren et al. (2007)].

3.6.3 Leaky waves and trapped waves

Bound long waves travel with the wave groups forcing them from deeper water toward the shoreline where bed-slope effects can no longer be ignored. We relax the problem further by allowing the two incident waves to be obliquely incident. The short wave surface elevation is then given by (in complex notation):

$$\eta_s(x,y,t) = \mathrm{Re}\left\{a_1 \exp i(\omega_1 t - \int k_{1,x} dx - k_{1,y} y) + a_2 \exp i(\omega_2 t - \int k_{2,x} dx - k_{2,y} y)\right\} \quad (3.143)$$

where the wave amplitudes, a_i, are subject to shoaling and wave breaking which can be calculated with the wave energy balance, eq. 2.19. The combined surface elevation can be expressed again as a rapid oscillation of a slowly varying wave amplitude A:

$$\eta_s(x,y,t) = \mathrm{Re}\left\{A(x,y,t) \exp i \left(\frac{\Sigma\omega}{2}t - \frac{\int \Sigma k_x dx}{2} - \frac{\Sigma k_y}{2} y\right)\right\} \quad (3.144)$$

The corresponding wave energy:

$$E(x,y,t) = \mathrm{Re}\left\{\frac{1}{2}\rho g A^2(x,y,t)\right\} = \bar{E} + \mathrm{Re}\left\{\hat{E} \exp i(\Delta\omega t - \Delta k_y y)\right\} \quad (3.145)$$

with the mean wave energy given by:

$$\bar{E} = \frac{1}{2}\rho g \left(a_1^2 + a_2^2\right) \quad (3.146)$$

and the energy modulation by:

$$\hat{E} = \rho g a_1 a_2 \exp i\left(-\int \Delta k_x dx\right) \quad (3.147)$$

can be used to construct the radiation stresses:

$$\begin{aligned} S_{xx} &= \left(n + n\cos^2\theta - 0.5\right) E = \bar{S}_{xx} + \mathrm{Re}\left\{\hat{S}_{xx} \exp i(\Delta\omega t - \Delta k_y y)\right\} \\ S_{xy} &= \left(n\cos\theta\sin\theta\right) E = \bar{S}_{xy} + \mathrm{Re}\left\{\hat{S}_{xy} \exp i(\Delta\omega t - \Delta k_y y)\right\} \\ S_{yy} &= \left(n + n\sin^2\theta - 0.5\right) E = \bar{S}_{yy} + \mathrm{Re}\left\{\hat{S}_{yy} \exp i(\Delta\omega t - \Delta k_y y)\right\} \end{aligned} \quad (3.148)$$

where the wave angle θ corresponds to the mean wave angle made up of the two short wave components. The wave forcing is now fully described by the radiation stress gradients:

$$-F_x = \frac{\partial S_{xx}}{\partial x} + \frac{\partial S_{yx}}{\partial y}$$
$$-F_y = \frac{\partial S_{yy}}{\partial y} + \frac{\partial S_{xy}}{\partial x}$$
(3.149)

The wave forcing thus consists of a mean part, which is responsible for the mean set-up/set-down (see Section 3.5.2) and longshore current (see Section 3.5.1), and a time-varying part associated with the wave group modulation which forces the bound long wave, travelling with the wave groups.

For the case of obliquely incident short waves considered here the bound long wave direction is given by:

$$\theta_b = \arctan\left(\frac{\Delta k_y}{\Delta k_x}\right) \quad (3.150)$$

which is typically different from the mean wave direction of the two short wave components forcing the bound long wave. The alongshore wave number is constant for an alongshore uniform beach and the cross-shore wave number can be evaluated with the linear dispersion relation:

$$\Delta k_x(x) = \int_{x_0}^{x} k_{x,2} - k_{x,1} dx = \int_{x_0}^{x} k_2 \cos\theta_2 - k_1 \cos\theta_1 dx \quad (3.151)$$

where the individual wave incidence angles of the two short waves are obtained with Snel's law, eq. 2.23. Bound long waves get released at breaking where the wave group modulation disappears as the short waves dissipate and (partially) reflect at the shore line travelling towards deeper water as free infragravity waves ([Battjes et al. (2004)], [van Dongeren et al. (2007)]). This means that the direction of the outgoing free long wave is given by:

$$\theta_f = \arctan\left(\frac{\Delta k_y}{k_{f,x}}\right) \quad (3.152)$$

where the cross-shore wave number is obtained from the (long wave) dispersion relation:

$$k_{f,x} = \sqrt{\frac{\omega^2}{gh} - \Delta k_y^2} \quad (3.153)$$

Calculating the incidence angles of the incoming bound and returning free long wave for two short wave incident components with wave frequencies of 0.11 Hz and 0.09 Hz and corresponding incidence angles of 24.5 degrees and 15 degrees at 15 m water depth, shows that the refraction of free infragravity waves is significantly stronger (see left panel of Figure 3.19), in this case leading to trapping of the

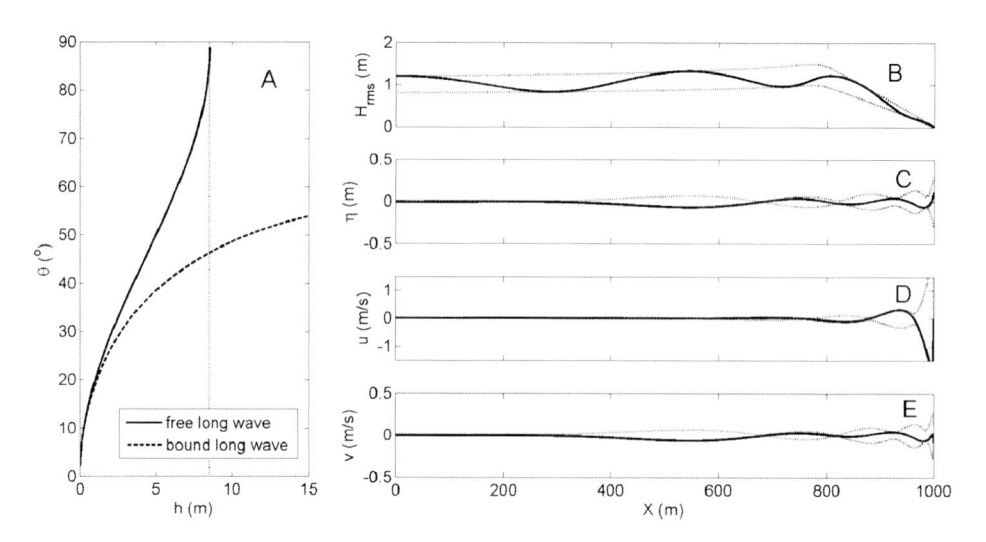

Fig. 3.19 Panel A: Example of refractive trapping of returning free long wave (solid line) after release of the bound long wave (dashed line) within the surf zone calculated for an alongshore uniform profile with a bed slope of 0.015. The turning point indicated by the dotted line. Panel B: Snapshot and envelope of the wave height modulation for two short wave components with $f_1= 0.09$ Hz and $f_2= 0.11$ Hz and $\theta_1 = 15^o$ and $\theta_1 = 24.5^o$ at the offshore boundary. Panel C: Corresponding long wave surface elevation, cross-shore velocity (panel D) and alongshore velocity (panel E).

returning free long wave (Matlab-code to calculate the long wave refraction has been included as **refraclong.m**).

The turning point is given by:

$$\theta_f = \arctan\left(\frac{\Delta k_y}{k_{f,x}}\right) = 90^0 \qquad (3.154)$$

which upon substitution of eq. (3.153) gives:

$$h_{turning} = \frac{\Delta \omega^2}{g \Delta k_y^2} \qquad (3.155)$$

Hence, due to the stronger refraction of the free infragravity waves, not all reflected infragravity waves propagate to the deeper water, but some refract back to the shore where reflection takes place again [Herbers et al. (1995)]. The infragravity waves that make it out to deeper water are called leaky waves, whereas the trapped waves are called edge waves [Ursell (1952)]. In the case the short wave forcing matches the edge wave dispersion relation, resonance can occur which will be discussed next.

3.6.4 Edge wave resonance

To solve for the long wave surface elevation the momentum equations now have to include the alongshore variation in the short-wave forcing and infragravity response:

$$\frac{\partial u}{\partial t}+u\frac{\partial u}{\partial x}+v\frac{\partial u}{\partial y}-\frac{\partial}{\partial x}D_h\frac{\partial u}{\partial x}-\frac{\partial}{\partial y}D_h\frac{\partial u}{\partial y}-fv+g\frac{\partial \eta}{\partial x}=\frac{F_x}{\rho h}-\frac{\tau_{bx}}{\rho h} \quad (3.156)$$

$$\frac{\partial v}{\partial t}+u\frac{\partial v}{\partial x}+v\frac{\partial u}{\partial y}-\frac{\partial}{\partial x}D_h\frac{\partial v}{\partial x}-\frac{\partial}{\partial y}D_h\frac{\partial v}{\partial y}+fu+g\frac{\partial \eta}{\partial y}=\frac{F_y}{\rho h}-\frac{\tau_{by}}{\rho h} \quad (3.157)$$

which can be combined with the continuity equation:

$$\frac{\partial \eta}{\partial t}+\frac{\partial hu}{\partial x}+h\frac{\partial v}{\partial y}=0 \quad (3.158)$$

to yield the linearized long wave equation (e.g. [Mei and Benmoussa (1984)]).

$$\frac{-1}{g}\frac{\partial^2 \eta}{\partial t^2}+h\frac{\partial^2 \eta}{\partial x^2}+\frac{dh}{dx}\frac{\partial \eta}{\partial x}+h\frac{\partial^2 \eta}{\partial y^2}=\frac{1}{\rho g}\left(\frac{\partial F_x}{\partial x}+\frac{\partial F_y}{\partial y}\right) \quad (3.159)$$

and the right hand side is known from eq. (3.149). Next we introduce periodic solutions for the surface elevation:

$$\eta = \mathrm{Re}\left\{\hat{\eta}\exp i(\Delta\omega t-\Delta k_y y)\right\} \quad (3.160)$$

where the complex valued $\hat{\eta}$ represents both the cross-shore varying amplitude and phase of the long waves. Substitution in the long wave equation gives:

$$\frac{d^2\hat{\eta}}{dx^2}+\frac{1}{h}\frac{dh}{dx}\frac{d\hat{\eta}}{dx}+\left(\frac{\Delta\omega^2}{gh}-\Delta k_y^2-\frac{i\Delta\omega\mu}{gh}\right)\hat{\eta}=\frac{1}{\rho gh}\left(\frac{d^2\hat{S}_{xx}}{dx^2}-i\Delta k_y\frac{d\hat{S}_{xy}}{dx}+\Delta k_y^2\hat{S}_{yy}\right) \quad (3.161)$$

to which a linear bottom friction term has been added ([Reniers et al. (2002)]) given by:

$$\mu = C_f \tilde{u}_{rms} \quad (3.162)$$

with C_f the friction coefficient and u_{rms} the average near-bed orbital velocity associated with the short waves.

[Schaffer (1993)] derived the analytical solutions for this equation on a profile with constant slope connected to a horizontal shelf. For an arbitrary bottom profile this forced long wave equation can be solved numerically, subject to a zero flux boundary condition at the shore line and a combination of an incoming bound long wave and outgoing free long wave offshore (see [Reniers et al. (2002)] for details). The corresponding velocities can be obtained from the momentum equations upon substitution of the long wave surface elevation:

$$\hat{u} = \frac{-g\frac{d\hat{\eta}}{dx}-\hat{F}_x}{i\Delta\omega+\mu} \quad (3.163)$$

$$\hat{v} = \frac{ig\Delta k_y \hat{\eta} - \hat{F}_y}{i\Delta\omega + \mu} \tag{3.164}$$

An example of the computed long wave response to two incident short waves with wave frequencies of 0.11 Hz and 0.09 Hz, wave amplitudes of 0.1 m and 0.5 m, and respective incidence angles of 25 degrees and 15 degrees at 15 m water depth is shown in the right panel of Figure 3.19 (Matlab code to calculate the bi-chromatic short wave transformation and corresponding long wave surface elevation and velocities has been included as **infragravity1d.m**). The wave energy modulation shows decreasing cross-shore length scales due to short-wave shoaling. Inside the surfzone the modulation disappears as the short waves break (panel B of Figure 3.19), and consequently the long wave forcing ceases thus releasing the bound long waves. Offshore of the turning point the long wave motion is primarily associated with the (small) progressive bound long wave (panel C). Inside of the turning point a cross-shore standing infragravity is observed, as can be inferred from the anti-nodes in the infragravity surface elevation and velocity envelopes (panel C-E of Figure 3.19), consistent with a trapped long wave.

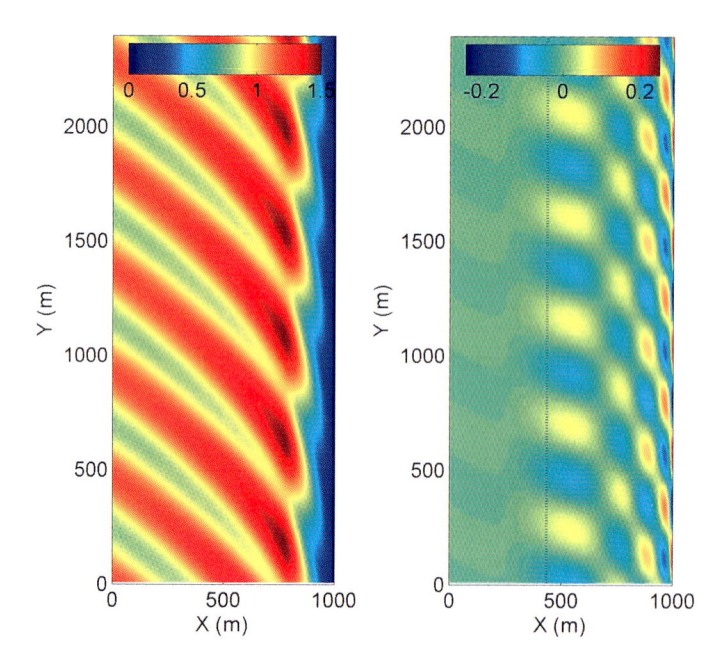

Fig. 3.20 Left panel: Snapshot of the modulated wave height in m denoted by the colour bar. Right panel: corresponding surface elevation in m denoted by the colour bar. Turning point for trapped long wave indicated by the dashed black line.

A snapshot of the wave energy (left panel of Figure 3.20) shows that the energy is modulated in the alongshore with a wave length of approximately 450 m which corresponds to the alongshore difference wave number, $\Delta k_y = 0.038$ (rad/m), made up by the two short wave components. Also, the energy modulation incidence angle is significantly larger than the mean angle of the incident waves (consistent with Figure 3.19), i.e. the bound long wave travels at much larger incidence angle than the two short wave components forcing it. Beyond the turning point there is a significant increase in the long wave surface elevation with an alongshore moving chess-board pattern associated with the cross-shore standing long wave (right panel of Figure 3.20).

Non-forced solutions to eq. (3.161) are present in the form of edge waves, i.e. free long waves trapped to the coast [Eckart (1951)] relation is given by:

$$\omega_e^2 = (2n_e + 1)gk_{y,e} \tan \beta_s \qquad (3.165)$$

where n_e is the edge wave mode number and β_s is bottom slope and the edge wave surface elevation is given by [Eckart (1951)]:

$$\hat{\eta}_e = \exp(-k_{y,e}x)L_n(2k_{y,e}x) \qquad (3.166)$$

where L_n is the Laguerre polynomial. An example of the surface elevation for the edge wave modes 0 through 2 for a fixed $k_{y,e}$ of 0.02 (rad/m) are shown in Figure 3.21. For a fixed alongshore edge-wave number, the cross-shore extent increases with mode number, where the mode number corresponds to the number of zero crossings.

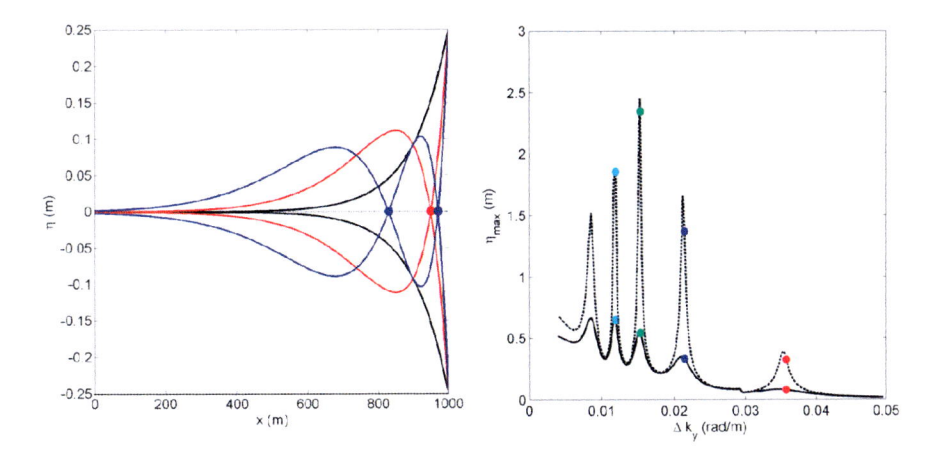

Fig. 3.21 Left panel: Edge wave surface elevation envelope for zero mode (black), first mode (ref) and second mode (blue) calculated with eq. 3.167 for an alongshore edge wave number of 0.01 (rad/m). Right panel: Maximum surface elevation at the water line with a friction factor of 0.001 (dashed lines) and 0.005 (solid line) calculated with eq. 3.161 for increasing difference wave angle between two short wave components. Expected resonant edge wave frequencies indicated by the dots based on the edge wave dispersion relation, eq. 3.165.

In the case the wave group forcing matches the edge wave dispersion relation the infragravity response becomes resonant, leading to large amplitude edge waves. This is examined by using the model to plot the maximum surface elevation at the water line as function of the infragravity alongshore wave number where we keep the incidence angle of the first short wave component at 15 degrees and gradually increase the incidence angle of the second short wave component from 15 degrees to 70 degrees. The peaks in the infragravity response coincide with the theoretical edge wave frequencies as derived by [Eckart (1951)]. This holds for the case with little damping, $C_f = 0.001$ and increased damping, $C_f = 0.005$. Note that the response is very narrow. This means that the wave-group forcing has to be exact to obtain a resonant response. The latter is unlikely to occur in the field as wave conditions change continuously resulting in a smoother infragravity response (e.g. [Huntley et al. (1984)], [Oltman-Shay et al. (1989)]).

The bi-chromatic approach can be used to predict the full infragravity response to directionally spread short waves by considering all possible interactions between the individual frequency-directional components within the full spectrum and has shown considerable skill in predicting the infragravity wave climate ([Reniers et al. (2002)], [Reniers et al. (2010b)], [MacMahan et al. (2010b)]).

3.6.5 *Very Low Frequency motions*

In the following we will use the same model to examine the presence of Very Low Frequency motions. Considering again a bi-chromatic wave train made up of two incident waves with frequencies of 0.099 Hz and 0.101 Hz, wave amplitudes of 0.1 m and 0.5 m, and respective incidence angles of 12.5 degrees and -12.25 degrees. This yields a slowly alongshore propagating circulation pattern consisting of transient rips (see Figure 3.22). The alongshore propagation speed is determined by:

$$c_{y,vlf} = \frac{\Delta \omega}{\Delta k_y} \qquad (3.167)$$

which is approximately -0.4 m/s for the present case. Note that the VLF motion is only generated at the location of wave breaking, i.e. within the surf zone.

The wave-group induced VLF velocity response within the surf zone has been verified in the laboratory [Fowler and Dalrymple (1990)] and in the field [MacMahan et al. (2010b)]. In the field, the presence of a frequency-directional incident wave field allows for a wide range of interactions between the different short wave components leading to a spectrum of VLFs, although there are clear limitations on both the maximum VLF-frequency and alongshore wave numbers (see [MacMahan et al. (2010b)] for details).

VLFs reside in the vortical region of the $f - k_y$-spectrum which is defined by:

$$k_y < \frac{\omega}{\sqrt{gh}} \qquad (3.168)$$

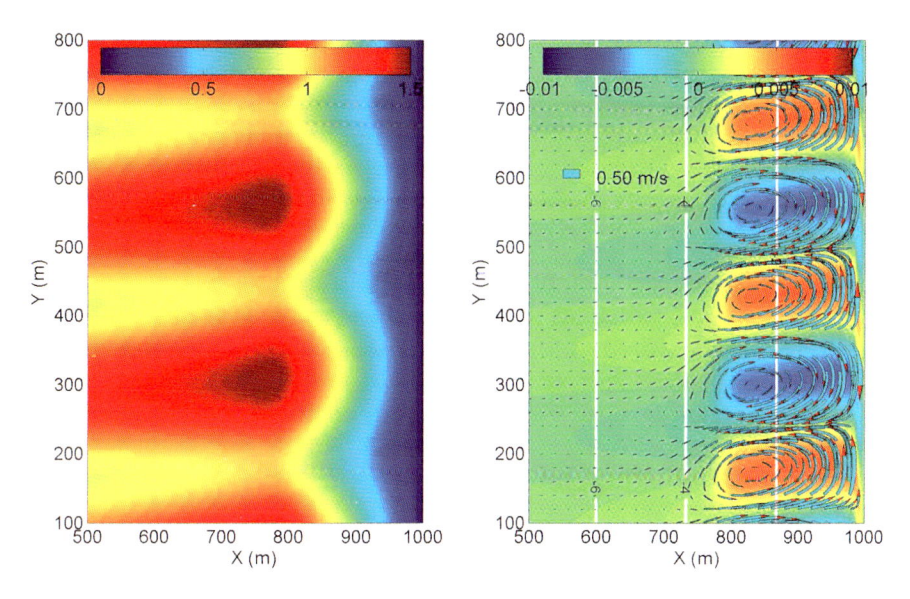

Fig. 3.22 Left panel: Snapshot of the modulated wave height in m denoted by the colour bar. Right panel: Corresponding vorticity field in s-1 indicated by the color bar with the super-imposed velocity transport vectors integrated over 300 s showing horizontal circulations corresponding to transient rip currents propagating along the beach with a velocity of -0.4 m/s. Velocity magnitude corresponds to the width of the arrows (see velocity scale on the left).

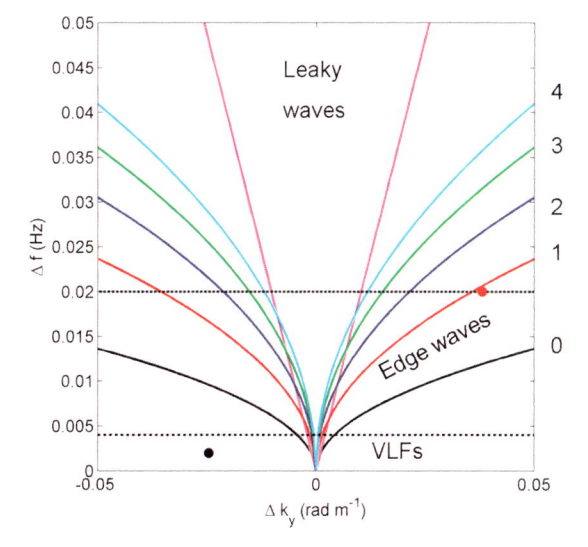

Fig. 3.23 f-k_y distribution with leaky wave regime enclosed by magenta lines, discreet edge wave dispersion curves (with corresponding mode numbers to the right) for a alongshore uniform bathymetry with 0.015 slope and 15 m offshore depth. VLF-regime is below the 0.004 Hz frequency cut-off (dash-dotted line). Alongshore edge wave resonance numbers for $\Delta f = 0.02$ Hz at occur intersection of dashed line and edge wave curves (see right panel of Figure 3.21 for infragravity model response). Δf-Δk_y positions for trapped wave calculation (red dot, see Figure 3.20) and VLF calculation (black dot, see Figure 3.22)

Edge waves also reside in this region as discreet curved dispersion ridges (see Figure 3.23) encompassing the area between the zero-mode egde wave and the boundary between the leaky waves and vortical motions. Given the fact that the VLF-velocity response is strongly dependent on the difference frequency (residing in the denominator in eq. 3.163), with a strong (weak) response for small (larger) frequency differences, their energy is restricted to frequencies less than 0.004 Hz ([MacMahan et al. (2010b)]).

To calculate the generation of VLFs over arbitrary bathymetry and including potential non-linear interactions between the wave group generated VLFs a more comprehensive model based on the non-linear shallow water equations is required as discussed in Section 10.5.

3.7 Vertical structure of the current

In this chapter we will briefly discuss the vertical structure of tide, wind and wave-driven currents in the coastal area. We will, for simplicity, assume that the vertical structure does not depend much on inertial terms and therefore we consider only (quasi-)stationary situations. Also, we assume that horizontal advection and diffusion terms can be neglected as we consider undisturbed flows in large areas, rather than flows in the vicinity of coastal structures or other disturbances.

3.7.1 Tide (or slope) driven current profile

In the case of tide-driven currents or currents just driven by a water level slope, the relevant terms in the horizontal momentum balance (taken in longshore or y-direction) are shown below:

$$\frac{\partial v}{\partial t} + u\frac{\partial v}{\partial x} + v\frac{\partial v}{\partial y} + w\frac{\partial v}{\partial z} + f_{cor}u = \frac{\partial}{\partial x}\left(\nu_h \frac{\partial v}{\partial x}\right) + \frac{\partial}{\partial z}\left(\nu_v \frac{\partial v}{\partial z}\right) - \frac{1}{\rho}\frac{\partial p}{\partial y} + \frac{w_y}{\rho} \quad (3.169)$$

Using the hydrostatic pressure assumption, we get:

$$\frac{\partial}{\partial z}\left(\nu_v \frac{\partial v}{\partial z}\right) = g\frac{\partial \eta}{\partial y} \quad (3.170)$$

Since $g\frac{\partial \eta}{\partial y}$ is not a function of z, we can write:

$$\nu_v \frac{\partial v}{\partial z} = -g\frac{\partial \eta}{\partial y}z \quad (3.171)$$

The left-hand side is equal to the shear stress divided by the density, which means that the shear stress has a linear distribution from zero at the surface (since there is no forcing there) to $\rho g h \frac{\partial \eta}{\partial y}$ at the bottom. A typical vertical distribution of the eddy viscosity is a parabolic profile:

$$\nu_v = -\kappa v_* z \frac{(h+z)}{h} \qquad (3.172)$$

where κ is von Karman's constant and the shear velocity is:

$$v_* = \sqrt{\frac{\tau}{\rho}} = \sqrt{g h \frac{\partial \eta}{\partial y}} \Rightarrow \frac{\partial \eta}{\partial y} = \frac{v_*^2}{gh} \qquad (3.173)$$

Substituting (3.172) and (3.173) into (3.171) we get:

$$\frac{\partial v}{\partial z} = g \frac{\partial \eta}{\partial y} \frac{h}{\kappa v_*(h+z)} = \frac{v_*}{\kappa} \frac{1}{(h+z)} \qquad (3.174)$$

Subsequent integration of (3.174) to z produces:

$$v = \frac{v_*}{\kappa} \ln(h+z) + C \qquad (3.175)$$

with the usual assumption that $v = 0$ at $z=-h+z_0$, we get the logarithmic velocity distribution:

$$v = \frac{v_*}{\kappa} \ln(h+z) - \frac{v_*}{\kappa} \ln(h+z_0) = \frac{v_*}{\kappa} \ln \frac{h+z}{z_0} \qquad (3.176)$$

The logarithmic velocity profile shown in Figure 3.24 is calculated with eq. (3.176) for a water depth of 5 m, a bed roughness, k_s, of 0.01 m and a depth averaged velocity of 0.7 m/s (the procedure to compute the vertical profile is given by the example Matlab-code **vertprotide.m**). The latter is used to calculate the shear velocity and is related to the alongshore slope (eq. (3.173)).

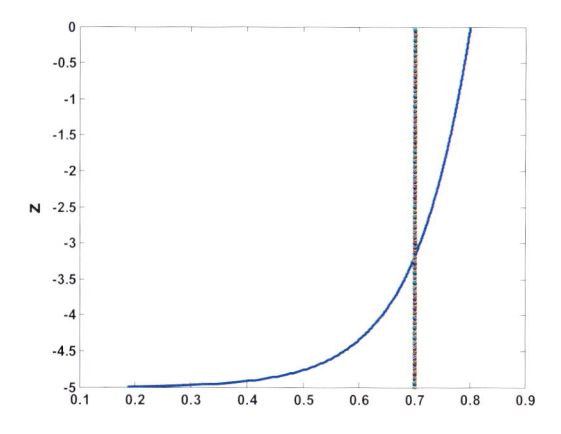

Fig. 3.24 Calculated vertical velocity profile for a depth-averaged velocity of 0.7 m/s.

3.7.2 Wind-driven current profile

In the absence of other forcing mechanisms and strong horizontal gradients the equations governing the vertical profile of the wind-driven current are the following:

$$\frac{\partial u}{\partial t} + u\frac{\partial u}{\partial x} + v\frac{\partial u}{\partial y} + w\frac{\partial u}{\partial z} - f_{cor}v = \frac{\partial}{\partial y}\left(\nu_h \frac{\partial u}{\partial y}\right) + \frac{\partial}{\partial z}\left(\nu_v \frac{\partial u}{\partial z}\right) - g\frac{\partial \eta}{\partial x}$$

$$\frac{\partial v}{\partial t} + u\frac{\partial v}{\partial x} + v\frac{\partial v}{\partial y} + w\frac{\partial v}{\partial z} + f_{cor}u = \frac{\partial}{\partial x}\left(\nu_h \frac{\partial v}{\partial x}\right) + \frac{\partial}{\partial z}\left(\nu_v \frac{\partial v}{\partial z}\right) - g\frac{\partial \eta}{\partial y}$$

(3.177)

On the open ocean, with the additional assumptions that the flow is stationary, the water surface is horizontal and the vertical viscosity is constant, this leads to the well-known Eckman spiral. This is easily derived using complex notation, where

$$s = u + i\,v \tag{3.178}$$

In other words, we specify the velocity s as a complex variable where u is the real part and v is the imaginary part. The advantage is that we can combine the two equations above into one that can be readily solved:

$$i\,f_{cor}\,s = \nu_t \frac{\partial^2 s}{\partial z^2} \tag{3.179}$$

The solution has the form:

$$s = s_0 \exp\left((a+bi)\,z\right) \tag{3.180}$$

which upon substitution this into eq. (3.179) gives:

$$a = b = 1/\delta, \quad \delta = \sqrt{\frac{2\nu_t}{f_{cor}}} \tag{3.181}$$

with the boundary condition:

$$\rho\nu_t \frac{\partial s}{\partial z} = \tau \tag{3.182}$$

Note that the shear stress τ is also a complex number, with the x-component being the real part and the y-component the imaginary part. We find the solution for the Eckman spiral:

$$s = \frac{\tau\delta}{\rho\nu_t(1+i)}\exp\left((1+i)\frac{z}{\delta}\right) \tag{3.183}$$

An example of the velocity distribution calculated with eq. (3.183) for an along-shore windstress of 1 Pa at a water depth of 120 m is shown in Figure 3.25 (the Matlab-code to solve for the velocity profile and generate the figure has been included as **vertprowind.m**).

The red arrow in Figure 3.25 indicates the wind shear stress direction, the purple arrows the velocity profile. The Coriolis force makes the velocity veer to the right in the northern hemisphere; by 45 degrees under the assumption of deep water, a constant viscosity and in the absence of closed boundaries. These assumptions are

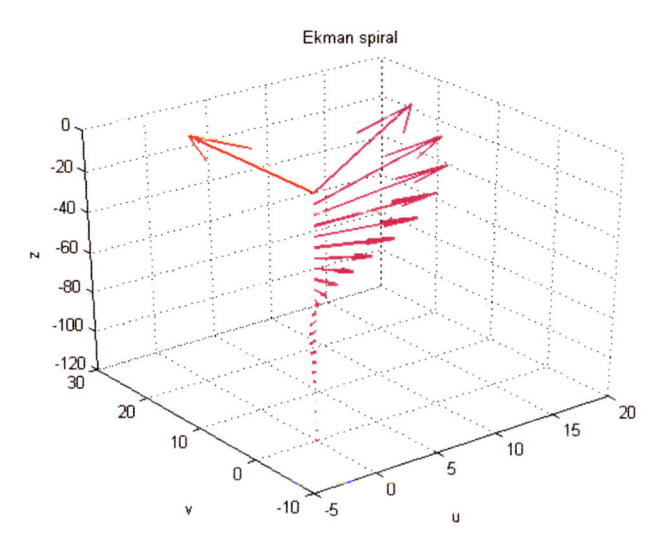

Fig. 3.25 Calculated vertical velocity distribution for an alongshore wind stress of 1 Pa at a water depth of 120 m with a constant turbulent eddy viscosity of 0.02 m2/s at a latitude of 51 degrees North.

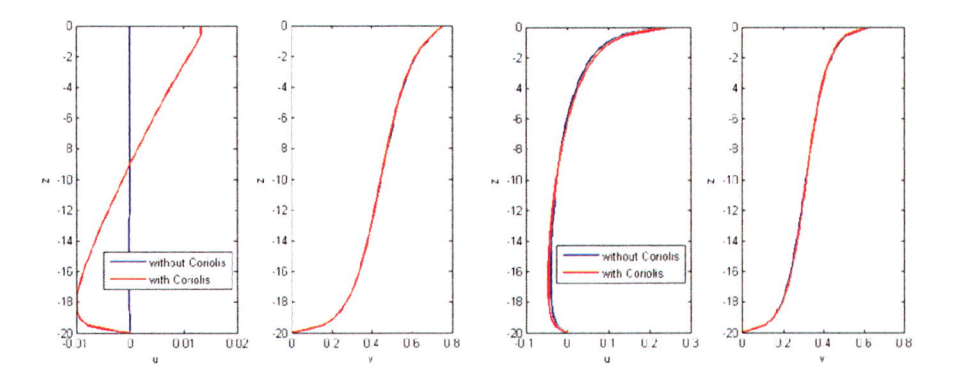

Fig. 3.26 Calculated cross-shore and alongshore velocity profiles with and without coriolis for a windspeed of 20 m/s in 20 m water depth. Left panels: Wind direction parallel to the coast. Right panels: Wind direction 45 degrees with respect to the shore normal in the direction of the coast.

rarely met, and even on the open ocean the angle is typically not more than 20 degrees (e.g. [Madsen (1978)]).

For coastal areas the depth is generally much less, typically we're interested in depths less than 20 m, and the presence of a closed coastal boundary makes a profound difference. The water surface cannot be assumed to be horizontal but will have a distinct slope in cross-shore (x-) direction. An analytical solution is not possible if we want to retain the Coriolis terms, but without too much trouble we can evaluate the importance of the Coriolis terms on the wind-driven flow profiles in shallow coastal areas. For a closed coast aligned with the y-axis we can still assume the longshore slope to be zero, but the cross-shore slope must be such that the depth-averaged cross-shore flow velocity becomes zero. We can achieve this by considering that there is a distance L between the vertical profile we consider and the coast, and that the depth and slope between the profile and the coast are constant. The equations to solve are:

$$
\begin{aligned}
\frac{\partial u}{\partial t} &= f_{cor} v + \frac{\partial}{\partial z}\left(\nu_v \frac{\partial u}{\partial z}\right) - g \frac{\partial \eta}{\partial x} \\
\frac{\partial v}{\partial t} &= -f_{cor} u + \frac{\partial}{\partial z}\left(\nu_v \frac{\partial v}{\partial z}\right) \\
\frac{\partial \eta}{\partial t} &= \frac{2h}{L} \int_{-h}^{0} u \, dz \\
\frac{\partial \eta}{\partial x} &= \frac{\eta}{L}
\end{aligned}
\qquad (3.184)
$$

We apply a parabolic viscosity profile similar to that in the previous section. Though not very efficiently, these equations when solved numerically converge to a stationary situation that allows us to evaluate the effect of the Coriolis terms in coastal areas. As an example we consider a wind shear stress of 1 Pa, roughly equivalent with a windspeed of 20 m/s, a depth of 20 m and an angle w.r.t. the shore normal of 90 and 45 degrees, respectively (see Figure 3.26).

We see that only in the case of purely alongshore wind, the cross-shore velocity with Coriolis effect is significantly different from that without it, but very small, in the order of 1 cm/s. The effect on the longshore velocity is negligible in both cases. This opens the way to relatively simple analytical solutions for the wind-driven current profile in shallow coastal areas, as we'll see in the next sections.

3.7.3 *Wind driven longshore current profile*

In the case of stationary, purely wind-driven longshore flow there is no alongshore pressure gradient, so all terms except the vertical shear stress gradient turn to zero:

$$
\frac{\partial v}{\partial t} + u\frac{\partial v}{\partial x} + v\frac{\partial v}{\partial y} + w\frac{\partial v}{\partial z} + f_{cor} u = \frac{\partial}{\partial x}\left(\nu_h \frac{\partial v}{\partial x}\right) + \frac{\partial}{\partial z}\left(\nu_v \frac{\partial v}{\partial z}\right) - \frac{1}{\rho}\frac{\partial p}{\partial y} + \frac{w_y}{\rho}
\qquad (3.185)
$$

In other words:

$$
\frac{\partial}{\partial z} \nu_v \frac{\partial v}{\partial z} = 0 \;\Rightarrow\; \nu_v \frac{\partial v}{\partial z} = \frac{\tau_{wy}}{\rho} \;\Rightarrow\; \frac{\partial v}{\partial z} = \frac{\tau_{wy}}{\rho \nu_v}
\qquad (3.186)
$$

The vertical shear stress distribution is uniform from top to bottom. Although the vertical viscosity distribution is likely to be different from that for a purely slope-driven flow, simulations using more sophisticated models show that it is still largely a parabolic distribution, mainly because the length scale reduces to zero both at the water surface and at the bottom. For convenience we will therefore keep the same distribution of the vertical viscosity as (3.172).

The result is then:

$$\frac{\partial v}{\partial z} = -\frac{\tau_{wy}}{\rho v_* \kappa z \frac{(h+z)}{h}} = -\frac{\tau_{wy}}{\rho v_* \kappa} \frac{(h+z-z)}{z(h+z)}$$

$$= -\frac{\tau_{wy}}{\rho v_* \kappa}\left(\frac{1}{z} - \frac{1}{h+z}\right)$$

(3.187)

This can be integrated to:

$$v = \frac{\tau_{wy}}{\rho v_* \kappa} \ln\left(\frac{h+z}{z}\right) + C$$

(3.188)

Again, with the assumption that $v = 0$ at $z=-h+z_0$, we get another logarithmic velocity distribution:

$$v = \frac{\tau_{wy}}{\rho v_* \kappa} \ln\left(\frac{h+z}{z}\right) - \frac{v_*}{\kappa} \ln\left(\frac{z_0}{z_0-h}\right)$$

$$= \frac{\tau_{wy}}{\rho v_* \kappa} \ln\left(\frac{h+z}{z} \frac{z_0-h}{z_0}\right)$$

(3.189)

3.7.4 Wind-driven cross-shore current profile

In the cross-shore direction, there is a balance between the cross-shore component of the wind stress, the bed shear stress and the cross-shore slope term. Apart from these terms there is a contribution due to Coriolis, which is important in continental shelf currents as it induces large-scale upwelling and down welling, but is less important in shallow coastal areas. If we stick to the first three forces, we face the problem that the bed shear stress is now unknown, since it is a function of the – as yet unknown – velocity near the bed. We need an additional condition to solve this problem, which is that the depth-averaged cross-shore velocity is zero. Another complication is, that since the bed shear stress is not yet known, we do not know the vertical viscosity yet. We will assume here, that the viscosity profile has the same parabolic shape and that a representative shear velocity is governed by the wind shear stress.

We can now derive the cross-shore wind-driven current profile as follows. We start with the vertical shear stress distribution:

$$\tau_{xz} = \tau_{wx} + \frac{\tau_{wx} - \tau_{bx}}{hz}$$

(3.190)

This states simply that the shear stress varies linearly between the bed shear stress (still unknown) and the cross-shore wind shear stress component. Since:

$$\nu_v \frac{\partial u}{\partial z} = \frac{\tau_{xz}}{\rho}$$

(3.191)

we can write, with the parabolic viscosity distribution in (3.172):

$$\begin{aligned}
\frac{\partial u}{\partial z} &= \frac{\tau_{xz}}{\rho \nu_v} = -\frac{\tau_{xz}}{\rho \kappa v_* z \frac{(h+z)}{h}} = -\frac{\tau_{xz}}{\rho \kappa v_*} \frac{h}{z(h+z)} \\
&= -\frac{\tau_{xz}}{\rho \kappa v_*} \frac{(h+z-z)}{z(h+z)} = -\frac{\tau_{xz}}{\rho \kappa v_*}\left(\frac{1}{z} - \frac{1}{h+z}\right) \\
&= -\frac{1}{\rho \kappa v_*}\left(\frac{\tau_{wx}}{z} - \frac{\tau_{wx}}{h+z} + \frac{\tau_{wx}-\tau_{bx}}{h+z}\right) = -\frac{1}{\rho \kappa v_*}\left(\frac{\tau_{wx}}{z} - \frac{\tau_{bx}}{h+z}\right)
\end{aligned} \qquad (3.192)$$

This expression contains two terms that have z in the denominator, and after integration the result is a combination of logarithmic functions:

$$u = -\frac{1}{\rho \kappa v_*}\left(\tau_{wx} \ln z - \tau_{bx} \ln(h+z) + C\right) \qquad (3.193)$$

Using the bed boundary condition $u=0$ at $z=-h+z_0$ we find:

$$u = -\frac{1}{\rho \kappa v_*}\left(\tau_{wx} \ln \frac{z}{z_0-h} - \tau_{bx} \ln \frac{h+z}{z_0}\right) \qquad (3.194)$$

In this expression, the bed shear stress is still unknown. We can solve this from the condition that the depth-integrated velocity in cross-shore direction must be zero. After integrating (3.194) we get:

$$\begin{aligned}
\frac{1}{h}\int_{-h+z_0}^{0} u\, dz &= -\frac{1}{\rho \kappa v_*}\left(\tau_{wx}\ln\frac{z}{z_0-h} - \tau_{bx}\ln\frac{h+z}{z_0}\right) = \\
&-\frac{1}{\rho \kappa v_* h}\left[\tau_{wx}\left(\frac{\frac{z}{z_0-h}\ln\frac{z}{z_0-h} - \frac{z}{z_0-h}}{\frac{1}{z_0-h}}\right) - \tau_{bx}\left(\frac{\frac{h+z}{z_0}\ln\frac{h+z}{z_0} - \frac{z}{z_0}}{\frac{1}{z_0}}\right)\right]_{-h+z_0}^{0} = \\
&-\frac{1}{\rho \kappa v_* h}\left[\tau_{wx}\left(z\ln\frac{z}{z_0-h} - z\right) - \tau_{bx}\left((h+z)\ln\frac{h+z}{z_0} - z\right)\right]_{-h+z_0}^{0} = \\
&-\frac{1}{\rho \kappa v_* h}\left(-\tau_{bx}\left((h)\ln\frac{h}{z_0}\right) - \left(\tau_{wx}(-(z_0-h)) - \tau_{bx}(-(z_0-h))\right)\right) = \\
&-\frac{1}{\rho \kappa v_*}\left(\tau_{bx}\left(1 - \ln\frac{h}{z_0} - \frac{z_0}{h}\right) - \tau_{wx}\left(1 - \frac{z_0}{h}\right)\right) \approx -\frac{1}{\rho \kappa v_*}\left(\tau_{bx}\left(1 - \ln\frac{h}{z_0}\right) - \tau_{wx}\right)
\end{aligned} \qquad (3.195)$$

and setting the result equal to zero we find:

$$\tau_{bx} \simeq \frac{\tau_{wx}}{1 - \ln\frac{h}{z_0}} \qquad (3.196)$$

This very simple expression allows us to solve the vertical distribution of the cross-shore wind-driven current given by expression (3.192).

The value of v_* represents the shear velocity induced by both longshore and cross-shore currents. A reasonable approximation is that it is determined by the total wind shear stress:

$$v_* = \sqrt{\frac{|\tau_w|}{\rho}} = \sqrt{\frac{\sqrt{\tau_{wx}^2 + \tau_{wy}^2}}{\rho}} \qquad (3.197)$$

Under this approximation, we can directly solve the longshore velocity profile (from (3.189) and the cross-shore velocity profile (from (3.194)) for an arbitrary wind stress vector, using (3.196) to obtain the cross-shore component of the bed shear stress.

This computation procedure is illustrated in Figure 3.27 for a case with 45 degree wind shear stress with respect to the shore normal at a water depth of 5 m and a bed roughness k_s, of 0.05 m (Matlab-code illustrating the computational procedure is included as **vertprowindoblique.m**). Apart from the computed vertical, we also plot the mean velocity as it would follow from a depth-averaged approach. The main difference is in the upper part of the profile, where the shear stress is much higher in the case of wind-driven currents than for a slope-driven current with the same bed shear stress. In terms of depth-averaged current, the difference is rather small for this case.

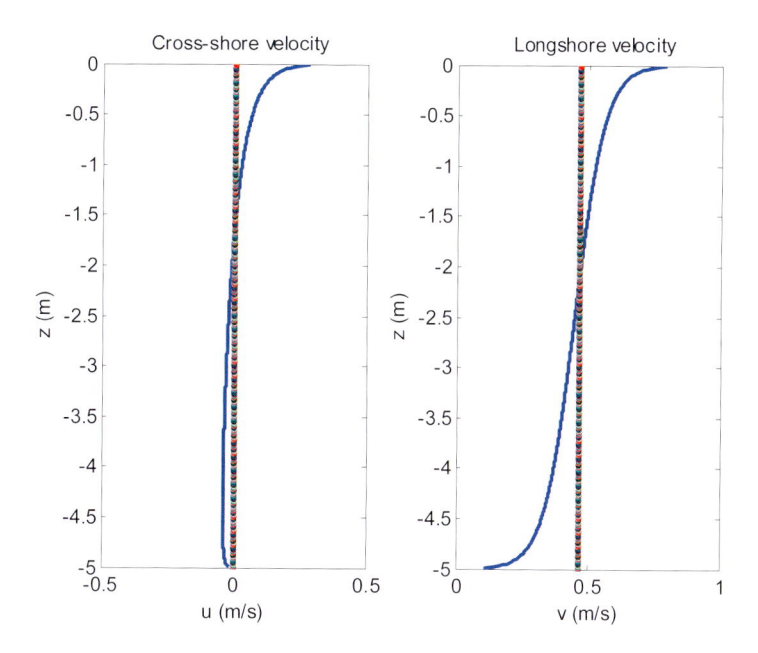

Fig. 3.27 Computed wind-driven velocity profiles for onshore wind at a 45 degree angle with the shore normal

3.7.5 Wave driven current profile

3.7.5.1 Traditional Eulerian approach

To a very reasonable approximation, the same procedure as for the wind-driven current can be followed to compute the wave-driven current profile. The forcing consists of a contribution due to the roller, which acts just like a surface stress, and a depth-invariant contribution which merely modifies the wave setdown/setup. The longshore forcing can be computed from the roller dissipation as:

$$\tau_{sy} = \frac{D_r}{C} \sin(\vartheta) \qquad (3.198)$$

The longshore wave-driven current profile is then given by eq. (3.189). To a good approximation, the cross-shore forcing is given by:

$$\tau_{sx} = \frac{D_r}{C} \cos(\vartheta) \qquad (3.199)$$

An additional point of consideration is the wave-induced mass flux, which now has to be taken into account in the cross-shore mass balance. The depth-averaged cross-shore current velocity is now balanced by the onshore mass flux. This leads to a modification of eq. (3.196):

$$-\frac{1}{\rho \kappa v_*} \left(\tau_{bx} \left(1 - \ln \frac{h}{z_0} \right) - \tau_{wx} \right) + u_{stokes} = 0 \Rightarrow$$
$$\Rightarrow \tau_{bx} = \frac{\tau_{wx} + \rho \kappa v_* u_{stokes}}{\left(1 - \ln \frac{h}{z_0}\right)} \qquad (3.200)$$

We can use this result in eq. (3.194) to obtain the cross-shore wave-driven current profile. This simple approach can be expected to produce reasonable results within the surf zone.

So far we have neglected effects of boundary layer streaming [Longuet-Higgins (1953)], which can lead to slightly shoreward velocities near the bottom outside the breaker zone. In [Reniers et al. (2004b)], a vertical profile model is presented and compared against detailed field data at Duck, NC. In this model, boundary layer streaming is accounted for by including a near-bottom layer with modified shear stress distribution. Also, the turbulent viscosity profiles are modified in order to increase the viscosity near the surface in case of wind- or wave-driven currents. Their model, which can still be solved analytically, has shown considerable skill in representing the observed velocity profiles in longshore and cross-shore directions.

3.7.5.2 *Generalized Lagrangian Mean approach*

A disadvantage of the 'Eulerian' approach presented above is that the vertical distribution of the Stokes drift cannot be taken into account; only the depth-integrated Stokes drift, the total mass flux, is taken into account. In [Walstra et al. (2000)] a rather simple approach based on Generalised Lagrangean Mean (GLM) theory is implemented in a 3D flow model, and is shown to give good results both inside and outside the surf zone. In this approach we consider the mass and momentum balance in terms of the GLM (lagrangian mean) velocity:

$$u_L = u_E + u_{stokes} \qquad (3.201)$$

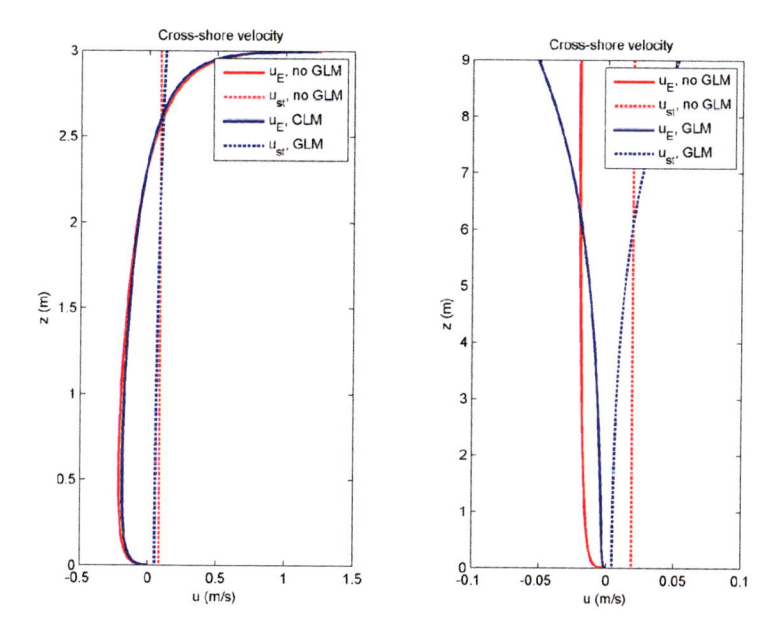

Fig. 3.28 Vertical distribution of wave-driven velocity for normally incident waves of 1 m root mean square wave height at 3 m and 9 m water depth respectively.

In case of a closed coast boundary, the depth-averaged GLM velocity is zero. The Stokes velocity is now allowed to have a vertical distribution according to [Phillips (1977)]:

$$u_{\text{stokes}} = \frac{1}{2}\left(\frac{\pi H_{rms}}{L}\right)^2 C \frac{\cosh(2k(z+h))}{(\sinh(kh))^2} \cos(\theta) \qquad (3.202)$$

In the longshore uniform case the horizontal momentum balance reduces to:

$$\nu_v \frac{\partial u_L}{\partial z} = \frac{\tau_{xz}}{\rho} \qquad (3.203)$$

At the bottom, the Eulerian velocity must equal zero, so the GLM velocity must be equal to the Stokes drift at the bottom. From this condition the vertical profile follows:

$$u_L = -\frac{1}{\rho \kappa v_*}\left(\tau_{wx} \ln \frac{z}{z_0 - h} - \tau_{bx} \ln \frac{h+z}{z_0}\right) + u_{\text{stokes},z=-h} \qquad (3.204)$$

We can now work out the unknown bottom shear stress by imposing that the depth-averaged GLM velocity is zero:

$$-\frac{1}{\rho \kappa v_*}\left(\tau_{bx}\left(1 - \ln \frac{h}{z_0}\right) - \tau_{wx}\right) + u_{\text{stokes},z=-h} = 0 \Rightarrow \\ \tau_{bx} = \frac{\tau_{wx} + \rho \kappa v_*(u_{\text{stokes},z=-h})}{\left(1 - \ln \frac{h}{z_0}\right)} \qquad (3.205)$$

Now we can substitute the bed shear stress into eq. (3.204) to obtain the vertical profile of the GLM velocity. The Eulerian velocity is obtained by applying eq.

(3.201), in other words subtracting the Stokes drift profile from the GLM velocity profile. This yields quite a different profile from the one obtained by the Eulerian approach, especially for lower wave height over depth ratios.

The cross-shore and longshore velocity profiles computed according to both the Eulerain and GLM approaches for a given wave height $H_{rms}=1$ m and water depths of 3 and 9 m respectively are shown in Figure 3.28 (Matlab-code to calculate the velocities and generate Figure 3.28 is included as **returnflow.m** for further clarification). In the shallow water case, both approaches lead to very similar results; in deeper water however, there is a distinct difference and the GLM approach leads to the 'inverted Stokes' profiles regularly found outside the surf zone ([Monismith et al. (2007)], [Lentz et al. (2008)]).

3.8 3D Wave-driven currents on a non-uniform coast

An example of three dimensional flow over variable bathymetry is presented in the following. These calculations have been performed with Delft3D ([Lesser et al. (2004)], [Walstra et al. (2000)]) in which the wave energy equation for narrow banded waves (eq. (2.19)):

$$\frac{\partial E_w}{\partial t} + \frac{\partial}{\partial x}\left(E_w c_g \cos(\theta_m)\right) + \frac{\partial}{\partial y}\left(E_w c_g \sin(\theta_m)\right) = -D_w - D_f$$

and the roller energy equation (eq. (2.31)):

$$\frac{dE_r}{dt} = \frac{\partial E_r}{\partial t} + \frac{\partial E_r c \cos\theta_m}{\partial x} + \frac{\partial E_r c \sin\theta_m}{\partial y} = D_w - D_r$$

are used to calculate the wave-breaking related surface stresses (eq. (3.26))

$$\tau_{sx} = \frac{D_r}{c}\cos\theta_m$$

$$\tau_{sy} = \frac{D_r}{c}\sin\theta_m$$

and the corresponding body force, (eq. (3.27)):

$$-F_{w,x} = \left(\frac{\partial S_{xx}}{\partial x} + \frac{\partial S_{yx}}{\partial y}\right) + \tau_{s,x}$$

$$-F_{w,y} = \left(\frac{\partial S_{yy}}{\partial y} + \frac{\partial S_{xy}}{\partial x}\right) + \tau_{s,y}$$

where the radiation stresses are obtained from eq. (3.24) and (3.25). Next the short-wave averaged non-linear 3D shallow water equations (eq. (3.17)):

$$\frac{\partial u}{\partial t} + u\frac{\partial u}{\partial x} + v\frac{\partial u}{\partial y} + w\frac{\partial u}{\partial z} - f_{cor}v = \frac{\partial}{\partial x}\left(\nu_h \frac{\partial u}{\partial x}\right) + \frac{\partial}{\partial y}\left(\nu_h \frac{\partial u}{\partial y}\right) + \frac{\partial}{\partial z}\left(\nu_v \frac{\partial u}{\partial z}\right) - \frac{1}{\rho}\frac{\partial p}{\partial x} + \frac{F_{w,x}}{\rho h}$$

$$\frac{\partial v}{\partial t} + u\frac{\partial v}{\partial x} + v\frac{\partial v}{\partial y} + w\frac{\partial v}{\partial z} + f_{cor}u = \frac{\partial}{\partial x}\left(\nu_h \frac{\partial v}{\partial x}\right) + \frac{\partial}{\partial y}\left(\nu_h \frac{\partial v}{\partial y}\right) + \frac{\partial}{\partial z}\left(\nu_v \frac{\partial v}{\partial z}\right) - \frac{1}{\rho}\frac{\partial p}{\partial y} + \frac{F_{w,y}}{\rho h}$$

$$\frac{\partial Uh}{\partial x} + \frac{\partial Vh}{\partial x} + \frac{\partial \eta}{\partial t} = 0$$

$$p = p_a + \int_z^\eta \rho g dz$$

$$\frac{\partial u}{\partial x} + \frac{\partial v}{\partial y} + \frac{\partial w}{\partial z} = 0$$

with $\vec{u}$ including the Stokes drift, are solved for with the bottom friction defined by the parameterization by [Soulsby et al. (1993)] of [Fredsoe (1984)], and the turbulent eddy viscosity calculated from a k-ε model with inclusion of wave-breaking generated turbulence [Walstra et al. (2000)].

The model is run in wave-group resolving mode were the time-varying incident wave energy was obtained with the single summation random phase model approach outlined in Section 2.7. Here we present time averaged results, i.e. averaged over many wave groups, of the three dimensional flow field. The vertical grid spacing consists of 10 percent increments of the total water depth (including the tidal elevation, mean setdown/setup, and infragravity elevations). The model domain is approximately 1 km alongshore and extends 700 m offshore, hence the boundaries are well away from the centre area which is presented here. Cross-shore grid spacing varies between 15 m offshore to approximately 5 m within the surf zone and a fixed 10 m grid spacing in the alongshore direction. The spacing is chosen to:

- adequately represent the bathymetric features
- resolve the wave group related wave and fluid motions.

Model coefficients for bottom friction, n (Manning coefficient), wave breaking, γ, and roller dissipation, β, are set at 0.02, 0.45, and 0.1. More details on the modeling can be found in [Reniers et al. (2009)].

The bathymetry at this site is characterized by year-round present shore-connected shoals cut by rip-channels [MacMahan et al. (2005)], i.e. a transverse-bar system [Wright and Short (1984)]. The rip channel spacing is approximately 125 m (panel A of Figure 3.29). Wave breaking occurs predominantly over the shallower shoals resulting in lower wave heights over the shoals and higher waves within the rip channel. Wave shoaling around the 3 m depth contour is apparent with higher waves opposite the rip channel locations. Comparisons of predicted wave heights with observations with in-situ instruments showed a skill of 0.85, explaining 85 percent of the variance at this location, for the constant γ value. The body force associated with the wave transformation shows a persistent offshore forcing outside of the surf zone due to wave shoaling. Inside the surf zone the body force is small and mostly related to alongshore variations in the wave height pushing the water toward the centre of the shoals (panel A of Figure 3.29).

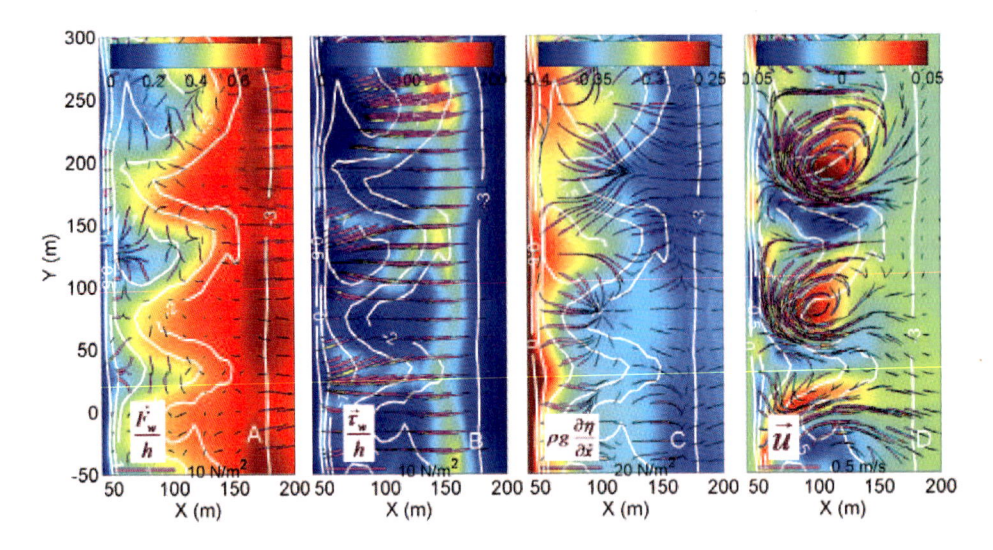

Fig. 3.29 Panel A: Root mean square wave height (indicated by the color bar in m) and corresponding body force given by the arrows. Panel B: Corresponding roller energy (indicated by the color bar in J/m2) and wave breaking related surface shear stress indicated by the arrows. Panel C: Mean water elevation (indicated by the color bar in m) and related pressure gradients indicated by the arrows). Panel D: Vorticity (indicated by the color bar in s-1) and surface (magenta arrows) and near-bed (cyan arrows) Eulerian velocities resulting from the combined wave and pressure gradient forcing. The arrow thickness corresponds to magnitude indicated by the scales at the bottom of each panel. Depth contours in m are indicated by the white lines.

The wave transformation clearly reflects the underlying bathymetry, with waves focusing on the shallow shoals and diverging from the deeper rip channels. As wave breaking predominantly occurs over the shoals the roller energy is highest offshore of the shoals (panel B in Figure 3.29). The corresponding surface stress is aligned with the incident wave direction demonstrating the focusing of the wave energy on the shoals (panel B of Figure 3.29). The wave-breaking related shear stress forcing, eq. (3.26), is clearly dominant over the body forcing within the surf zone, pushing the water towards the shore again centred on the shoals (compare panels A and B in Figure 3.29).

As a result the mean water level inside the surf zone is elevated over the shoals with respect to the rip channel locations. This leads to pressure gradients which push the water from the shoals towards the deeper rip channels (see panel C of Figure 3.29), thereby creating strong vortical circulations (see panel D of Figure 3.29). Outside the surfzone, the pressure gradient associated with the set down, cancels the wave forcing induced by the shoaling waves, and no spurious flows develop. The increased wave height, outside the surf zone, opposite the rip-channels coincides with the presence of offshore directed currents resulting in wave-current interaction (see Section 2.9).

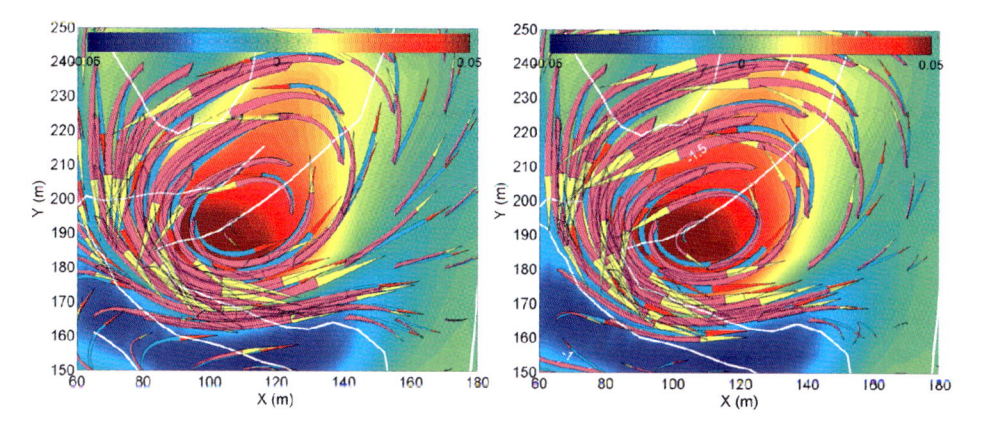

Fig. 3.30 Left panel: Detailed view of vorticity (indicated by the color bar in s^{-1}) and surface (magenta arrows) and near-bed (cyan arrows) Eulerian velocities. The arrow thickness corresponds to velocity magnitude with the scale given at the bottom of panel D in Figure 3.29. Depth contours in m are indicated by the white lines. Right panel: Equivalent to the left panel but now showing the GLM flow field

The vertical variation of the flow is limited within the surf zone but can be significant at the outer edge of the surf zone where the near-bed flow is typically more offshore directed than the surface flow (see panel D in Figure 3.29). This is expected to lead to significant differences for the transport of suspended and bed-load material.

The GLM and Eulerian mean flow velocities are very similar (compare panels in Figure 3.30). A detailed look at the flow velocities at the outer surf zone shows that the Eulerian flow field (left panel of Figure 3.30) generally has stronger offshore directed velocities. The alongshore velocities are only moderately affected (as Stokes drift acts in the predominantly onshore wave direction only), and as a result, the GLM velocities (right panel of Figure 3.30) exhibit a stronger rotation (i.e., smaller radius) and at times opposite flow directions compared with the Eulerian velocities. This in turn has important consequences for the transport of floating matter and the exchange of surfzone material with the inner shelf [Reniers et al. (2009)].

Note that a comparison of the time-averaged predicted surface velocities with GPS drifter-inferred mean velocities ([MacMahan et al. (2010a)]) showed good correspondence with a model skill of O(0.6) for both the mean surface velocity magnitude and direction. In view of the potential effects of small bathymetric errors [Plant et al. (2009)] this is considered a good match.

Chapter 4

Sediment transport

4.1 Introduction

Sediment transport is the essential link between the waves and currents and the morphological changes. It is a strong and non-linear function of the current velocity and orbital motion, the sediment properties such as grain diameter and density and the small-scale bed features often lumped together in the parameter 'bed roughness'. Typically transport is subdivided into bed load transport, which takes place just above the bed and reacts almost instantaneously to the local conditions, and suspended load transport, which is carried by the water motion and needs time or space to be picked up or to settle down.

The complexities of sediment transport are staggering, and one can easily spend a lifetime studying a particular aspect of, say, cross-shore transport over rippled beds or grain-grain interactions in a sheet flow layer. A vast amount of work has been carried out and will continue to be done in order to unravel the principles of sediment transport and to develop practically applicable formulations. It is a major challenge to morphologists to use as much of this knowledge as possible without being bogged down by the thought that we have to understand every detail before we go on to morphology. Even with the present state of knowledge we can make useful morphological models, because there are some general trends that are robust and lead to unambiguous morphological effects:

(1) Sand tends to go in the direction of the *near-bed* current.
(2) If the current increases, the transport increases by some power greater than 1.
(3) On a sloping bed transport tends to be diverted downslope.
(4) The orbital motion stirs up more sediment and thus increases the transport magnitude.
(5) In shallow water, the wave motion becomes asymmetric in various ways, which leads to a net transport term in the direction of wave propagation or opposed to it.

In this section, we will describe the general structure of sediment transport models applied in morphodynamic models, and discuss various transport mechanisms, without going into the details of the transport models, as there are many books about this subject, e.g. [Van Rijn (1993)], [Nielsen (1992)], [Fredsoe and Deigaard (1992)].

4.2 Suspended transport

Suspended sediment transport is different from bed load transport in that it does not react instantaneously to changes in the flow or wave conditions, but indirectly, through changes in the concentration field. In an accelerating flow, the concentration will typically be lower than the equilibrium concentration, which is the concentration that would occur for stationary and uniform conditions, because the sediment has to be picked up and transported upwards by turbulent dispersion. When the flow decelerates or the waves are reduced, there is more sediment in suspension than the flow can support and sediment settles out. Therefore we must first solve the 3D distribution of the sediment concentration before we can obtain the transports we are interested in.

4.2.1 3D Advection-diffusion equation for sediment

The distribution of sediment in suspension is governed by the 3D advection-diffusion equation:

$$\frac{\partial c}{\partial t} + u\frac{\partial c}{\partial x} + v\frac{\partial c}{\partial y} + (w - w_s)\frac{\partial c}{\partial z} - \frac{\partial}{\partial z}(\varepsilon_s \frac{dc}{dz}) - \frac{\partial}{\partial x}(\varepsilon_h \frac{\partial c}{\partial x}) - \frac{\partial}{\partial y}(\varepsilon_h \frac{dc}{dy}) = 0 \quad (4.1)$$

Here c is the concentration, w_s is the fall velocity, ε_s is the vertical dispersion coefficient and ε_h is the horizontal dispersion coefficient. This equation can be used at various time- and space scales, from detailed intra-boundary layer models and vortex-resolving ripple-induced transport to slowly-varying tide- wind- or wave driven current-induced transport. The horizontal and vertical dispersion coefficients must reflect the processes that are not explicitly resolved, and tend to be higher as the time- and space-scales considered are larger. As an example, in very detailed models of transport over bottom ripples every vortex shed by the ripples is resolved and the transport is dominated by the advection terms; diffusion only plays a minor role. If we go to a much coarser approach where the ripples are represented by an average roughness, the net advection transport by these turbulent vortices is represented as a gradient-type diffusive flux.

4.2.1.1 Bottom boundary condition

Near a sandy bed the sediment concentration reacts quickly to the bed shear stress, and various formulations exist for the near-bed concentration, e.g. [Zyserman and Johnson (2002)] and [van Rijn (1984)]. The reference concentration is given at a specified height, either related to the grain size or to the bed roughness. As described in [Lesser et al. (2004)], the flux of sediment between the bed and the flow can be approximated by:

$$S_z = -\varepsilon_s \frac{\partial c}{\partial z} - w_s c \qquad (4.2)$$

where the concentration gradient can be approximated by:

$$\frac{\partial c}{\partial z} \approx \frac{c(z_{ref} + \Delta z) - c_{ref}}{\Delta z} \qquad (4.3)$$

When the shear stress increases, the reference concentration c_{ref} will increase; this will lead to a more negative concentration gradient, and hence to a positive flux of sediment from the bed into the water column; when the shear stress decreases, the settling flux term will be dominant.

4.2.1.2 Sediment transport through a section

The horizontal flux of sediment through a section is given by integrating the advective and dispersive terms over the vertical:

$$\begin{aligned} S_{sus,x} &= \int_{-h}^{\eta} c u \, dz + \int_{-h}^{\eta} \varepsilon_h \frac{\partial c}{\partial x} \, dz \\ S_{sus,y} &= \int_{-h}^{\eta} c v \, dz + \int_{-h}^{\eta} \varepsilon_h \frac{\partial c}{\partial y} \, dz \end{aligned} \qquad (4.4)$$

Note that both concentration and velocity vary with depth and time, and that important net contributions remain after averaging, e.g. in the case of transport by waves, where we can decompose the velocity and concentration in a time-average part and an oscillatory part. We then get:

$$\begin{aligned} \int_t \int_{-h}^{\eta} c u \, dz dt &= \int_{-h}^{0} \bar{c} \bar{u}_L \, dz + \int_t \int_{-h}^{\eta} \tilde{c} \tilde{u} \, dz dt \\ \int_t \int_{-h}^{\eta} c v \, dz dt &= \int_{-h}^{0} \bar{c} \bar{v}_L \, dz + \int_t \int_{-h}^{\eta} \tilde{c} \tilde{v} \, dz dt \end{aligned} \qquad (4.5)$$

Clearly we have a contribution by the mean concentration, which is transported by the lagrangean mean velocity, and the oscillatory term, which is non-zero because concentration and velocity timeseries can be correlated. We will discuss this effect further in the section about wave skewness and asymmetry (see Section 4.4.1).

4.2.1.3 Equilibrium suspended transport

In cases where the bathymetry, sediment properties and flow vary slowly, the sediment transport can be assumed to be in equilibrium with the flow. Considering how sediment transport depends on the flow and sediment properties under such (quasi) uniform conditions already provides very useful general trends. We can derive the concentration profile for this case from the general advection-diffusion equation:

$$\frac{\partial c}{\partial t} + u \frac{\partial c}{\partial x} + v \frac{\partial c}{\partial y} + (w - w_s) \frac{\partial c}{\partial z} - \frac{\partial}{\partial z}(\varepsilon_s \frac{dc}{dz}) - \frac{\partial}{\partial x}(\varepsilon_h \frac{\partial c}{\partial x}) - \frac{\partial}{\partial y}(\varepsilon_h \frac{dc}{dy}) = 0 \quad (4.6)$$

Leaving out all non-stationary and non-uniform terms this equation reduces to:

$$w_s c + \varepsilon_s \frac{\partial c}{\partial z} = 0 \tag{4.7}$$

with the general solution:

$$c(z) = c_a \exp\left(-\int_a^z \frac{w_s}{\varepsilon_s} \partial z\right) \tag{4.8}$$

The shape of the concentration profile is linked to the distribution of the dispersion coefficient, if we assume the fall velocity to be constant with z. For a constant dispersion coefficient we get a simple logarithmic distribution;

$$c(z) = c_a \exp\left(-\frac{w_s}{\varepsilon_s} z\right) \tag{4.9}$$

for a parabolic distribution over the vertical we get the well-known Rouse profile:

$$c(z) = c_a \left(\frac{z+h}{z} \frac{a-h}{a}\right)^{-\frac{w_s}{\kappa u_*}} \tag{4.10}$$

The dispersion coefficient for fine-grained sediment is close to the turbulence viscosity, but not equal; usually a factor β_ν is applied to represent the ratio between the dispersion coefficient and the turbulence viscosity. This factor may depend on the fall velocity and on the flow conditions. Various empirical formulations have been suggested, see e.g. [Van Rijn (1993)] or [Van de Graaff (1988)].

For the case of combined current and waves various empirical distributions of the dispersion coefficient have been suggested. [Van Rijn (1993)] suggests a parabolic-constant distribution, with a constant dispersion coefficient in the wave boundary layer and a parabolic distribution above it.

In sophisticated 3D models, the dispersion coefficient is the product of a two-equation turbulence model, such as the $k-\varepsilon$ model applied in Delft3D [Lesser et al. (2004)]. In that case, appropriate source and sink terms for bottom- and surface- generated wave-induced turbulence have to be added, as for instance given by [Walstra et al. (2000)].

4.2.2 2DH Advection-diffusion equation for sediment

In many applications the horizontal variations are much more important than vertical non-uniformities, and a depth-averaged approach is justified. In that case we can apply the depth-averaged advection-diffusion equation:

$$\frac{\partial h\bar{c}}{\partial t} + \bar{u}\frac{\partial h\bar{c}}{\partial x} + \bar{v}\frac{\partial h\bar{c}}{\partial y} - \frac{\partial}{\partial x}\left(\varepsilon_h \frac{\partial h\bar{c}}{\partial x}\right) - \frac{\partial}{\partial y}\left(\varepsilon_h \frac{dh\bar{c}}{dy}\right) = S \tag{4.11}$$

The source/sink term S represents the exchange with the bottom and must be considered with some care. [Galappatti and Vreugdenhil (1985)] derived expressions for this source term:

$$S = \frac{h(\bar{c}_{eq} - \bar{c})}{T_s} \tag{4.12}$$

where $\bar{c}_{eq}$ is the equilibrium depth-averaged concentration and T_s is a typical timescale, which can be expressed as:

$$T_s = T_{sd} \frac{h}{w_s} \tag{4.13}$$

The dimensionless factor T_{sd} depends on the ratio of shear velocity to fall velocity, and can be approximated for the case of currents only, e.g. according to [Wang (1992)]. For wave-current combinations and strongly non-stationary situations [Reniers et al. (2004a)] propose a simple constant value of T_{sd}, in the order of 0.1, which was based on limited observations of the response of concentration to varying short wave energy in large-scale wave flume tests. Here we will propose a somewhat more physics-based but still rather simple approach. We start from the observation that in analytical solutions according to [Hjelmfelt and Lenau (1970)], profiles that are out of equilibrium tend to rotate around the near-bed concentration. We can represent this by multiplying the fall velocity by a factor α, which reduces the concentrations when it is greater than 1 and increases them when it is less than 1. For the situation with a constant dispersion coefficient over the vertical, as in eq. (4.9), we can relate the depth-averaged concentration to the reference concentration as follows:

$$\bar{c} = \frac{1}{h} \int_{-h}^{0} c\, dz = \frac{c_a}{h} \int_{-h}^{0} \exp\left(-\frac{\alpha w_s}{\varepsilon_s}(z+h)\right) dz$$

$$-\frac{\varepsilon_s}{\alpha w_s} \frac{c_a}{h} \left[\exp\left(-\frac{\alpha w}{\varepsilon_s}(z+h)\right)\right]_{-h}^{0} \approx \frac{\varepsilon_s}{\alpha w_s} \frac{c_a}{h} \approx \frac{1}{\alpha} \bar{c}_{eq} \Rightarrow \tag{4.14}$$

$$\alpha \approx \frac{\bar{c}_{eq}}{\bar{c}}$$

Though we cannot integrate the concentration as easily for the case of a Rouse profile, we have found by numerical integration that we can approximately use the same relationship to relate depth-averaged concentration to the equilibrium depth-averaged concentration. We can use this to estimate the source-sink term S, which follows from the balance between the upward diffusive flux and the downward settling flux:

$$\varepsilon_s \frac{\partial c}{\partial z} + \alpha w_s c_a = 0 \Rightarrow \varepsilon_s \frac{\partial c}{\partial z} + w_s c_a + (\alpha - 1) w_s c_a = 0 \Rightarrow$$
$$S_c = -\varepsilon_s \frac{\partial c}{\partial z} - w_s c_a = (\alpha - 1) w_s c_a = \left(\frac{\bar{c}_{eq}}{\bar{c}} - 1\right) w_s c_a \tag{4.15}$$

Here, $\bar{c}_{eq}$ is a function of c_a through the Rouse profile, which depends on the ratio between fall velocity and the shear velocity and on the bed layer thickness a and water depth h. For situations where c_a goes to zero the sedimentation must still continue; in tis case the term wc_a must be replaced by $w\bar{c}$. For the situation where sand is picked up from a clean bed, the initial concentration is set to $\bar{c} = c_a a/h$. With these modifications the formulation for the source term is:

$$S_c = \left(\frac{\bar{c}_{eq}}{\max(\bar{c}, c_a a/h)} - 1\right) w_s \max(c_a, \bar{c}) \tag{4.16}$$

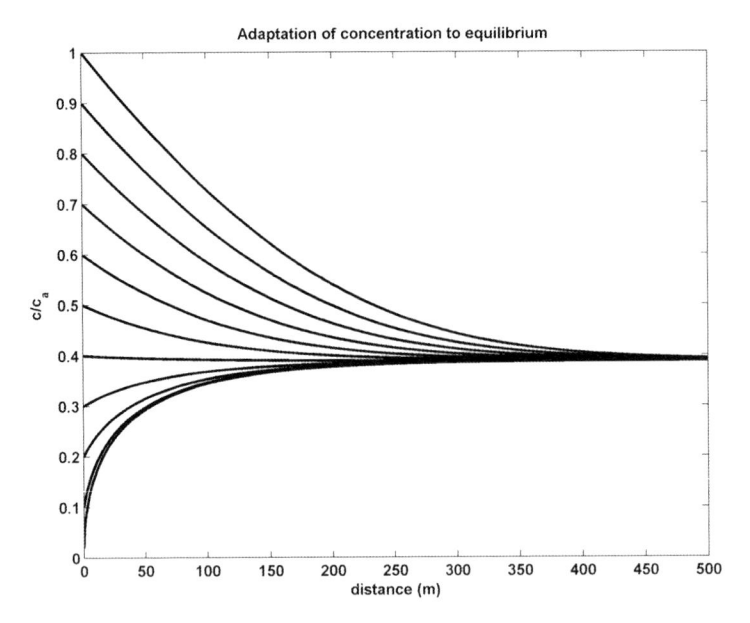

Fig. 4.1 Adaptation of depth-averaged concentration to equilibrium, u=2 m/s; C=65 m$^{1/2}$/s; w_s=0.01 m/s; a=0.1 m, h=5 m

In Figure 4.1 an example is shown of how the depth-averaged concentration adapts from different initial conditions to the equilibrium concentration. Clearly adaptation from a zero or small concentration is much faster than that from a high concentration; in the first case, there is a strong upward flux by turbulence, whereas in the second case the sedimentation is limited by the fall velocity.

4.3 Bed load and total load transport formulations

4.3.1 *Current-only situation*

Bed load transport, which takes place in a thin layer above the bed, can always be assumed to react directly to local flow conditions. Most bed load transport formulations contain a number of the following concepts:

- The bed shear stress by the flow acting on the sediment grains. It is often expressed in dimensionless form as the Shields parameter:

$$\theta = \frac{\tau}{\rho g \Delta D_{50}} \quad (4.17)$$

where τ is the bed shear stress, ρ the water density, g the acceleration of gravity, $\Delta = (\rho_s - \rho)/\rho$ is the relative sediment density, and D_{50} is the median grain diameter. The dimensionless shear stress reflects the balance between lifting

forces, which are proportional to shear stress and grain surface, and gravity, which is proportional to the relative density, g and the grain volume.
- The critical shear stress or critical Shields parameter for initiation of motion;
- Bed load transport in the direction of near-bed flow, as a function of the Shields parameter (minus the critical Shields parameter) to some power.
- Bed slope effects in the direction of the flow and in transverse direction.
- For rippled beds, the bed shear stress is often divided into form drag (because of the ripples) and skin friction (exerted directly on the sand grains), where the bed load transport is generally taken to be a function of the skin friction only.

A general form of bed load / total load transport formulations is given by:

$$S_b \sim \sqrt{\Delta g D_{50}^3} \theta^{b/2} \left(m\theta - n\theta_{cr}\right)^{c/2} \left(1 - \alpha \frac{\partial z_b}{\partial s}\right) \qquad (4.18)$$

A number of bed load transport formulae are captured by this formulation, e.g. [Meyer-Peter and Muller (1948)] *(c=3,b=0)*, [van Rijn (1984)] *(b=0,c=3-4)*. The coefficient m represents a ripple efficiency factor, which depends on the ratio of skin friction to form drag, and n may represent a factor for hiding and exposure in graded sediments.

4.3.2 Waves plus current

If sediment transport is already difficult, highly empirical and inaccurate for current-only situations, the situation with waves is much worse. Waves interact with the current in modifying the bed shear stress, the bed ripples, the sediment mobility and the near-bed current transporting the sediment.

After efforts to adapt bed-load transport formulas for current-only to combined current-wave situations by adapting the dimensionless shear stress (e.g. [Van de Graaff and Van Overeem (1979)]) most researchers have resorted to developing formulations directly fitted against as many data points as they could get hold of. Recent examples are [Ribberink (1998)], [Soulsby (1997)], [Gonzalez-Rodriguez and Madsen (2007)]. We will use the latter in most of our example models as it is quite simple in setup and has a behaviour very similar to the full expressions by [Van Rijn (1993)].

4.3.2.1 Soulsby-van Rijn formula

In many of the examples furtheron we will use the so-called Soulsby-van Rijn formula ([Soulsby (1997)]), because it has a number of attractive features:

- It is a very simple expression;
- It is easy to implement;
- It is reasonably close to Van Rijn's full formulations;
- It gives clear insight in mechanisms;
- It considers bed load and suspended load separately;

- It has combined current and wave effects;
- It has a critical velocity;
- It has a bed slope effect.

[Soulsby (1997)] developed the formulation by finding a suitable form of the equation that could be fitted to the numerical results of van Rijn 's 1993 model. The formulations are as follows:

$$\begin{aligned} S_{bx} &= A_{cal} A_{sb} u \xi \\ S_{by} &= A_{cal} A_{sb} v \xi \\ S_{sx} &= A_{cal} A_{ss} u \xi \\ S_{sy} &= A_{cal} A_{ss} v \xi \end{aligned} \quad (4.19)$$

where A_{cal} is a user-defined calibration factor, A_{sb} is a bed-load multiplication factor:

$$A_{sb} = 0.05 h \left(\frac{D_{50}/h}{\Delta g D_{50}} \right)^{1.2} \quad (4.20)$$

and A_{ss} is a suspended load multiplication factor:

$$A_{sb} = 0.012 D_{50} \frac{D_*^{-0.6}}{(\Delta g D_{50})^{1.2}} \quad (4.21)$$

The dimensionless grain diameter D_* is given by:

$$D_* = \left[\frac{g \Delta}{\nu^2} \right]^{1/3} D_{50} \quad (4.22)$$

where ν is the kinematic viscosity. The term ξ is a general multiplication factor that governs the power of the transport relation, determines the relative effects of current and waves, and includes a critical velocity:

$$\xi = \left(\sqrt{u^2 + v^2 + \frac{0.018}{C_f} U_{rms}^2} - U_{cr} \right)^{2.4} \quad (4.23)$$

Here:

$$C_f = \left[\frac{\kappa}{\ln(h/z_0) - 1} \right]^2 \quad (4.24)$$

and:

$$U_{cr} = \begin{cases} 0.19 D_{50}^{0.1} \log_{10}(4h/D_{90}) & , D_{50} \leq 0.5 \, mm \\ 8.5 D_{50}^{0.6} \log_{10}(4h/D_{90}) & , 0.5 < D_{50} \leq 2 \, mm \end{cases} \quad (4.25)$$

The root mean square orbital velocity, U_{rms}, can be computed by eq. 2.28.

4.4 Wave-driven transport

Waves contribute to both onshore and offshore sediment transport. Wave-driven on-shore transport is the primary mechanism in restoring the beach after severe storm-related erosion, bringing the sand back onshore. It is responsible for the relatively steep beach profiles typically observed during the summer and is essential in explaining onshore bar migration during moderate wave conditions. The dominant contributions to the wave-related onshore sediment transport are associated with the non-linear wave shape of the incident waves, the wave-induced streaming in the wave boundary layer and the Stokes drift. Wave-driven offshore transport is predominantly related to the return flow compensating the onshore directed Stokes drift and the phase coupling between wave groups and accompanying (bound) long waves. The various mechanisms will be discussed in the following showing its relevance and ways to include it in coastal modeling.

4.4.1 Wave skewness and asymmetry

As discussed in the section on wave modeling, in shallow water, but prior to breaking, both bound higher and lower harmonics are generated through triad interactions. The presence of higher harmonics, which are initially in phase with the primary waves generating them, results in more peaked wave crests and flatter troughs [Stokes (1847)]. A relatively simple way of representing the corresponding orbital velocity is given by [Elgar and Guza (1985)]:

$$u_\infty(t) = \text{Re}\left(A_{rms}\omega \sum_{m=0}^{N} \frac{1}{2^m} \exp i\left[(m+1)\omega t + m\Phi\right]\right) \quad (4.26)$$

where A_{rms} is related to the measured near-bed free stream velocity by:

$$A_{rms} = \sigma(u_\infty)\frac{\sqrt{2}}{\omega_p} \frac{1}{\sqrt{1 + \sum_{m=1}^{m=N}\left(\frac{1}{2^m}\right)^2}} \quad (4.27)$$

to match the variance in the analytical velocity signal with the observed variance. Generating a signal with $T_p = 8$ s, $\sigma(u_\infty) = 0.5$ m/s, N=10 and $\Phi = 0$, i.e. all harmonics are in phase, yields a Stokes-like wave signal (left panel of Figure 4.2). For a constant phase shift of 90° a saw tooth wave shape is obtained (right panel of Figure 4.2).

The motion of the sediment at the bed is a function of the near-bed velocity within the wave boundary layer, whereas velocity measurements are generally obtained outside of the wave boundary layer. Hence the free stream velocity, i.e. outside the boundary layer has to be translated into a near bed velocity. This involves an attenuation of the signal and phase shift. Given the near bed velocity the shear stress can be calculated, or the corresponding Shields number, to calculate

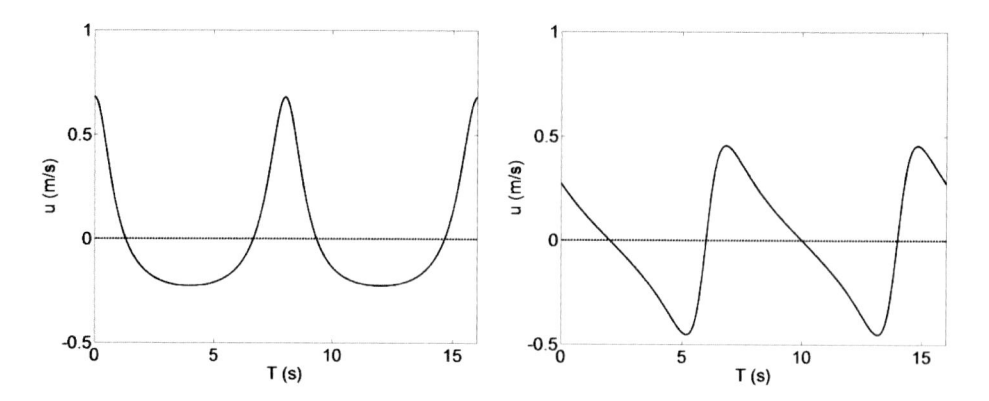

Fig. 4.2 Left panel: Stokes like free stream velocity generated with eq. (4.26) for Tp = 8 s, $\sigma(u_\infty)$ m/s, N=10 and $\Phi = 0$. Right panel: Similar but for a saw tooth shape free stream velocity obtained with $\Phi = 90°$

the sediment stirring. The general expression for the shear stress at the bed is given by:

$$\tau(t) = \rho C_f |u_\infty(t)| u_\infty(t) \qquad (4.28)$$

ignoring the phase shift within the wave boundary layer. If the shear stress exceeds a critical shear stress sediment is set into suspension, which is subsequently transported by the flow velocity. In absence of a mean flow the near-bed sediment transport then becomes proportional to:

$$S_{S_k} = K_s u^3 \qquad (4.29)$$

where K_s is a calibration coefficient. The third order velocity moment is closely related to the wave skewness:

$$S_k = \frac{<u^3>}{\sigma^3(u)} \qquad (4.30)$$

Hence, depending on the wave shape, this then leads to a net wave-averaged sediment transport in the direction of wave propagation, which has its maximum for Stokes waves (i.e. maximum skewness) and is negligible for saw-tooth shape waves (see Figure 4.3). This typically leads to maximum onshore transport at the offshore side of the bar and minimal transport on top of the bar. By including the phase shift the dynamics change as can be seen in the following.

Starting with the saw tooth free stream velocity signal (right panel of Figure 4.3) a constant phase shift of $30°$ (positive number corresponding to leading velocity signal in the wave boundary layer) is introduced to calculate the nearbed velocity response within the wave boundary layer (ignoring attenuation):

$$u_b(t) = \mathrm{Re}\left(A_{rms}\omega \sum_{m=0}^{N} \frac{1}{2^m} \exp i \left[(m+1)\omega t + m\Phi + \varphi_\tau \right] \right) \qquad (4.31)$$

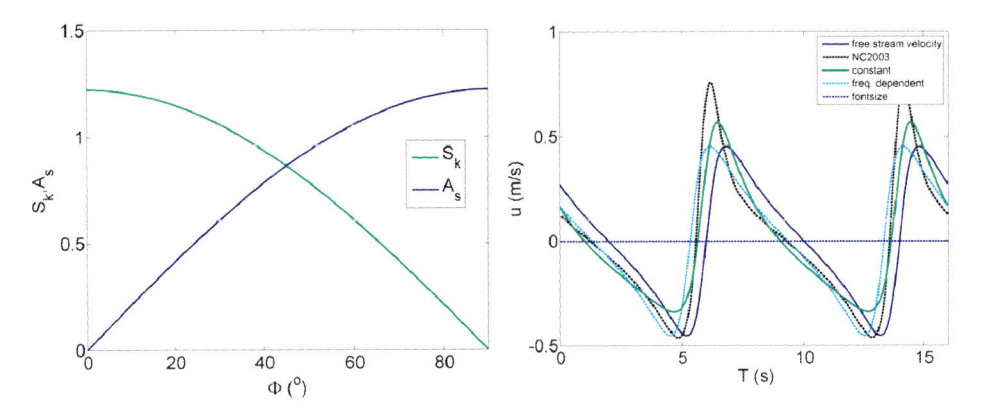

Fig. 4.3 Left panel: Velocity skewness, eq. 4.30, and asymmetry, eq. (4.36), calculated as function of the phase shift in eq. (4.26). Right panel: near bed velocity response corresponding to a constant phase shift (green), a frequency dependent phase shift (cyan), [Nielsen and Callaghan (2003)] (black dashed), and the corresponding free stream velocity as a reference (blue).

The introduction of the constant phase shift transforms the saw tooth velocity signal into a mixed skewed/asymmetric velocity signal (right panel of Figure 4.3), resulting into a positive sediment transport at locations where the waves are breaking. This transformation from a saw-tooth velocity signal to a more skewed signal was demonstrated by [Henderson et al. (2004)], whom observed a near frequency independent phase shift within the boundary layer using a detailed numerical model to calculate sediment transport under irregular waves. One of their key findings was that the near-constant phase shift resulted in the transformation of saw-tooth waves into partially skewed nearbed velocity profiles. This in turn could explain the onshore motion of the bar as observed during Duck 94 [Gallagher et al. (1996)]. Their findings are not fully consistent with existing analytic approximations to calculate the flow velocity within the wave boundary layer, which show a frequency dependent phase shift ([Mei (1989)], his Section 8.7.1).

In fact, if we introduce a frequency dependent phase shift according to:

$$u_b(t) = \mathrm{Re}\left(A_{rms}\omega \sum_{m=0}^{N} \frac{1}{2^m} \exp i\left[(m+1)\omega t + m\Phi + (m+1)\varphi_\tau\right]\right) \quad (4.32)$$

we obtain a time-lagged saw-tooth velocity profile that is identical to the free stream velocity (see right panel of Figure 4.3), and subsequently the net sediment transport according to eq. (4.30) is negligible again.

Outside the wave boundary layer the velocity acceleration is predominantly related to the pressure gradient:

$$\frac{\partial u}{\partial t} + u\frac{\partial u}{\partial x} + v\frac{\partial u}{\partial y} + w\frac{\partial u}{\partial z} - f_{cor}v = \frac{\partial}{\partial y}\left(\nu_h \frac{\partial u}{\partial y}\right) + \frac{\partial}{\partial z}\left(\nu_r \frac{\partial u}{\partial z}\right) - \frac{1}{\rho}\frac{\partial p}{\partial x} \quad (4.33)$$

Since both coriolis force and viscous terms can be neglected outside the wave boundary layer. Upon the assumption that the non-linear terms are small and can therefore be ignored we find a balance between flow acceleration and horizontal pressure gradient. It has been shown that these wave-induced pressure gradients can lead to sediment transport in the way of plug flow ([Madsen (1975)], [Sleath (1999)], [Foster et al. (2006)]) and coarse particle transport ([Drake and Calantoni (2001)], [Terrile et al. (2006)]). [Hoefel and Elgar (2003)] accounted implicitly for the pressure gradient contribution using the fluid acceleration as a proxy (note that at times the pressure gradient can deviate significantly from the fluid acceleration [van Thiel de Vries et al. (2008)]. Their approach thereby assumes that the phase shift is in fact frequency dependent in the form described by eq. (4.32), i.e. a shape conserving translation of the free stream velocity to the bed, as they used the free stream acceleration to calculate the pressure gradient induced transport ([Hoefel and Elgar (2003)]):

$$S_{as} = K_a \left(\frac{<a_\infty^3>}{\sigma^2(a_\infty)} - \text{sgn}[<a_\infty^3>]a_{crit} \right) \quad (4.34)$$

which can in turn be written as a function of the wave asymmetry:

$$S_{as} = K_a \left(\sigma(a) A_s - \text{sgn}[<a_\infty^3>]a_{crit} \right) \quad (4.35)$$

where:

$$A_s = \frac{<a^3>}{\sigma^3(a)} \quad (4.36)$$

Hence, in this case the wave averaged sediment transport is again a function of the wave shape, but now with zero transport for skewed waves and maximum transport for the saw-tooth wave shape (left panel of Figure 4.3).

An alternative contribution to the wave-shape related sediment transport is given by [Nielsen (1992)] who suggests that the wave boundary layer thickness under asymmetric waves is thinner under the advancing wave crest than under the wave trough, resulting in higher shear stresses under the wave crest. This effect has been included by incorporating the acceleration into the shear stress formulation ([Nielsen and Callaghan (2003)]):

$$\tau = \rho c_f \left[u_\infty \cos\varphi_\tau + \frac{1}{\omega_p} \frac{\partial u_\infty}{\partial t} \sin\varphi_\tau \right]^2 \quad (4.37)$$

Note that the corresponding near bed velocity signal is not only of different shape (see right panel of Figure 4.3) but also has increased variance with respect to the free stream velocity signal due to the simplified transformation involving the peak radial frequency only ([Terrile et al. (2009)]).

So the inclusion of the phase shift, i.e. either frequency independent or dependent, determines the (maximum) bed shear stress under a skewed/asymmetric wave and the need for an explicit contribution of the pressure gradient to explain the onshore motion of bars under moderate conditions. At this point it remains to be seen which expression is the most appropriate given the lack of direct evidence of the flow velocities within the wave boundary layer over a mobile bed.

Note that additional complexity is present due to the vertical structure of both the sediment concentration and wave and flow-dynamics, and the qualitative analysis presented above is for the near-bed response on a flat bed only. In the presence of ripples or fine sediment the sediment transport direction may actually be opposite from the wave direction due to this vertical variation (e.g. [Ribberink and Chen (1993)], [Janssen et al. (1998)]). As a result the various wave-shape related transport contributions require a calibration coefficient (see section 7.2).

Also note that in the analytical expressions presented above, the amplitudes of the higher harmonics are independent of the local conditions, whereas in reality they depend on the local wave height, wave period and water depth and up-wave conditions (e.g. [Eldeberky and Battjes (1995)], [Beji and Battjes (1993)]). These non-linear amplitude and phase evolution over variable topography can be well predicted with intra-wave modeling (e.g. [Dingemans (1997)] and references therein).

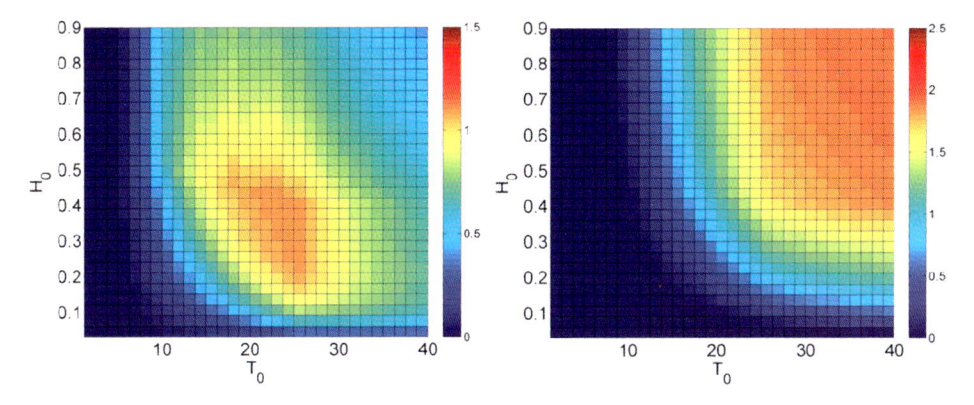

Fig. 4.4 Near-bed velocity Skewness calculated with stream function theory ([Rienecker and Fenton (1981)]) in combination with empirical phase shift (eq. (4.38)) as function of the normalized wave height and normalized wave period (eq. (4.39)). Right panel: similar for the corresponding asymmetry.

In absence of an intra-wave evolution model, the approach outlined by [Van Thiel de Vries (2009)] can be used where the weights for the higher frequencies are calculated with stream function theory (e.g. [Rienecker and Fenton (1981)]) and the phase is estimated from empirical formulations (e.g. [Ruessink and van Rijn (2011)]):

$$\Phi = \frac{\pi}{2} - \frac{\pi}{2}\tanh\left(0.64/U_r^{0.60}\right) \quad (4.38)$$

where the Ursell number is given by: $U_r = \frac{3}{8}\frac{kH}{(kh)^3}$. Performing these computations for a wide range of wave conditions and defining the non-dimensional wave height and non-dimensional wave period as:

$$H_0 = \frac{H}{h} \quad T_0 = T\sqrt{\frac{g}{h}} \quad (4.39)$$

the corresponding nearbed velocity skewness and asymmetry can be tabulated (Figure 4.5). Given the local wave height, wave period and water depth the corresponding S_k and A_s are easily obtained. This is demonstrated for a case of normally incident waves with 8 second peak period on a single barred beach where the wave transformation has been calculated with the wave energy balance, eq. (2.19) (upper panel of Figure 4.5).

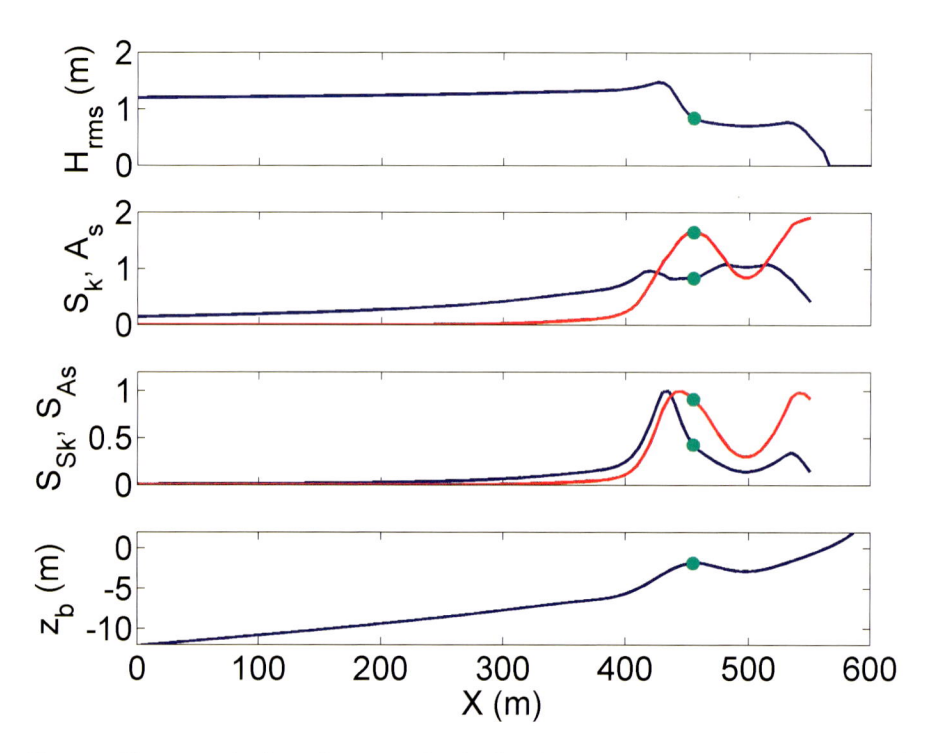

Fig. 4.5 Upper panel: Cross-shore wave height distribution. 2nd panel: corresponding S_k (blue) and A_s (red) obtained from tabulated data (see Figure 4.4). 3rd panel: Skewness (blue) and asymmetry-related (red) sediment transports (normalized by their respective maxima). Lower panel: Single barred bottom profile and position of the bar crest indicated by green dots.

Given the velocity skewness and asymmetry the corresponding intra-wave transports related to skewness (eq. (4.29)) and asymmetry (eq. (4.35)) can then easily be evaluated with:

$$S_{sk} = K_s <u^3> = K_s S_k \sigma^3(u) = K_s S_k u_{rms}^3 \qquad (4.40)$$

and:

$$S_{as} = K_a \left(\sigma(a) A_s - \text{sgn}[A_s] a_{crit} \right) \cong K_a \left(\omega_p u_{rms} A_s - \text{sgn}[A_s] a_{crit} \right) \qquad (4.41)$$

respectively (Figure 4.5), where u_{rms} is calculated with linear wave theory from the local wave height, wave period and water depth. The results are in line with the findings by [Elgar et al. (1997)] that the wave skewness and its corresponding transport has its maximum offshore of the bar crest (2nd and 3rd panel of Figure 4.5), whereas the maximum asymmetry and corresponding transport is in the vicinity of the bar crest position.

4.4.2 *Lagrangian drift*

As discussed in section 3.2.3 the presence of waves also leads to a mass flux in the direction of wave propagation. All surface floating as well as suspended material present in the watercolumn is subject to the Stokes drift. However, it strongly depends on the inertia of the particles that are being transported. Supended particles with little inertia follow the orbital motion (e.g. suspended fines) and are transported by the wave induced Stokes drift. Coarser sediment is more susceptible to gravity, does not (fully) follow the fluid parcel paths, and will reside closer to the bed and thus be less affected by the Stokes drift. The contribution of the Stokes drift on the onshore sediment transport rate can be significant and of similar magnitude as the wave skewness induced sediment transport ([Henderson et al. (2004)], [Ruessink et al. (2007)]).

4.4.3 *Streaming*

As discussed earlier, the wave energy dissipation in the boundary layer due to bottom friction results in a net forcing in the direction of wave propagation [Longuet-Higgins (1953)]. This leads to a wave-averaged streaming in the direction of wave propagation as demonstrated below by considering harmonic unidirectional waves propagating in the x-direction over a horizontal bed. In that case the cross-shore wave-averaged momentum equation reduces to:

$$\frac{\partial \bar{u}}{\partial t} + \bar{u}\frac{\partial \bar{u}}{\partial x} + \bar{v}\frac{\partial \bar{u}}{\partial y} + \bar{w}\frac{\partial \bar{u}}{\partial z} - f_{cor}\bar{v} = \frac{\partial}{\partial x}\nu_h\frac{\partial \bar{u}}{\partial x} + \frac{\partial}{\partial y}\nu_h\frac{\partial \bar{u}}{\partial y} + \frac{\partial}{\partial z}\nu_v\frac{\partial \bar{u}}{\partial z} - \frac{1}{\rho}\frac{\partial \bar{p}}{\partial x} + \frac{w_x}{\rho} \quad (4.42)$$

where:

$$\frac{w_x}{\rho} = \frac{\partial <uw>}{\partial z} \quad (4.43)$$

and for constant eddy viscosity the time-averaged wave Reynold's stress is given by eq. (3.33). Vertical integration of eq. (4.42) then yields the steady streaming near the bed ([Longuet-Higgins (1953)]):

$$\bar{u}(z) = \frac{\omega^2 a^2 k}{4\sinh^2 kh}\left(3 - 2(\beta z + 2)e^{-\beta z}\cos(\beta z) - 2(\beta z - 1)e^{-\beta z}\sin(\beta z) + e^{-2\beta z}\right) \quad (4.44)$$

Due to the fact that the eddy viscosity is constant everywhere the streaming is also present in the interior flow although the forcing is limited to the wave boundary

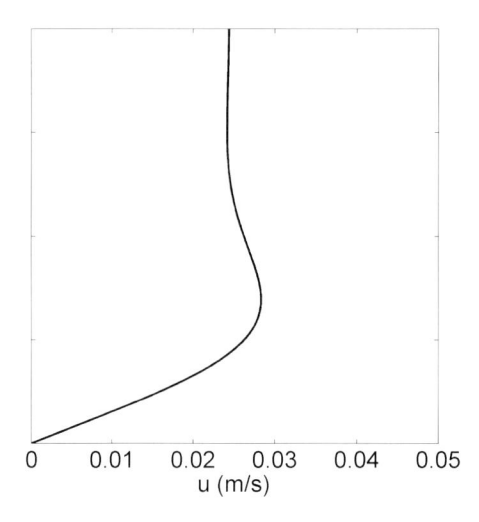

Fig. 4.6 Wave-averaged streaming velocity calculated with eq. (4.44) for a regular wave with amplitude $a = 0.4$ m and $T = 6$ s at 5 m water depth.

layer (Figure 4.6). Formulations based on time and depth varying eddy viscosity are given by [Trowbridge and Madsen (1984)].

Suspended material near the bed experiences both the Stokes drift and the Eulerian boundary layer streaming. Although the corresponding velocities are small, $O(\text{cm/s})$, the persistent flux in the wave direction results in significant transport potential over long durations (e.g. [Ruessink et al. (2007)]). Note that as waves start breaking on the beach slope the friction induced wave forcing within the wave boundary layer is countered by a larger pressure gradient driving the return flow (see Section 4.5). Hence the steady streaming is only relevant for non-breaking waves.

4.4.4 Wave group induced bound long waves

The randomness in the wave field is characterized by the presence of wave groups (Sections 2.7 and 3.6). To evaluate the sediment transport associated with wave group induced long waves we consider a wave field of normally incident waves with a H_{rms} of 1 m and a T_p of 6 s at the offshore boundary of a single barred profile (upper panel of Figure 4.7). The wave transformation is again calculated with the wave energy balance eq. 2.19. The parameterization by [Sand (1982)] then allows for a simple way[1] to calculate the corresponding (bound) long wave response:

$$H_b = -2Gh^2(\Sigma\omega, \Delta\omega, h)\frac{a^2}{h^2} \tag{4.45}$$

---

[1] More sophisticated methods are discussed in Section 3.6 and 10.5

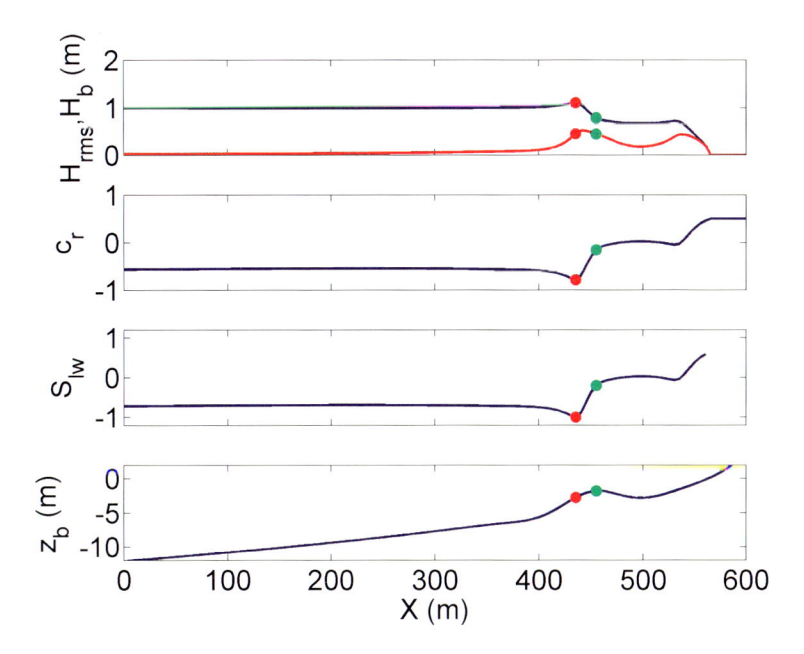

Fig. 4.7 Upper panel: Cross-shore root mean square wave height distribution for normally incident waves with a peak period of 6 seconds (blue line) and corresponding long wave response calculated with eq. 4.45. 2nd panel: corresponding correlation coefficient cr obtained from eq. 4.47. 3rd panel: S_{lw} scaled with maximum absolute value. Bottom panel: Single barred bottom profile and position of the bar crest (green dots) and location of maximum incident wave height (red dots).

where the transfer function Gh^2 is given by [Sand (1982)] and a^2 represents the short wave variance made up of two incident waves with different frequencies but equal amplitude calculated from [Roelvink and Stive (1989)]:

$$\frac{1}{8}H_{rms}^2(x) = a^2(x) + \frac{1}{8}H_b^2(x) \qquad (4.46)$$

where H_{rms} represents the combined incident and infragravity wave height. Note that without this constraint H_b becomes unrealistically large for shallow water. Setting the difference frequency to one fifth of the peak frequency and the sum frequency to two times the peak frequency the bound long wave height is calculated with eqs. (4.45) and (4.46). The calculated bound infragravity wave height shows a strong increase prior to wave breaking. Note that although the calculated infragravity variance may correspond reasonably well with observations it is not necessarily bound [2] to the incident wave groups as shown by [Roelvink and Stive (1989)]. To account for this, the phase coupling expressed by the phase lag, ϕ, between the wave groups and the long waves is estimated with an empirical function based on the results of [Roelvink and Stive (1989)]:

---

[2] Implying a 180° phase shift between wave groups and long wave

$$\phi = \pi c_r = \pi \left(0.5 - 0.9 \frac{H_{rms,x}^2}{H_{rms,o}^2}\right) \tag{4.47}$$

Once the long wave amplitude and phase lag are resolved the instantaneous surface elevation can be constructed at any point within the cross-shore (upper panel of Figure 4.8):

$$\eta = \eta_s + \eta_l = a(x)\cos(\omega_1 t) + a(x)\cos(\omega_2 t) + 0.5 H_b(x)\cos(\Delta\omega t + c_r \pi) \tag{4.48}$$

where the individual radial frequencies are given by:

$$\begin{aligned} \omega_1 &= \omega_p - \tfrac{\Delta\omega}{2} \\ \omega_2 &= \omega_p + \tfrac{\Delta\omega}{2} \end{aligned} \tag{4.49}$$

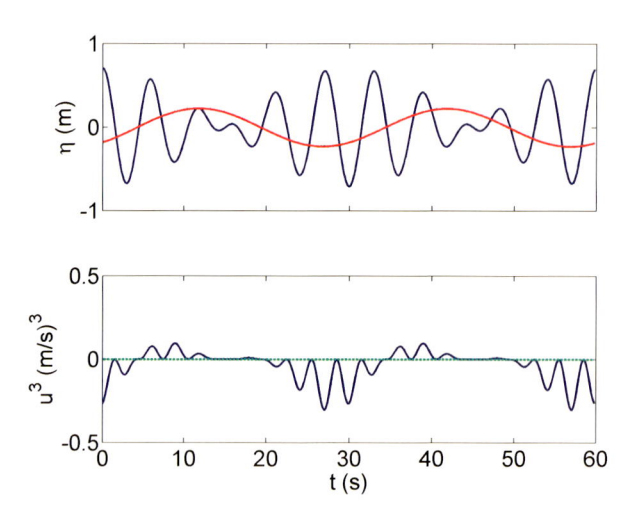

Fig. 4.8 Upper panel: Surface elevation of bichromatic wave (blue line) and corresponding long wave (red line) at position of maximum H_{rms} (Figure 4.7). Lower panel: Corresponding instantaneous third order moment of combined incident and long wave near bed velocity signal (eq., blue line).

To calculate the long wave related sediment transport we consider the corresponding velocity signal made up by the combined incident and long wave wave motions:

$$u(x,t) = u_s(x) + u_l(x) = \hat{u}_1(x)\cos(\omega_1 t) + \hat{u}_2(x)\cos(\omega_2 t) + \hat{u}_b(x)\cos(\Delta\omega t + c_r(x)\pi) \tag{4.50}$$

where the near-bed incident velocity amplitudes have been calculated with linear wave theory. Considering the cubic velocity moment to be a proxy the net bound long wave sediment transport contribution (averaged over the wave groups) is proportional to:

$$S_{lw} =< u^3 > \tag{4.51}$$

Given the fact that the higher waves within the wave group stir up more sediment than the lower waves (lower panel of Figure 4.8), the net-transport direction (averaged over the wave groups) is offshore where c_r is less than zero and onshore where c_r is greater than zero. As a result the net transport is generally offshore directed with a maximum offshore of the bar crest and the onshore transport is only present near the water line (Figure 4.7). Given the cross-shore phase-coupled long wave sediment transport distribution and the underlying profile the general effect is an offshore propagating and atenuation of the bar [Roelvink (1993)].

Note that free long waves also contribute to the sediment transport, however because they are not coupled to the incident waves there is no preferential net sediment transport direction and the transports act more or less as a diffusion operator limiting bar growth ([Roelvink (1993)], [Reniers et al. (2004a)]). Including the long wave bottom boundary layer flow does result in net sediment transport for free long waves ([Holman and Bowen (1982)]), however typically they are on order of magnitude smaller than the bound long wave transport.

4.5 Return flow

The mass transported onshore by breaking waves within the surf zone needs to return offshore to conserve mass. This is easily envisioned in a narrow laboratory wave flume where the alongshore variation in bathymetry is absent and the wave-related mass-flux can only be compensated within the vertical (i.e. the case of horizontal mass flux compensation is discussed in the next section). This mass-flux compensation results in approximately parabolic velocity profiles within the surf zone with maximum offshore velocities in the lower part of the water column and smaller or even onshore velocities near the surface (see Section 3.7.4).

Outside of the surf zone the maximum offshore directed velocities typically occur near the surface [Lentz et al. (2008)]. These types of distribution have been observed both in the lab [Roelvink and Reniers (1995)] and in the field (e.g. [Garcez Faria et al. (2000)], [Reniers et al. (2004b)]).

The sediment transport associated with the mean return flow constitutes the dominant contribution to the total sediment transport responsible for dune erosion during energetic conditions ([Steetzel (1993)],[van Thiel de Vries et al. (2008)]). It plays a similar role for the formation of sand bars during storm conditions as demonstrated in Figure 4.9, where the sediment transport associated with the return flow is calculated from the measured mean sediment concentration profiles and mean

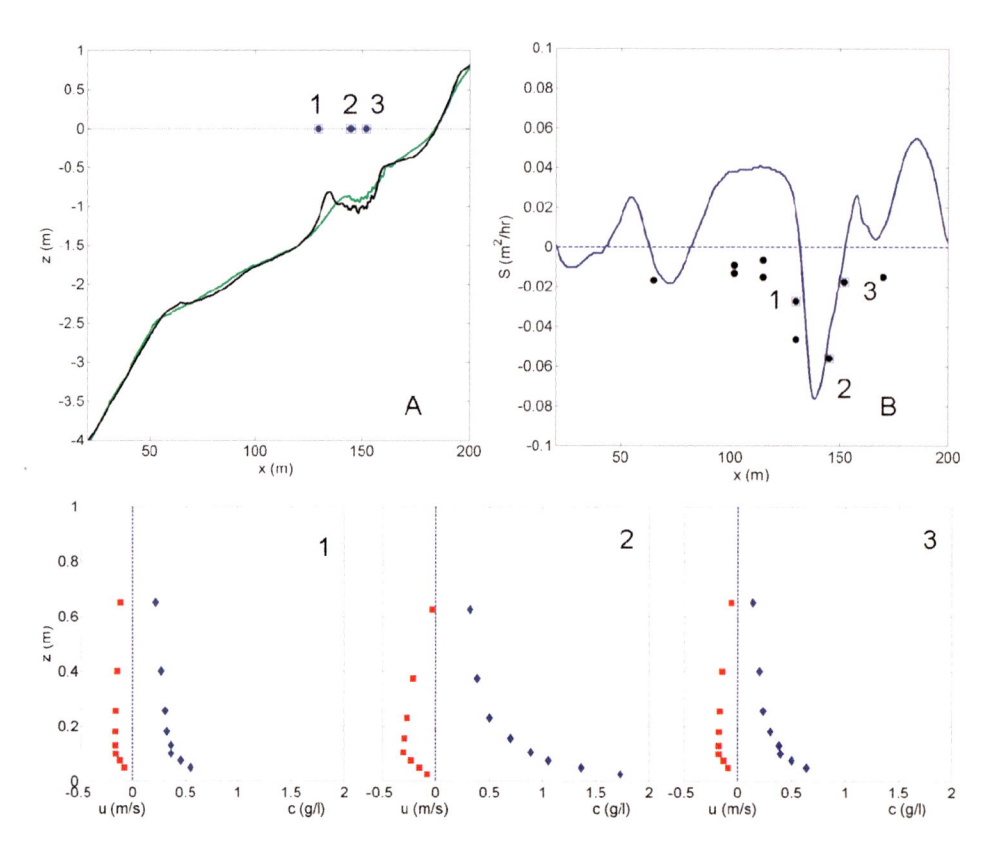

Fig. 4.9 Profile evolution (panel A, intial profile in green and final profile in black) and inferred sediment transport rates (blue line in panel B) for test 1B of the LIP11D experiments [Roelvink and Reniers (1995)]. Three lower panels show the measured hourly mean return flow velocities (red squares) and corresponding mean sediment concentrations (blue diamonds) at locations centred around a bar crest (indicated by the position numbers in panel A). Depth-integrated sediment transport rate estimates (black dots in panel B), using eq. (4.52), are compared to the inferred sediment transport rates (panel B).

return flow velocities:

$$\bar{S}_x = \int_{z=-h}^{z=0} \bar{u}\,\bar{c}\,dz \qquad (4.52)$$

which is compared to the sediment transport rates inferred from the observed bathymetric changes for test 1B (panel A of Figure 4.9), showing that in the vicinity of the bar crest where wave breaking is most intense the sediment transport is indeed dominated by the mean flow contribution (panel B in Figure 4.9). Away from the bar other transport components associated with the incident wave skewness and asymmetry as well as long waves become important or even dominant which explains why the inferred sediment transport rates can become positive, i.e. onshore directed, even though the return flow related transport is always offshore directed (panel B in Figure 4.9).

A consistent finding in the observations is that the return-flow velocities tend to have their maximum shoreward of the bar crest, resulting in a phase shift between the sediment transport rates and underlying bed profile. The physical mechanism causing this onshore shift is not well understood and typically ascribed to the breaker delay, i.e. the time a wave needs to break in response to local changes in the bottom profile [Roelvink et al. (1995)]. This effect is in addition to the roller transition (see Section 2.6.3).

4.5.1 Breaker delay

The effect of breaker delay can be introduced in several ways. [Roelvink et al. (1995)] introduced a delayed water depth in the wave transformation, where the dissipation due to wave breaking is calculated with a weighted depth:

$$h_d(x) = \frac{\int\limits_{x-\lambda}^{x} (\lambda - (x-\xi))h(\xi)d\xi}{\int\limits_{x-\lambda}^{x} (\lambda - (x-\xi))d\xi}$$

The integration distance, λ, is typically in the order of 1 (local) wave length. Alternatively the wave and or roller related mass flux can be delayed [Reniers et al. (2004a)] to push the maximum return flow velocity into the trough. Alternative expressions to the same effect are provided by [Dronen and Deigaard (2007)] and [van Rijn and Wijnberg (1996)].

4.6 Rip circulation cells

In the presence of strong alongshore variability in the surfzone bathymetry the wave-related mass flux compensation occurs predominantly in the horizontal through rip current circulations, in contrast to the vertical compensation expected for alongshore uniform coasts and laboratory flumes discussed in section 4.5. This is especially true for normally incident waves. Due the absence of strong off-shore directed pressure gradients (see Figure 3.29 in Section 3.8) the onshore flow velocity over the shallower shoals can attain velocities that are of similar order as the offshore directed (rip) currents within the rip current channels (e.g. [MacMahan et al. (2010b)] and are significantly larger than the return flow velocities observed for alongshore uniform coasts. The cross-shore transport capacity related to the mean flow, acting offshore in the channel and onshore over the shoals, thus becomes much larger for similar incident wave conditions compared with alongshore uniform coasts. An important consequence of this that an initially offshore located bar intersected by rip channels can quickly merge with the shore line forming a transverse bar.

A similar response can be expected in the presence of an shoreface nourishment of limited alongshore length, where onshore flows push the nourished sand onshore resulting a rapid progression of the nourishment towards the shore line. The efficiency of these shoreface nourishments are a function of their alongshore length, alongshore separation and also the wave climate, with increased onshore progression for decreasing incidence angles (i.e. maximum onshore motion for normally incident waves) [Koster (2006)].

The latter is associated with the fact that the flow becomes increasingly alongshore directed, resulting in elongation of alongshore interrupted bars/shoals ([Dronen and Deigaard (2007)], [Smit (2010)]). For persistent alongshore flows, e.g. tidal currents, this leads to coherent alongshore meandering bars as observed along many a coast [Van Enckevort and Ruessink (2003)]. For persistently cross-shore flows irregularly spaced rip channels are the norm ([Ranasinghe et al. (2000)], [Orzech et al. (2010)]).

Chapter 5

Morphological Processes

5.1 Introduction

In this chapter we will deal with some typical morphological processes that govern the behaviour of coastal areas.

5.2 Some principles

5.2.1 *Propagation of bed forms*

Morphological behaviour is characterized by a strong interaction between the changing bottom and the sediment transport that leads to these changes. Sometimes this leads to a propagating behaviour, as in the case of bottom disturbances in a flow; sometimes a more diffusive behaviour follows from the reaction of the transport to the bottom change, as in the case of coastline behaviour. In many cases the combined system of waves, currents, sediment transport and bottom change tends towards an equilibrium situation.

The governing equation for the change of bed level z_b is:

$$(1-\varepsilon)\frac{\partial z_b}{\partial t} + \frac{\partial S_{bx}}{\partial x} + \frac{\partial S_{by}}{\partial y} = D - E \qquad (5.1)$$

where ε is the porosity, S_{bx} and S_{by} are the bed-load transports in x- and y-direction, D is the deposition rate of suspended sediment and E is the erosion rate of suspended sediment. Very often the distinction between bed-load transport and suspended-load transport is not made in a morphological model, in which case the equation reduces to:

$$(1-\varepsilon)\frac{\partial z_b}{\partial t} + \frac{\partial S_x}{\partial x} + \frac{\partial S_y}{\partial y} = 0 \qquad (5.2)$$

with S_x and S_y the transports in x- and y-directions respectively.

A first observation we can make when looking at this equation is that a positive transport gradient leads to erosion; in other words, increasing transport in the transport direction leads to erosion, whereas decreasing transport leads to accretion.

This is easily explained by the fact that when the gradient is positive, more sediment leaves a certain control area than enters it.

Another important point is, that strong sediment transports in themselves do not lead to erosion or accretion, but only the *gradients* in the transport can do this. Of course the gradients typically scale with the magnitude of the transport, so it is not to say that the magnitude is not important, but we often find stable situations with large transports passing through the area.

As a first illustration of how bottom changes and sediment transport interact, let us look at the case of a hump-like disturbance in a channel. We assume a simple sediment transport formula, where transport is proportional to a power of the velocity.

$$S_x = au^b \tag{5.3}$$

If the hump is uniform across the channel, we very much simplify the hydrodynamics of the case, because the velocity then follows directly from the discharge per unit width, q_x and the water depth h:

$$u = \frac{q_x}{h} \tag{5.4}$$

Combining the last two equations we get a simple expression for the sediment transport:

$$S_x = a\left(\frac{q_x}{h}\right)^b = aq_x^b h^{-b} \tag{5.5}$$

where the transport is only a function of the water depth, as the coefficients and the discharge are constant along the channel. If we also assume that the water level reduction because of the accelerated flow over the hump is small, we can approximate the water depth h by $h = -z_b$ and write the sediment balance equation in 1D as:

$$(1-\varepsilon)\frac{\partial h}{\partial t} - \frac{\partial S_x}{\partial x} = 0 \tag{5.6}$$

Since transport is only a function of h, we can write the transport gradient as:

$$\frac{\partial S_x}{\partial x} = \frac{\partial S_x}{\partial h}\frac{\partial h}{\partial x} = -baq_x^b h^{-b-1}\frac{\partial h}{\partial x} = \frac{-bS_x}{h}\frac{\partial h}{\partial x} \tag{5.7}$$

Combining the last two equations we then get:

$$\frac{\partial h}{\partial t} + \frac{bS_x}{h(1-\varepsilon)}\frac{\partial h}{\partial x} = 0 \tag{5.8}$$

and we see that we now have a simple wave equation, with a celerity or characteristic velocity equal to:

$$c_x = \frac{bS_x}{h(1-\varepsilon)} \tag{5.9}$$

This describes the propagation of bed forms, where the propagation speed is proportional to the transport rate and the power in the transport formula, and inversely proportional to the bottom depth. The solution for this case is very simple using the method of characteristics: to get the profile at a certain time we simply apply a horizontal shift equal to the elapsed time multiplied by the celerity to each point on the hump:

$$x(h,t) = x(h,t_0) + c_x(h)\,(t-t_0) \qquad (5.10)$$

Because the celerity at the top of a hump will be larger than at the bottom of it, a hump will also deform, and steepen, as is illustrated in Figure 5.1 below. After a while the hump develops into a steep front, and then the method of characteristics breaks down as it would lead to an overhanging front. Of course this would be avoided by avalanching in reality. This behaviour has important consequences for numerical models: first, the bed celerity determines the time step that can be chosen in the numerical scheme and second, the numerical scheme must be capable of handling steep fronts.

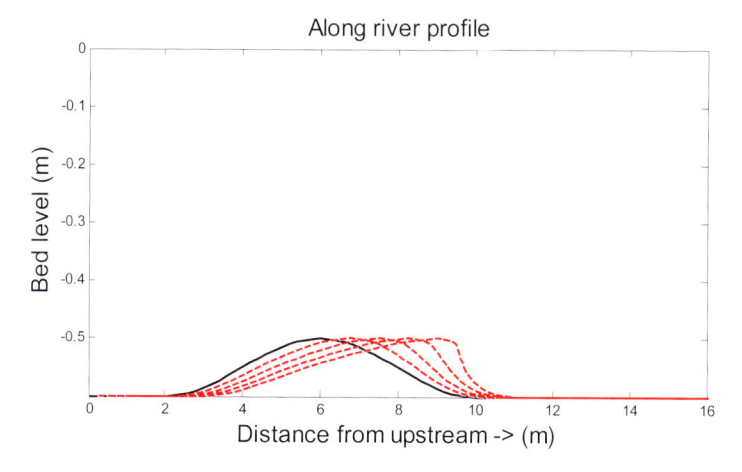

Fig. 5.1 Propagation of a hump in a river obtained with analytical solution using method of characteristics.

In Figure 5.2 we see the same method applied to a bottom depression, which can be seen as a dredged trench across the river. Here we see a typical behaviour, often observed in the field, where the trench migrates downstream and at the same time the upstream slope steepens into a front and the downstream slope flattens out.

5.2.2 *Equilibrium depth*

In cases where the current pattern is dominantly forced by the geometry, as is the case near coastal structures or at inlet mouths, we can get a reasonable first idea of the equilibrium depth of the channel by considering the sediment balance in combination with, again, a simple transport formula.

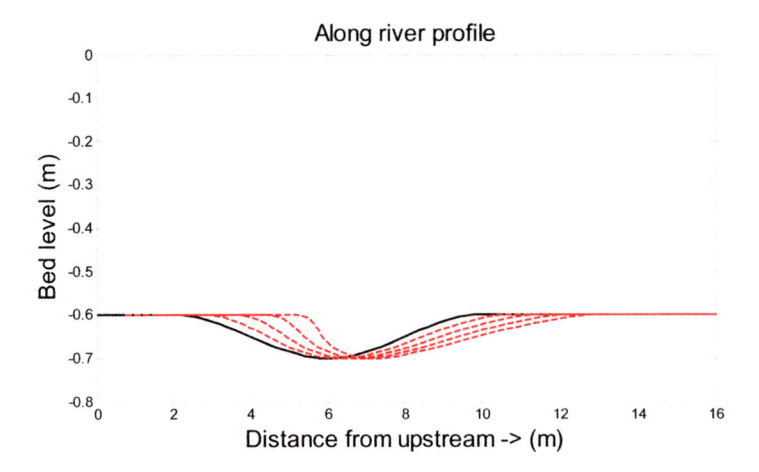

Fig. 5.2 Propagation of a bottom depression in a river obtained with analytical solution using method of characteristics.

Looking at Figure 5.3 below, we consider the flow and transport between the two streamlines indicated. In an equilibrium situation, the depth- and width-integrated discharge and sediment transport are constant, so:

$$B_2 u_2 h_2 = B_1 u_1 h_1 \qquad (5.11)$$

$$B_2 S_2 = B_1 S_1 \Rightarrow B_2 u_2^b = B_1 u_1^b \qquad (5.12)$$

$$\frac{h_2}{h_1} = \frac{B_1 u_1}{B_2 u_2} = \frac{B_1 u_1}{B_2 \left(\frac{B_1}{B_2}\right)^{1/b} u_1} = \left(\frac{B_1}{B_2}\right)^{1-1/b} \qquad (5.13)$$

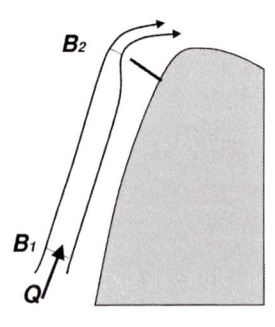

Fig. 5.3 Example of flow contraction around an obstruction

5.3 Open coasts

Open (sandy) coasts are primarily shaped by the waves and wave-driven currents. The waves continuously modify the shape of the cross-shore profile. This is seen most dramatically during storms, when dune erosion can create spectacular scarps and we see houses and resorts falling into the water. During calmer conditions, beaches can recover and generally a healthy beach profile varies around an equilibrium shape (e.g. [Dean (1973)]). In the following we will discuss the morpological mechanisms responsible for these change in profile shape. A distinction is made between cross-shore profile behavior on predominantly alongshore uniform coasts and coasts with significant alongshore variation in the nearshore bathymetry.

Wave-driven longshore currents move sand along the coast; where there are gradients in this longshore transport, this leads to structural changes in the sediment budget, where beaches may accrete or erode in the longer term. Typical examples that will be discussed are river deltas and coastline changes adjacent to breakwaters.

5.3.1 *Cross-shore profile behavior*

5.3.1.1 *Bar-dynamics*

At a coast with alongshore bars the cross-shore bottom profile responds to the incoming waves in a quasi-predictable way, with bars propagating offshore during erosive conditions and onshore during moderate wave conditions. This is the result of the cross-shore balance between onshore and offshore transport contributions related to both waves and currents (e.g. [Roelvink and Stive (1989)], [Roelvink and Broker (1993)], [Thornton et al. (1996)], [Hoefel and Elgar (2003)], [Henderson et al. (2004)], [Ruessink et al. (2007)]). Some of the underlying transport processes have been discussed in section 4.3, including wave skewness and asymmetry, boundary layer effects, streaming, bound infragravity waves and mean flow contributions. All of these processes are a function of the incident wave conditions and their interaction with the bottom profile. As a result of the continuously changing offshore wave conditions, which typically occur on a much faster time scale than the morphological time scale at which the beach responds, the bottom profile is never in equilibrium with the forcing conditions.

5.3.1.2 *Bar generation and propagation*

Although sand bars are ubiquitous it is often not clear what underlying mechanisms are responsible for their generation, propagation and decay. A key parameter in bar behaviour is the phase-shift between the sediment transport pattern and the underlying bottom profile. A positive, i.e., onshore, phase shift allows bars to grow while propagating. A zero phase shift results in propagation only, whereas a negative phase shift leads to damping of the bar. These effects can be readily demonstrated with the simplified assumption that on a undulating bed given by:

$$h = h_0 + \text{Re}\,(h_1 \exp(ikx - i\omega t)) \tag{5.14}$$

where h_0 is the undisturbed water depth and h_1 is the amplitude of the bed anomaly, the sediment transport is proportional to the local water depth (panel A Figure 5.4):

$$S = -S_0 + \alpha h_1 Re\,[\exp(ikx - i\omega t - i\varphi)] \tag{5.15}$$

with S_0 the mean offshore directed sediment transport and α is a proportionality factor with increased (decreased) offshore transport at smaller (larger) depth (panel B of Figure 5.4). Both the bed and sediment transport are allowed to vary harmonically in time and space as expressed by the frequency ω and wave number k. We also allow the frequency to be complex to represent both propagation and temporal growth or decay but keep the wave number real, hence no spatial growth or decay. The phase shift between the sediment transport pattern and the underlying bed is expressed by φ. Next, the bed elevation changes are obtained from the sediment balance equation:

$$\frac{\partial z_b}{\partial t} + \frac{\partial S}{\partial x} = -\frac{\partial h}{\partial t} + \frac{\partial S}{\partial x} = 0 \tag{5.16}$$

by substituting the expressions for the water depth 5.14 and the sediment transport 5.15 in the sediment balance 5.16. This gives the following relationship between frequency and bed undulation wave number:

$$i\omega h_1 \exp(ikx - i\omega t) = -i\alpha k h_1 \exp(ikx - i\omega t - i\varphi) \tag{5.17}$$

which reduces to:

$$\omega = -\alpha k \exp(-i\varphi) \tag{5.18}$$

After substitution in the expression for the water depth, eq. (5.14), the bed evolution is given by:

$$h = h_0 + \text{Re}\,(h_1 \exp(ikx + i\alpha k \exp(-i\varphi)t)) \tag{5.19}$$

For a zero phase shift, $\phi = 0$, the frequency is real and we retrieve a bed feature of constant shape propagating in the negative x-direction:

$$h = h_0 + \text{Re}\,(h_1 \exp(ikx + i\alpha k t)) \tag{5.20}$$

as shown in panel A of Figure 5.4 after some arbitrary time lapse t-t_0, where t_0 corresponds to the starting time.

Introducing a positive phase shift displaces the sediment transport pattern with the underlying bottom profile (compare panels B and D of figure 5.4) and the frequency now has a complex and real contribution resulting in the following expression for the bed evolution:

$$h = h_0 + \text{Re}\,(h_1 \exp(ikx + i\alpha k \exp(-i\varphi)t)) \tag{5.21}$$

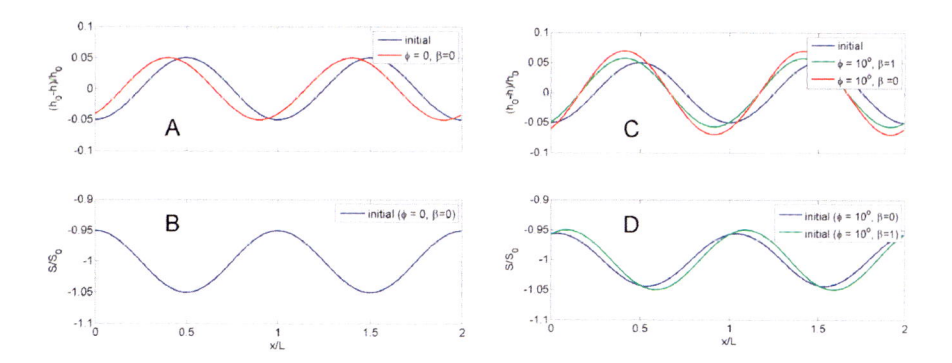

Fig. 5.4 Panel A: Initial normalized bed anomaly (blue line) and bed evolution after $t - t_0$ time lapse (red line) for a zero degree transport phase lag without bed slope related transport as function of normalized distance with L the anomaly wave length. Panel B. Initial normalized sediment transport for zero phase lag. Panel C. Initial normalized bed anomaly (blue line), bed evolution after t-t0 time lapse (red line) for a 10° transport phase lag without bed slope related transport (red line) and with bed slope transport included (green line). Panel D: Initial normalized sediment transport for 10° phase lag without (blue) and with bed slope related transport (green line).

which can be written as:
$$h = h_0 + \mathrm{Re}\left(h_1 \exp(ikx)\exp(iak\cos\varphi t)\exp(ak\sin\varphi t)\right) \quad (5.22)$$
corresponding to a growing and propagating bed feature (see panel C of Figure 5.4). The growth rate of the bed undulation is given by the argument of the third exponential term on the right hand side of eq. (5.22), i.e. $ak\sin\varphi$, which is a function of the wave number. Note that for a negative phase lag, $-180° < \varphi < 0°$, the argument of the exponential function becomes negative leading to a damping of all undulations. Also note that the propagation direction is a function of the phase lag, where a large positive phase lag, $\varphi > 90°$, results in an onshore propagation of growing bed undulations (not shown). For positive φ the largest growth rates are obtained for the shortest wave numbers. Without damping this will result in unlimited growth of the bed amplitudes yielding an unrealistic situation. However, as the bed profile becomes more undulated, bed-slope related transports become increasingly important, damping the short scale undulations, as will be shown next.

5.3.2 Bed-slope related transport

The sediment transport is extended with a bed-slope related term:
$$S = -S_0 + \alpha h_1 \mathrm{Re}\left[\exp(ikx - i\omega t - i\varphi)\right] + \beta \frac{\partial h}{\partial x} \quad (5.23)$$
with β again a proportionality factor. Upon substitution of eq. (5.14) this yields the following expression for the sediment transport:
$$S = -S_0 + \alpha h_1 \mathrm{Re}\left[\exp(ikx - i\omega t - i\varphi)\right] + \beta h_1 \mathrm{Re}\left[ik\exp(ikx - i\omega t)\right] \quad (5.24)$$

The addition of the slope related transport results in an increase in the down-slope sediment transport and an accompanying decrease in the up-slope sediment transport (compare green and blue lines in panel D of Figure 5.4). Substitution of the extended sediment transport expression into the sediment balance, eq. (5.16), yields:

$$\omega = -\alpha k \exp(i\varphi) - i\beta k^2 \qquad (5.25)$$

and for the bed evolution equation:

$$h = h_0 + \mathrm{Re}\left(h_1 \exp(ikx)\exp(i\alpha k \cos\varphi t)\exp(k\sin\varphi t)\exp(-\beta k^2 t)\right) \qquad (5.26)$$

where $\exp(-\beta k^2 t)$ now reduces the growth rate of the bed undulations for longer wave lengths (compare red and green lines in the upper right panel C of Figure 5.4) and inhibits the growth of short scale disturbances when $\beta k^2 > k \sin\phi$.

5.3.2.1 Long term bar behaviour

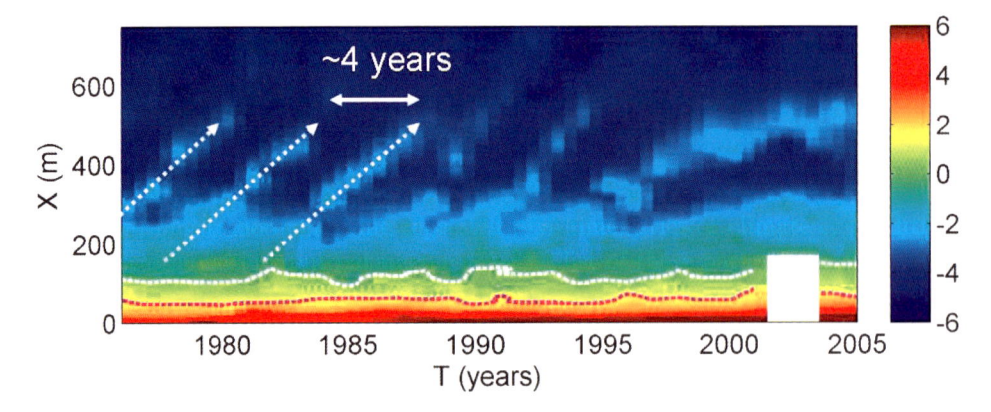

Fig. 5.5 Observations of bar evolution at Noordwijk, the Netherlands, obtained from the bi-annual Jarkus profile measurements. The anomolous bar response after 1999 is associated with a nourishment. Approximate shore line indicated by the white dashed line. Dune foot location indicated by the magenta dashed line.

The long term behaviour of a barred coast often displays a seaward cycling of the bars, where new bars are generated close to the shore line and offshore bars decay with time shown in Figure 5.5. The cycle-time varies with beach location, with rapid time scales in the order of 2 years for steeper beaches such as Duck, North Carolina, USA (e.g. [Plant et al. (1999)]) and much longer time scales for mildly sloping beaches such as $O(12)$ years at Egmond, The Netherlands ([Wijnberg (2002)]). These time scales are most likely related to the offshore forcing conditions and sand volume of the bars, where fat bars react much slower to the incident wave forcing than skinny bars ([Van Enckevort and Ruessink (2003)], [Smit et al. (2005)]). Within the yearly cycle there is typically still significant variation in bar position and bar shape subject to the variations in the incident wave conditions at the seasonal and shorter storm time scales ([Plant et al. (1999)]).

Table 5.1 Parameter settings for bottom profile eqs. 5.23-5.24

z_r	A_b	x_b	R_b	L_b	φ_b	A	b	T_b
6 m	2 m	150 m	50 m	100 m	-pi/4	1.4	0.38	4 yr

The mechanisms behind the cyclical behaviour of the bars can be examined by considering the sediment transport distribution that is required to change the barred profile. Starting with a barred profile given by [Bakker and De Vroeg (1988)]:

$$z_{b,mean} = z_r - A(x - x_r)^b \tag{5.27}$$

and:

$$z_b = z_{b,mean} - z_{b,anomaly} = z_{b,mean} - A_b \exp\left(-\left(\frac{x - x_b}{R_b}\right)^2\right) \cos(k_b x - \omega_b t + \varphi_b) \tag{5.28}$$

with the parameter values given in the Table 5.1, where $k_b = \frac{2\pi}{L_b}$ and $\omega_b = \frac{2\pi}{T_b}$. The profile has a clear bar-trough signature with the crest located approximately 120 m from the shore line (see Figure 5.6).

The corresponding transport gradients can be obtained from the sand balance, eq. (5.16):

$$\frac{\partial S_x}{\partial x} = -\frac{\partial z_b}{\partial t} = A_b \omega_b \exp\left(-\left(\frac{x - x_b}{R_b}\right)^2\right) \sin(k_b x - \omega_b t + \varphi_b) \tag{5.29}$$

which in turn can be integrated to give the time and spatially varying sediment transport rates (lower panel of Figure 5.6) responsible for the time-evolution of the barred profile. The local phase shift between the bar positions and the concurrent sediment transport pattern are examined by plotting the minima (maxima) in the sediment transport rate (bed-anomaly). Locations of positive spatial shift correspond to locations of growing bed features, as described above whereas locations with a negative spatial shift correspond to damping of the bed features. Based on the snapshot shown in Figure 5.6 we therefore expect the bed undulations near the shore line to grow, while they propagate offshore. While propagating through the surf zone the positive phase shift then gradually becomes negative leading to an attenuation of the bed anomalies.

This is further illustrated in a time-space plot of the bed evolution during a number of years (Figure 5.7), where the spatial shift can be inferred from by the distance between the black dashed line (bed maxima) and white dashed line (sediment transport minima). Bars are generated near the shore line at locations of positive phase shift, subsequently grow while traveling offshore. At the outer edge of the surf zone the phase shift becomes negative and subsequently the bars decay and disappear as they propagate offshore.

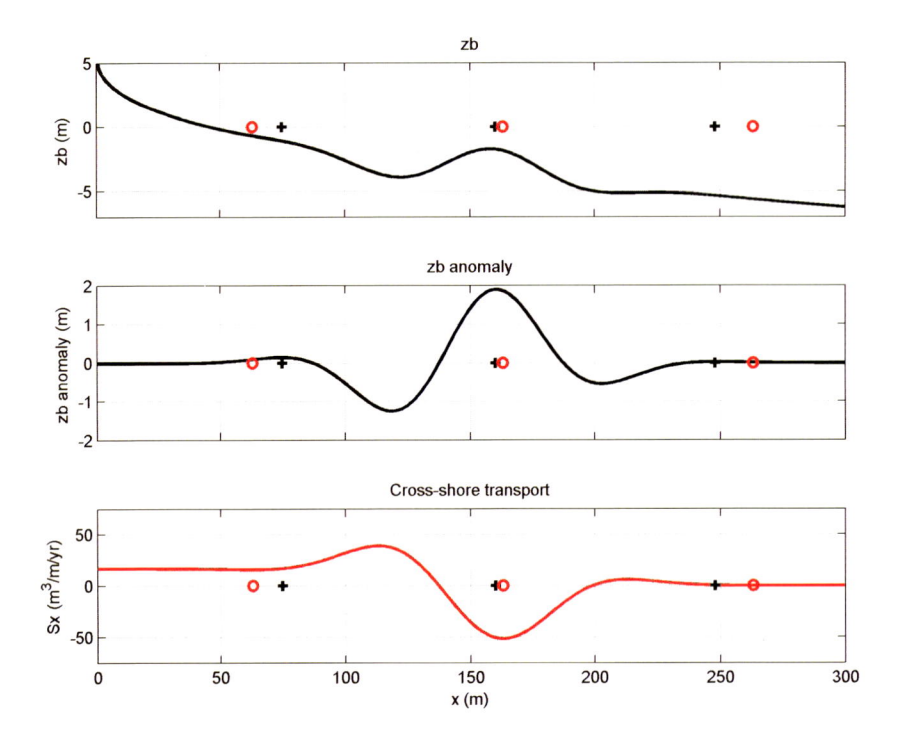

Fig. 5.6 Upper panel: Initial Bakker and de Vroeg bottom profile (eqs. (5.27) and (5.28)). Middle panel: Corresponding bed anomaly. Lower panel: Initial sediment transport distribution (obtained after integration of eq. (5.29)). The spatial shift between bed elevation and sediment transport is indicated by their respective maxima (plusses) and minima (circles).

5.3.3 *Dune erosion and overwash*

The coastal impacts and coastal vulnerability of sandy coasts to severe storms has become alarmingly clear with recent hurricane Ivan and Katrina landfalls along the gulf coast in the USA. And although existing tools to assess dune erosion under extreme storm conditions have been applied successfully along relatively undisturbed coasts ([Vellinga (1986)], [Steetzel (1993)], [Nishi and Kraus (1996)]), they are inadequate for more complex situations, for instance when the coast has significant alongshore variability such as in the case of barrier islands protecting the hurricane impact of the main-land coast. In that case the elevation, width and length of the barrier island, as well as the hydrodynamic conditions (surge level) of the back bay should be taken into account to assess the coastal response. Furthermore, these models should be able to represent not just the swash regime during the initial phase of dune erosion, but also during overtopping during the collision regime, the overwash regime and inundation ([Sallenger (2000)]).

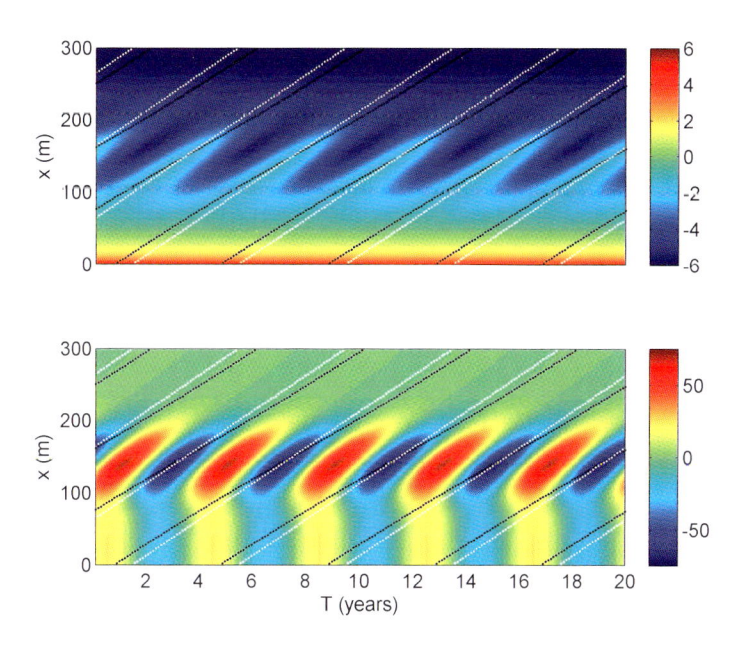

Fig. 5.7 Upper panel: Simulated bed evolution based on eq. (5.28) (bed level in m indicated by the color bar). Lower panel: Corresponding sediment transport distribution (obtained after integration of eq. (5.29)) (sediment transport in m^3/m/yr indicated by the colorbar). Phase shift between bed anomalies and sediment transport indicated by their respective maxima (black dots) and minima (white dots).

Swash motions during storm conditions are generally dominated by infragravity waves generated by the grouped short waves (see Section 3.6) due to the fact that the short waves break within the surfzone and infragravity waves persist [Raubenheimer and Guza (1996)]. The infragravity swash consists of a combination of leaky and trapped infragravity waves (see Section 3.6.3) as observed by e.g. [Huntley et al. (1984)]. The sediment transport processes in the surf and swash zone are complex, with sediment stirring by a combination of short-wave and long-wave orbital motions, currents and wave breaking-induced turbulence. However, intra-wave sediment transports due to wave asymmetry and wave skewness (see Section 4.4.1) are known to be relatively minor compared to long-wave and mean current contributions (e.g. [van Thiel de Vries et al. (2008)]) (see also Section 4.4.5). As a result we can use a relatively simple and transparent formulation, such as Soulsby - Van Rijn ([Soulsby (1997)]), in a short-wave averaged but wave-group resolving model of surf zone processes (see Section 4.3.2). This formulation has been applied successfully in describing dune erosion ([Roelvink et al. (2009)]), overwash ([McCall et al. (2010)]), and the generation of rip channels ([Damgaard et al. (2002)] and [Reniers et al. (2004a)]).

An important element in describing dune erosion is the slumping of dry sand and the subsequent transport by the swash waves and return flow: without it the upper beach scours down and the dune erosion process slows down considerably. One-dimensional (cross-shore) models such as DUROSTA ([Steetzel (1993)]) focus on the underwater offshore transport and obtain the supply of sand by extrapolating these transports to the dry dune. [Overton and Fisher (1988)], [Nishi and Kraus (1996)] focus on the supply of sand by the dune based on the concept of wave impact. Both approaches rely on heuristic estimates of the runup and are well suited for 1D application but difficult to apply in a horizontally 2D setting.

$$\left|\frac{\partial z_b}{\partial x}\right| > m_{cr} \tag{5.30}$$

Here we apply an avalanching mechanism, where we account for the fact that saturated sand moves more easily than dry sand, by introducing both a critical wet- and dry slope with typical values for $m_{cr,dry} = 1$ and $m_{cr,wet} = 0.3$ (see eq. (5.30). As a result slumping is predominantly triggered by a combination of infragravity swash run-up on the previously dry dune face and the (smaller) critical wet-slope ([Roelvink et al. (2009)]).

During the overwash regime the flow is dominated by low-frequency motions on the time-scale of wave groups, carrying water over the dunes. This onshore flux of water is an important landward transport process where dune sand is being deposited on the island and within the shallow inshore bay as overwash fans (e.g. [Leatherman et al. (1977)], [Wang and Horwitz (2007)]). This requires the inclusion of the wave-group forcing of low frequency motions in combination with a robust momumtum-conserving drying/flooding formulation (such as [Stelling and Duinmeijer (2003)]) and concurrent sediment transport and bed-elevation changes. Example applications for dune erosion and overwash are presented in Sections 10.3 and 10.4 respectively.

5.3.4 *Rip channel dynamics*

Similar to sand bars, rip channels display (quasi-) periodic behaviour in space and time. Their distribution is often characterized by their quasi-rhithmic alongshore spacing. Their temporal evolution is coupled to the beach state as outlined by [Wright and Short (1984)], ranging from alongshore uniform beaches, i.e. without rip channels, to shore attached rip features. Recently the dynamics of sand bars, shoals and rip channels have been observed with video observation techniques [Holman and Stanley (2007)]. The video captures photographic images of breaking waves that show up as a region of high light intensity. Averaging over a number of images results in a time exposure, showing areas of frequent wave breaking as light and areas of non-breaking waves as being dark (see left panel of Figure 5.8). As a result the light intensity works as a proxy for the underlying water depth, with high intensities for relatively shallow areas and low intensities for the deeper parts. This is confirmed by overlaying the measured depth contours on the time exposure (left panel of Figure 5.8). This clearly shows the shoals (light areas) and deeper rip-channels (dark areas with the surf zone). By tracking the changes in intensity over time a good picture of the nearshore morphological response to the local hydrodynamics can be obtained. These video observations show that alongshore quasi-rhithmic features are ubiquitous on beaches around the globe (e.g. [Lippmann and Holman (1989)], [Ranasinghe et al. (2000)], [Van Enckevort and Ruessink (2003)], [Turner et al. (2006)]). The general idea is that alongshore varying bathymetry

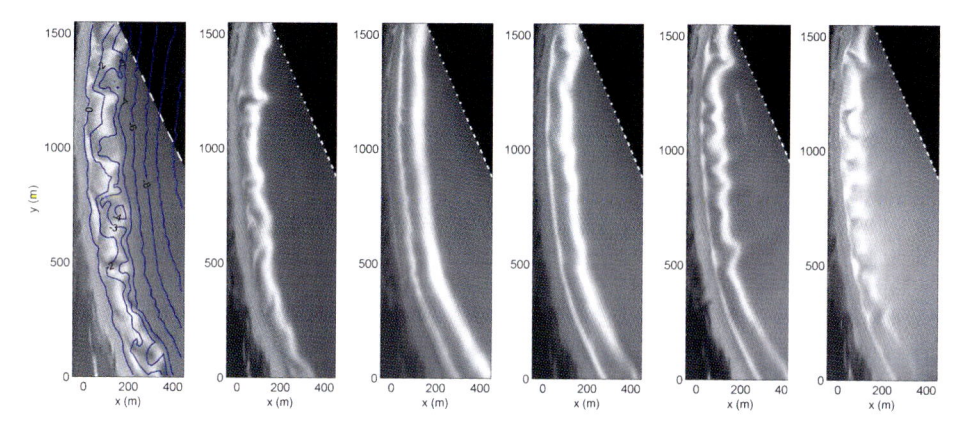

Fig. 5.8 Panel sequence from left to right. Panel A: Time exposure of wave breaking [Holman and Stanley (2007)] overlain with depth contours obtained from a bathymetric survey during RDEX [Reniers et al. (2001)] with darker (lighter) areas corresponding to rips (shoals). Panels B-F: A sequence of time exposures showing an upstate transition to a reset with an alongshore uniform bar and through (Panel C) and subsequent downstate transition to shore attached shoals and rips at Palm Beach, Australia (courtesy of Graham Symonds).

disappears during storm conditions as has been observed for a number of occasions were the high waves transform the alongshore variable bathymetry into a more or less alongshore uniform beach (compare panels B and C of Figure 5.8). However, the underlying process that is responsible for the transition is presently not understood, even though this is a very strong signal. Recent modeling efforts by [Smit (2010)] suggest this may in a significant part be related to wave breaking turbulence. The longshore current may also play an important role by stretching and translating the morphological features thus creating an alongshore uniform beach. During quiescent times after the rest event the morphology typically starts displaying rithmic features again which gradually develop into rip-channels and shoals. Continued quiescent conditions lead to then shore-attached shoals and rip-channel formation (last panel of Figure 5.8).

The generation of a quasi-rhithmic beach can be explained by the inherent instability of a barred beach to normally incident waves [Hino (1974)], where small perturbations develop into crescentic bar patterns as amply demonstrated by [Falques et al. (2000)]. The effect is illustrated with a numerical model computation using the depth-averaged shallow water equations with the Soulsby van Rijn sediment transport formula (see Section 4.3.2) for normally incident monochromatic waves of

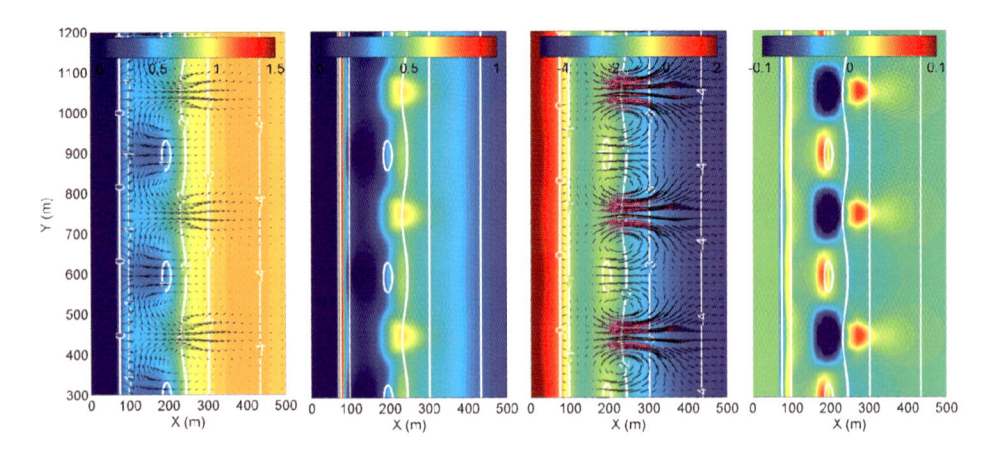

Fig. 5.9 Panel sequence A-D from left to right. Positive feedback mechanism for normally incident waves propagating over bed undulations superimposed on an alongshore uniform beach (depth contours given by solid white lines in (m)). Panel A: computed wave height (indicated by the color bar in (m)) and flow field (indicated by the arrows). Panel B: Corresponding depth-averaged sediment concentration (color bar in (g/l)). Panel C: Resulting sediment transport pattern (indicated by the arrows) and depth (color bar in (m)). Panel D: Calculated bed changes (color bar in m) demonstrating the positive feedback resulting in deeper channels and shallower shoals (calculations performed with Delft3D).

1 m wave height and a peak period of 10 s. The initial bathymetry has of an offshore bar with a 1m alongshore modulation in height with a length scale of 300 m (see Figure 5.9). The elevation change associated with the bed perturbations leads to an excess radiation stress over the shallower parts of the bar and a counter circulation through the deeper part of the bar crest (see panel A in Figure 5.9). Wave breaking at the offshore side of the shoals leads to elevated sediment concentrations (panel B in Figure 5.9) thereby pushing sand onshore over the shoal towards the coast (panel D in Figure 5.9). In the deeper rip channels the increased offshore directed velocities result in an offshore directed sediment transport (panel C in Figure 5.9) deepening the rip channels with subsequent accretion further offshore. This leads to a positive feedback and subsequent amplification of the bed perturbations (Figure 5.9). The initial alongshore spacing is a function of many variables, including bar position, bar shape, type of wave breaking, turbulent diffusion, grain size, as assessed from numerical modeling (e.g. [Dodd et al. (2003)], [Damgaard et al. (2002)], [Dronen and Deigaard (2007)], [Reniers et al. (2004a)], [Calvete et al. (2007)], [Castelle et al. (2010)]).

The origin of the initial bed perturbations varies. The bathymetry in the nearshore is never truly uniform and undulations are the norm. These may result from pre-existing bathymetric variability that was not erased during the reset-event. Alternatively initial undulations can be induced by quasi-steady circulation patterns forced by groups of breaking waves [Reniers et al. (2004a)], standing edge waves [Holman and Bowen (1982)] and wave-current interaction [Dalrymple (1975)].

5.3.5 Plan shape evolution

Fig. 5.10 Aerial view of the coastline near showing smoothly arching coastline sections.

When we look at sandy coastal shapes and evolution on the length scale of kilometres to tens of kilometres, the dominant feature is the smoothly arching coastline, as in the example above. In such situations we can make some assumptions about the coastal behaviour that make it possible to describe it in a relatively simple way:

- The depth contours in the shallow area are almost straight and parallel, so that the effect of their curvatures on the refraction of waves can be neglected.
- The longshore variations of the coastline are only gradual, so that the longshore transport is always in equilibrium with the wave conditions.
- The shape of the cross-shore profile is more or less the same everywhere, and when the coastline shifts, it remains the same;

Based on the first and second assumptions, we can consider the cross-shore profile to be locally uniform, so that we can use one-dimensional wave and roller energy balances to derive the cross-shore variation of wave energy and radiation stresses, for given offshore wave conditions. From these we can compute longshore velocity profiles and longshore sediment transport distributions over the profile. After integrating the longshore transport rate over the surf zone we then get the total longshore transport for the given wave condition. In this 'process-based' approach we can still consider the effects of the particular shape of the cross-shore profile. Alternatively, we can apply a total longshore transport formula such as the CERC equation or the Kamphuis formula to compute the longshore transport directly from the incident wave conditions and some sediment properties.

Under the third assumption we can derive a simplified sediment balance equation. If we orientate the x-axis along the shoreline and the y-axis pointing offshore, we can derive the longshore sediment balance as follows. We consider a stretch of coastline that moves a distance Δy, because of a difference in sediment transport

ΔS_x acting over a time period Δt.

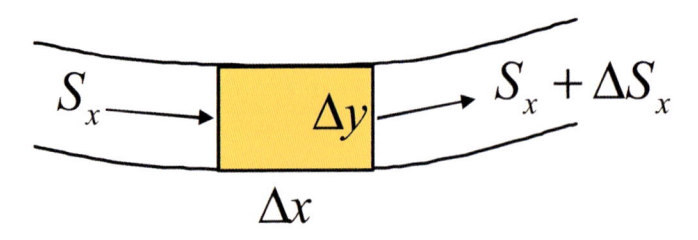

Fig. 5.11 Plan view principle sketch coastline model

The change in cross-sectional area over this time is ΔA, so:

$$\Delta V = \Delta A \Delta x = d\Delta y \Delta x = -\Delta S_x \Delta t \Rightarrow \frac{\Delta y}{\Delta t} = -\frac{1}{d}\frac{\Delta S_x}{\Delta x} \quad (5.31)$$

$$\lim_{\Delta x \to \infty} \Rightarrow \frac{\partial y}{\partial t} = -\frac{1}{d}\frac{\partial S_x}{\partial x}$$

This simple equation relates the rate of change of the coastline to the gradient of longshore sediment transport. An important aspect here is the profile height d, which spans the vertical distance over which the profile moves uniformly, typically from the dune top to the so-called *depth of closure* h_c.

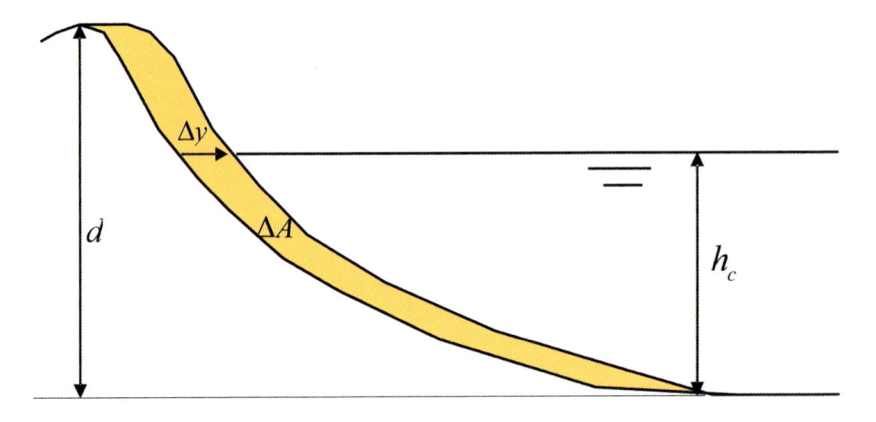

Fig. 5.12 Principle sketch sediment balance with uniform profile shift

A vital next step in creating a coastline model is now to relate the longshore transport to the coast orientation φ_c. Here we can either use a profile model and evaluate the total longshore transport for different values of the coast orientation, or use a simplified total longshore transport model. In either case the result is a so-called $S-\varphi$ relation, which gives the transport as a function of the coast orientation. Typically such an $S-\varphi$ relation looks like this:

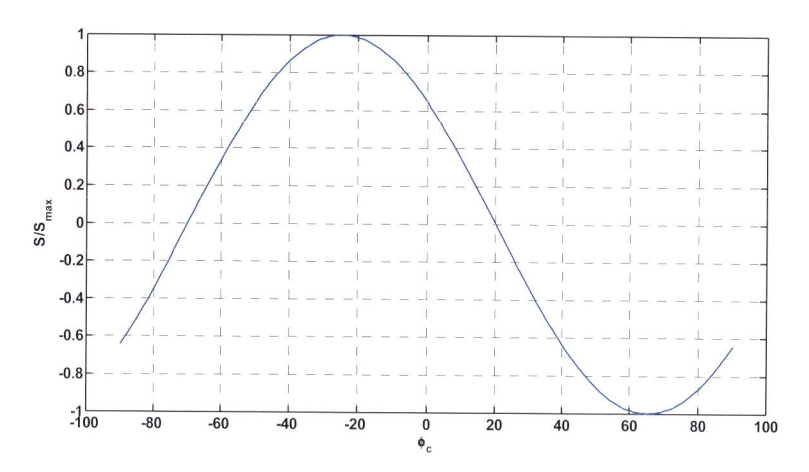

Fig. 5.13 Alongshore sediment transport normalized by it's maximum as a function of the orientation of the coast for a mean incidence angle of 20 degrees.

In this example the mean wave direction is 20 deg, which leads to zero transport for a coast orientation of 20 deg. For larger coast angles the transport is in negative x direction, for smaller coast angles in positive direction. The maximum transport occurs close to a wave direction at 45 deg. to the shore normal.

For relatively small wave angles relative to the coast, the curve can be approximated by a straight line:

$$S_x \approx -s_x \varphi_c + S_{x,0} = -s_x \arctan \frac{\partial y}{\partial x} + S_{x,0} \approx -s_x \frac{\partial y}{\partial x} + S_{x,0} \qquad (5.32)$$

where s_x is the so-called *coastal constant* and $S_{x,0}$ is the transport for $\varphi_c = 0$. The longshore transport gradient is then easily obtained as:

$$\frac{\partial S_x}{\partial x} = -s_x \frac{\partial^2 y}{\partial x^2} \qquad (5.33)$$

Combining this with eq. (5.31) we get the Pelnard-Considere's [Pelnard-Considere (1954)] equation:

$$\frac{\partial y}{\partial t} = \frac{s_x}{d} \frac{\partial^2 y}{\partial x^2} \qquad (5.34)$$

This diffusion equation has solutions in the form of the error function, and can explain a range of coastal behaviour, such as:

- Accretion and erosion next to a groin or harbour
- Development of a river delta
- Longshore dispersion of a nourishment

5.3.5.1 Accretion and erosion next to groin

The solution for this case is given by the following function:

$$y_* = \left[\exp\left(-x_*^2\right) - x_*\sqrt{\pi}\left(1 - erf(x_*)\right)\right] sign(x) \qquad (5.35)$$

$$x_* = \frac{|x|}{\sqrt{4at}}, \quad y_* = \frac{y}{\sqrt{4at}}\frac{\sqrt{\pi}}{\varphi'}, \quad a = \frac{s_x}{d} = \frac{S_\infty}{\varphi' d}$$

where φ' is the angle of wave incidence.

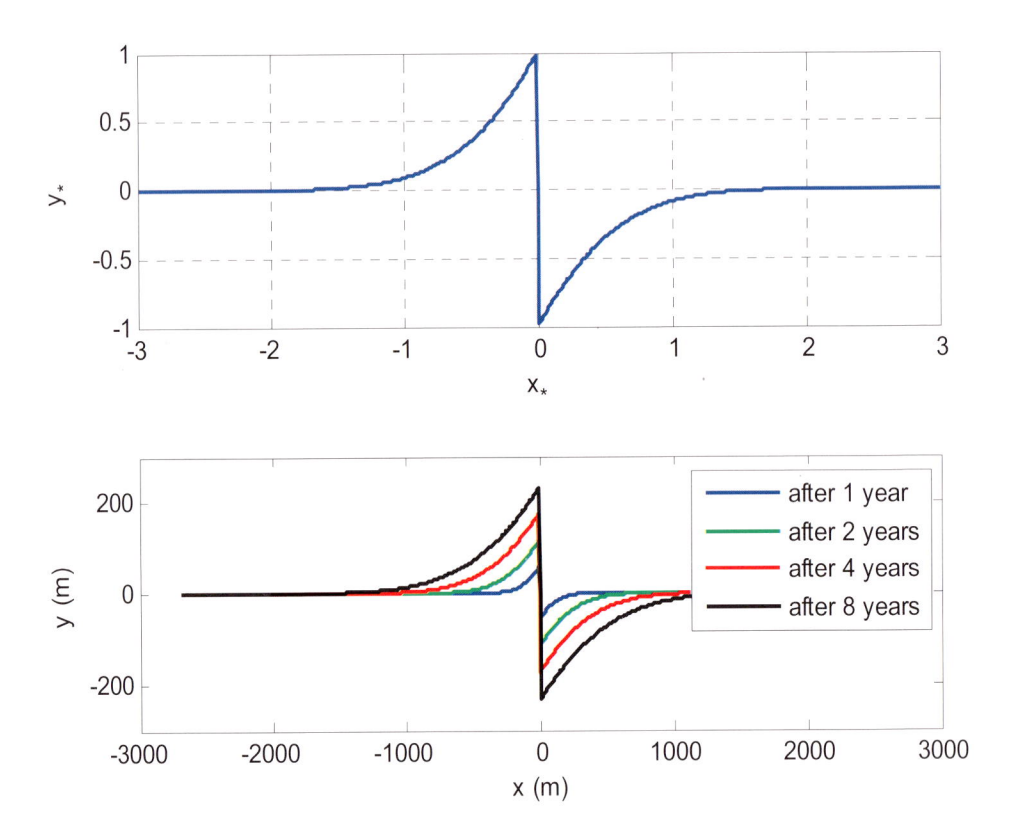

Fig. 5.14 Solution of Pelnard-Considere for the case of a groin; dimensionless shape (top panel) and example dimensional evolution for the case of sx=500,000 m3/yr/rad, wave angle 20 deg.

The shape of the function between square brackets is shown in Figure 5.14. It is 1 at the groin and goes to 0 at infinity. Clearly it scales with the square root of time and a. The accretion or erosion of the beach immediately next to the groin is given by:

$$y(x_0) = \varphi' \sqrt{\frac{4at}{\pi}}, \quad a = \frac{s_x}{d} = \frac{S_\infty}{\varphi' d} \tag{5.36}$$

In the lower panel of Figure 5.14 the evolution in time is shown in dimensional form, for $a = 100,000$ m^2/yr. We now clearly see that for a quadratic increase of the time, the dimensions of the accretion and erosion patterns in x- and y-direction both increase linearly.

5.3.5.2 Development of a growing river delta

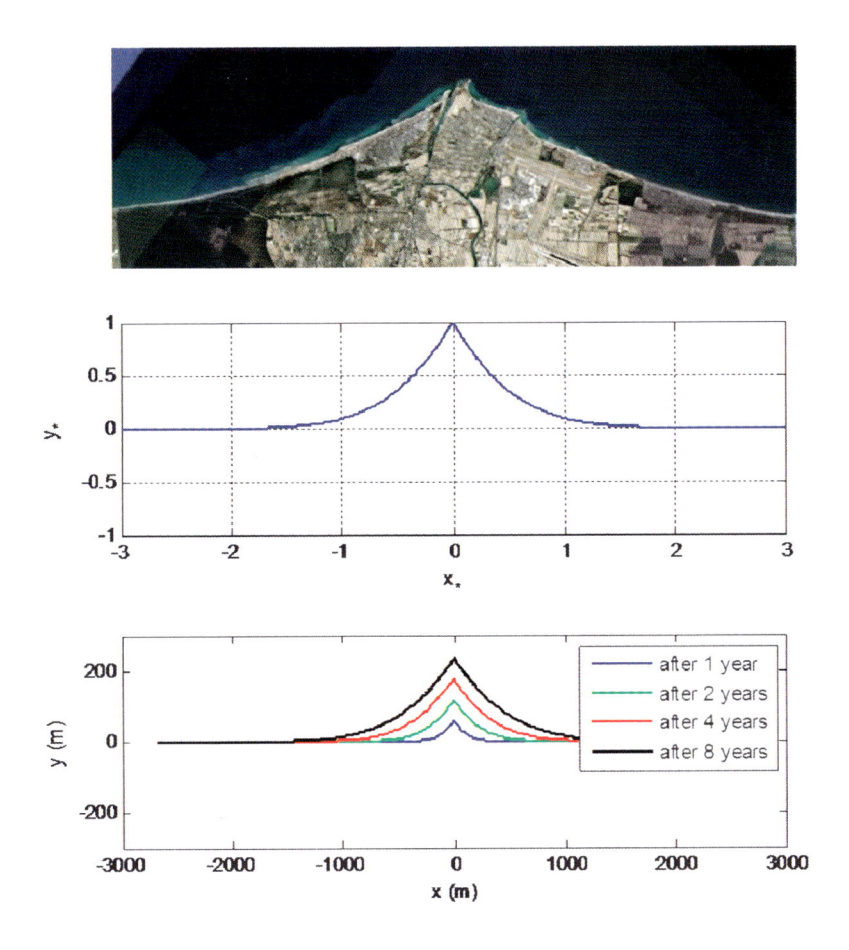

Fig. 5.15 Solution of Pelnard-Considere model for river delta; top panel: dimensionless solution; bottom panel: dimensional example for a river output of 350,000 m3/yr and a coastal constant of 500,000 m3/yr/rad

The same type of solution can be applied to the case of a growing river delta, by considering that the sediment transport is split between both sides of the delta. The boundary condition, for perpendicularly incident waves, is then that at both ends of the river mouth:

$$|\varphi_0| = \frac{1}{2}\frac{S_{river}}{s_x} \qquad (5.37)$$

The shape of the solution is shown in Figure 5.15. The development is very much like the previous case, but the erosive part has been flipped around. From eq. (5.33) it follows that the inclination of the delta at the mouth depends on the balance between the river output and the coastal constant, which indicates how well the waves can redistribute the sediment along the shore.

A good example is that of the Tiber delta near Rome, Italy. As in many developed countries, deforestation in the past led to large soil erosion and from this high river output of sediments. At present, many rivers have now been dammed and reforestation and other measures to control soil erosion have been put in place, and as a result many of these deltas are eroding, since the waves still tend to transport sand away from the mouth.

5.3.5.3 Longshore dispersion of a beach nourishment

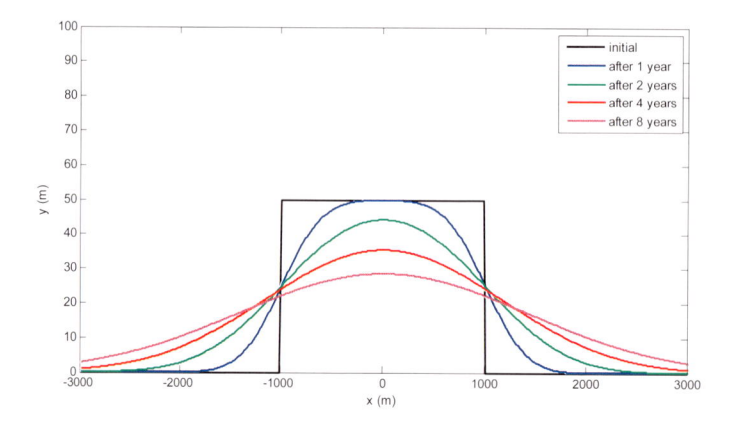

Fig. 5.16 Alongshore dispersion of a beach nourishment subject to waves at small incidence angles.

For the longshore dispersion of a beach nourishment with an initially rectangular plan view shape, the following solution is found (e.g. [Rijkswaterstaat (1988)]):

$$y = \frac{B}{2}\left[erf\left(\frac{L/2 - x}{\sqrt{4at}}\right) + erf\left(\frac{L/2 + x}{\sqrt{4at}}\right)\right] \qquad (5.38)$$

Here B is the initial width of the nourishment and L the length of the nourishment. According to this solution, the shape of the nourishment smoothens out in a symmetrical fashion (see Figure 5.16). Note that the direction of the waves (under the assumption of small wave angle) does not affect the solution, nor does the nourishment propagate along the coast under these conditions.

5.3.5.4 *High-angle instability*

When dominant waves approach the coast at an angle of more than approximately 45 degrees, the coastline evolution may become unstable: the maximum longshore transport rate occurs at an angle of around 45 degrees, so if the angle at a given point is higher than that of its upstream neighbour, sand will accumulate there until the maximum transport is reached again. [Ashton and Murray (2006)] elegantly show through coastline modeling that this can explain spectacular coastal shapes such as sand waves, flying spits and capes.

5.3.5.5 *Mechanisms for propagating behaviour*

From the classical coastline theory it follows that the coastline response to disturbances is purely diffusive, However, it has often be noted that certain features had a propagating character, which cannot be easily reconciled with this theory. Examples of this are the large-scale sand waves along the Dutch coast [Verhagen (1989)] and the propagation of an 'erosion shadow' along the beaches of Gold Coast, Australia, after blocking of sediment by an extension of the Tweed River training walls.

[Falques and Calvete (2004)] analyse instability and propagation character of coastline disturbances, and identify the effect of 2D refraction as a possible cause of propagating behaviour, because it can enhance the transport at the horns relative to that in the bays.

In general, we get a propagating behaviour if the longshore transport not only depends on $\partial y / \partial x$, but also on y itself. This can be readily seen by assuming that transport varies linearly as a function of the cross-shore position of the coastline, with a gradient of $1/Bd$:

$$S_x \approx \left(-s_x \frac{\partial y}{\partial x} + S_{x,0}\right)\left(\frac{y-y_0}{B} + 1\right) \Rightarrow$$
$$\frac{\partial S_x}{\partial x} = \frac{1}{B_d}\frac{\partial y}{\partial x}\left(-s_x \frac{\partial y}{\partial x} + S_{x,0}\right) - \left(\frac{y-y_0}{B} + 1\right) s_x \frac{\partial^2 y}{\partial x^2} \approx \frac{S_{x,0}}{B_d}\frac{\partial y}{\partial x} - s_x \frac{\partial^2 y}{\partial x^2} \quad (5.39)$$

Assuming a disturbances that is small relative to B_d we then get:

$$\frac{\partial y}{\partial t} + \frac{S_{x,0}}{d\,B_d}\frac{\partial y}{\partial x} = \frac{s_x}{d}\frac{\partial^2 y}{\partial x^2} \quad (5.40)$$

which clearly leads to an alongshore propagation of the disturbance at a celerity of approx. $S_{x,0}/(d\,B_d)$. This can have important consequences for the longshore spreading of nourishments, because there are several possible mechanisms through which longshore transport may depend on the absolute coastline position, for instance:

- The presence of groins, which are more effective on a sediment-starved beach than when they are covered with sand;
- The presence of a beach wall or revetment; again, on a sediment-starved beach the transport rate may be much less because waves expend their energy on rocks rather than on the beach.

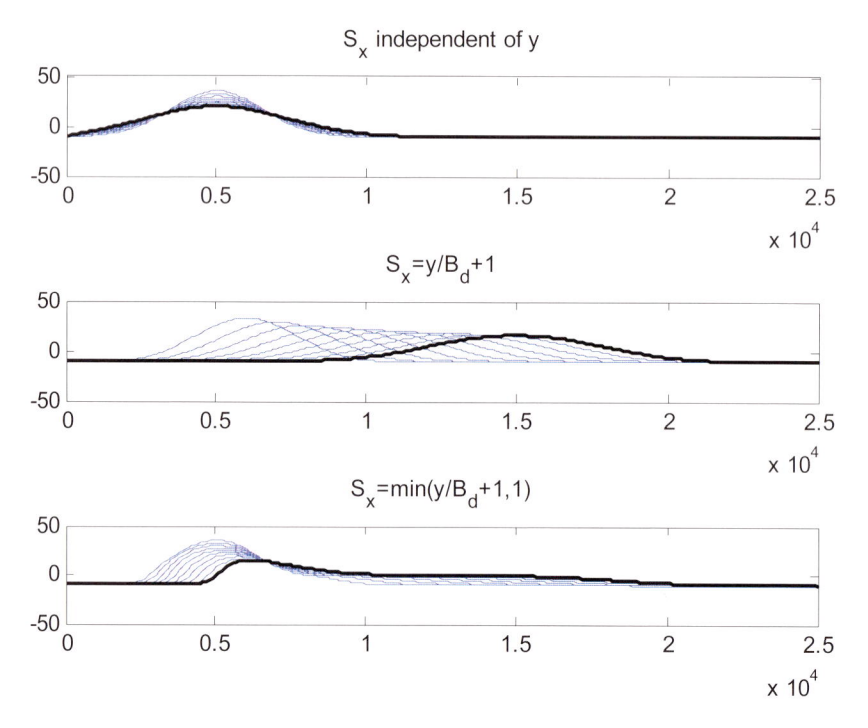

Fig. 5.17 Example of development of coastline after nourishment, where a) longshore transport S_x is independent of coastline position y; b) transport is a linear function of y, where B_d is 50 m; c) transport is a linear function of y, where B_d is 50 m and S_x is undisturbed when $y > 0$. In all cases initial coastline position before nourishment is $y=-10$ m.

In Figure 5.17, an example simulation is shown for a) no dependence of transport on coastline position; b) a linear dependence, c) a linear dependence with a maximum. Clearly, there is a major difference in coastline behaviour: in cases b) and c) the nourishment propagates or spreads out along a much larger stretch of coast. The Matlab-code to calculate the coastline evolution and plot the results has been included as **nourishment.m**.

5.4 Tidal inlets and Estuaries

5.4.1 *Ebb and flood tidal delta formation*

Tidal inlets often develop after breaching of a barrier during an extreme event such as a storm or hurricane. During a sufficiently strong breaching process the height of the barrier reduces to well below mean high water, allowing the tidal flow to flood and empty the back barrier. The tidal flows associated with this initial situation are very strong, in the order of several m/s, and quickly start to scour out the main channel in the gorge and to build up a flood delta (on the inside) and an ebb delta (on the outside). In the beginning, both ebb and flood delta are built up as rather uniform radial deposition lobes; created by the sudden reduction of flow velocities as the flow diverges. After a certain period these lobes, especially on the flood delta, become so shallow that the flow resistance increases and water level differences over the lobes increase. This then leads the flow to either be diverted and to create a new lobe, or to incise into the lobe if there is no such escape. In this way, a system of branching channels is created on the flood delta, which on the whole expands further and further into the back barrier until the lobes reach the basin edges and the system can be considered well-developed.

A similar end situation can be the result of a very different evolution. When low-laying areas are flooded during sea level rise, former river channels can become tidal channels and form the same kind of channel patterns and ebb deltas as in tidal basins that developed after breaching. In this case, however, the system has to be exporting in order to arrive at the same morphology as in the case before, where large amounts of sediment were imported into the lagoon that was breached.

In the meantime, the ebb delta typically develops less far into the sea than the flood delta into the bay, simply because the sea area is usually deeper and it takes much more sand to build out into sea than into the shallow back bay.

While the processes on the flood delta are entirely dominated by the tidal motion, those on the ebb delta may be strongly affected by waves. Waves on the ebb delta influence the sediment transport and deposition in various ways.

- They tend to increase the bed shear stress, leading to higher concentrations than for the case without waves. When combined with flood flow, this may lead to more import of sand, but combined with ebb flow it can lead to deposition further out to sea and thus to a more extended, deeper ebb delta.
- When the ebb delta is shallow enough, the waves drive a circulation pattern, typically leading to onshore flows over the shoals that escape through the channels, similar to the rip current case.
- In case of obliquely incident waves, the longshore current and transport feeds sand towards the inlet, which may push the ebb channel into the down-wave direction; besides, the oblique waves also lead to a longshore component in the circulation patterns that tends to shift the shoals in the down-wave direction.

- Finally, asymmetry and skewness of the wave orbital motions leads to (generally onshore directed) transports.

5.4.1.1 Relation with tidal propagation

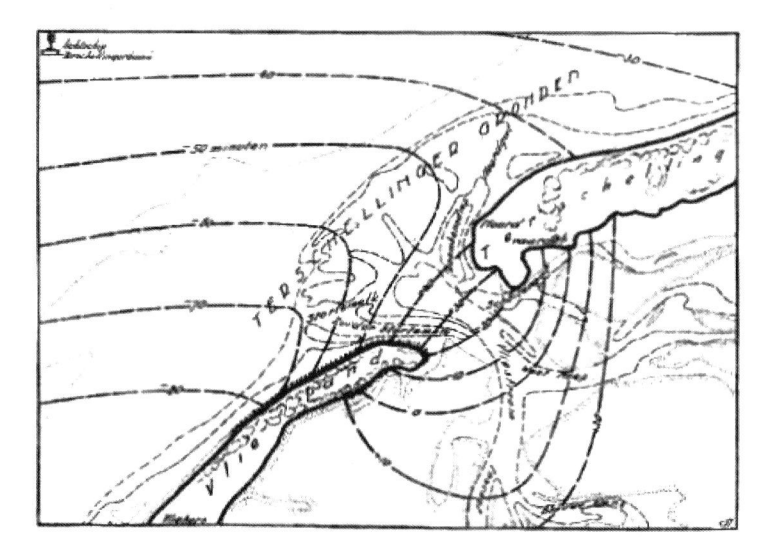

Fig. 5.18 Lines of equal tidal phase and channel pattern in Vlie inlet. from Van Veen (1936)

The direction of the tidal wave plays an important role in determining the orientation of the ebb delta channels. If the tide comes in perpendicular to the coast (as on the open ocean), a more or less symmetrical evolution can be expected, with a main ebb channel in the middle and flood channels on the sides. However, along shallow seas where the tide propagates alongshore, the phase differences across the inlet lead to an asymmetric development. [van Veen (1936)] with his "motorisch vermogen" (Dutch, probaly best translated as "moving power") theory estimated the most likely positions of channels by looking for the maximum water level gradient between pairs of points in the inlet mouth. The idea behind this is that if for some reason (for instance the asymmetrical tide propagation) large water level gradients occur, they will create strong currents over the area where this happens, which will be able to move sand out of the way and create a new channel. Strong water level gradients are therefore also an indication of the creation of a shortcut channel. [Sha and Van den Berg (1993)] investigated this concept further and showed the effect of the tidal basin in determining the current patterns and channel preferences. As we'll show in Chapter 10, this mechanism is clearly confirmed by modern numerical models. The idea is nicely illustrated in Figure 5.18, taken from [van Veen (1936)], which shows the main channel where the iso-phase lines are closest, viz. where the water level gradients are largest.

5.4.2 *Equilibrium relations*

Since the late nineteenth century, when especially in the US engineers started to modify tidal inlets in order to improve the navigability, there has been a lot of research aimed at finding practical relationships that can help in predicting the impacts of changes to the inlets and in choosing the dimensions of inlet works. Such relations offer a simple alternative to costly and complex physical or numerical models and were available long before such models could be dreamed of.

5.4.2.1 *Cross-sectional area vs. tidal prism*

There is a large body of literature concerning the dependence of the cross-sectional area of the gorge of a tidal inlet on the tidal prism. [LeConte (1905)] first suggested a linear relationship between tidal prism P and minimum cross-section A_c, based on observations on some inlets in California. Later, [O'Brien (1931)],[O'Brien (1969)] extended the observations to 28 inlets and arrived at a small modification of the relationship:

$$A_c = CP^n \qquad (5.41)$$

with C and n empirically determined coefficients; O'Brien found a value of 0.85 for n.

Following the work of [O'Brien (1931)], [O'Brien (1969)] and others, [Jarrett (1976)] integrated data of the former studies, included new measurements and assigned different values for 'C' and 'n' depending on the location in the United States with specific tidal conditions (Pacific, Gulf or Atlantic) and the character of the inlet (i.e. unjettied or with a single jetty and with two jetties).

Based on an analysis of UK estuaries [Townend (2005)] improved the fit to the [O'Brien (1931)] relationship by including the ratio of estuarine length and the tidal wavelength. [Hume and Herdendorf (1993)] verified the P/A relationship for 16 estuary inlets in New Zealand with a different type of geological background, ranging from barrier enclosed basins to embayments with a volcanic bathymetry. They concluded that the P/A relationship is obeyed, although they distinguished, similar to [Townend (2005)], different values for C and n for different types of estuaries. [Powell *et al.* (2006)] verified the P/A relationship for tidal entrances all around Florida; in Figure 5.19 their data is redrawn with their optimum relationship of $P=6.25.10^{-5}\ A_c^{1.00}$. It is clear that, even when not considering some outliers, the scatter is still huge, given that it is plotted on log-log paper.

It is interesting to note that although A_c is supposed to be a function of P, the relationship is usually plotted with A_c on the x-axis. Maybe this is because for a given inlet configuration, P is a direct function of A, for purely hydrodynamic reasons. This is related to [Escoffier (1940)]'s famous curve, which relates the cross-sectionally mean velocity amplitude to the cross-sectional area, from which we can easily obtain the tidal prism as a function of the cross-section.

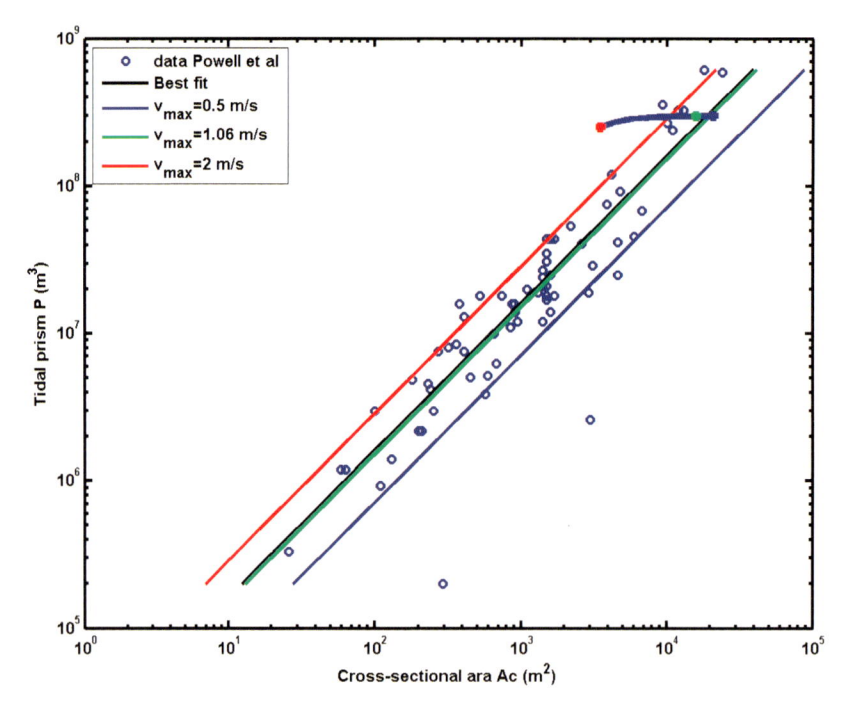

Fig. 5.19 Tidal Prism vs. cross-sectional area, data (circles) from [Powell *et al.* (2006)]; black line: best fit; colored lines: eq. (5.44) with different values of velocity amplitude; thick blue line: example evolution of an inlet through time.

If we consider the same idealized inlet as in Section 3.3.5, we can derive the Escoffier curve from eq. (3.81):

$$v_{\max} = \frac{A}{A_c}\frac{-\omega}{\sqrt{1+(\omega/\mu)^2}}\widehat{\eta}_{out}, \quad \mu = \frac{A_c h}{A}\frac{g}{\lambda L_{gorge}}, \quad \lambda \approx \frac{\pi}{4}C_f \widehat{u} \quad (5.42)$$

We have to specify the basin area, the cross-sectional area, the tidal frequency and a schematized length of the gorge. In this expression we have neglected entrance and exit losses, so the length of the gorge should be taken on the large side. We can remove the depth h from the equation by specifying a fixed width/depth ratio B/h; we then get:

$$h = \sqrt{A_c/(B/h)} \quad (5.43)$$

In Figure 5.20 we see this relation plotted for a basin area of 150 km², a gorge lenth of 4000 m, a width/depth ratio of 100 and a tidal amplitude of 1 m; besides, we show the sensitivity to the gorge length and width/depth ratio separately, Clearly, for small cross-sectional areas, the velocity amplitude is sensitive to the gorge length and the width/depth ratio, as these strongly influence the friction and thereby the

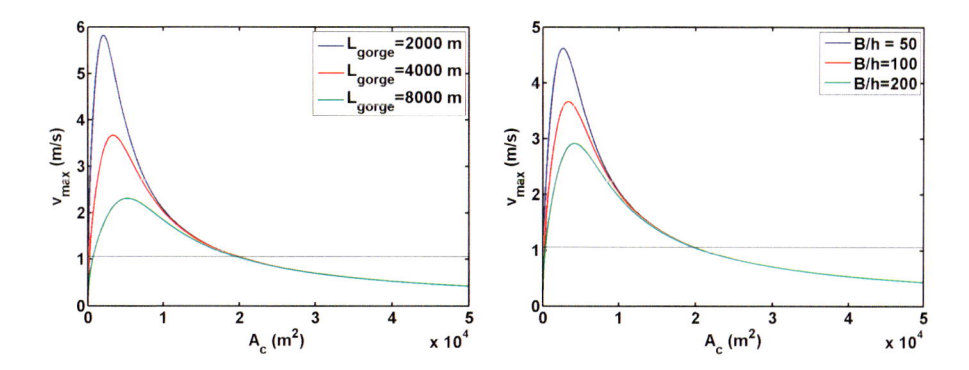

Fig. 5.20 Escoffier curves based on eq. 5.42, for different values of gorge length (left panel) and wifth/depth ratio (right panel). $A-150$ km^2; $C_f-0.004$, tidal amplitude 1 m.

damping of the tidal amplitude inside the basin and the tidal prism; for larger cross-sectional area, the velocity maximum is not sensitive to any of these parameters, as the tide will no longer be damped and the tidal prism is constant.

The tidal prism based on the same relationship can be expressed as:

$$P = \frac{2v_{\max} A_c}{\omega} = \frac{2A\widehat{\eta}_{out}}{\sqrt{1+(\omega/\mu)^2}} \qquad (5.44)$$

It follows then that for large A_c the friction parameter μ increases and the denominator in this expression goes to 1; therefore P will not be a function of A_c anymore, but only depends on the basin area and the tidal amplitude. From the middle part of this equation we readily see that if the velocity amplitude is given, the tidal prism depends linearly on the cross-sectional area A_c and the tidal angular frequency $\omega = 2\pi/T_{tide}$. It has often been noted that the velocity amplitude tends towards values around 1 m/s (3.5 ft/sec or 1.06 m/s) according to [LeConte (1905)], and if we apply this value we get almost exactly the best-fit value given by [Powell et al. (2006)]. In Figure 5.19 we have plotted P/A relations based on velocity amplitudes of 0.5 m/s, 1.06 m/s and 2 m/s respectively, and we see that the middle value comes close to the best fit and that most data points lie within the range given by the other two.

We can give a possible explanation for this behavior by following what happens with a tidal inlet after breaching. Initially, the tidal velocity amplitude will be very large, as the cross-sectional area will be small. This will lead to large sediment transports from the gorge area, both during ebb and during flood. These transports will mainly be deposited on the flood and ebb delta, and the total deposition rate will be close to the mean of the absolute value of the transport in the gorge, as little transport will be returned to the gorge. This is schematically represented in Figure 5.21 in panels (a) and (b). After a while, the ebb and flood deltas will be

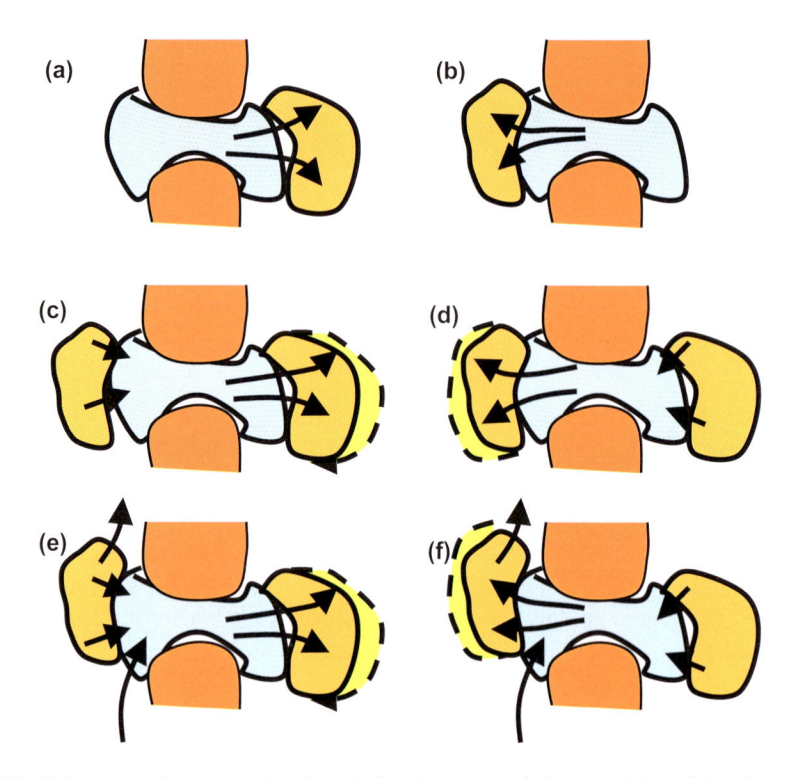

Fig. 5.21 Schematic of growth of ebb and flood deltas. a) flood, initial; (b) ebb-initial; (c) flood, after some development; (d) ebb, idem; (e) flood, with wave-driven transport; (f) ebb, with wave-driven transport

developed to such an extent that significant transport will also be directed towards the gorge, and the loss from the gorge will only be a fraction of the gross total transport; see panels (c) and (d). This means that the excavation of the inlet by the tidal current will slow down and the build-out of ebb and flood delta will continue at a decreasing rate. As there will always be some transport across the steep delta fronts, this process will continue for a long time with a slowly increasing inlet cross-section, while the tidal prism, after an initial increase, will remain more or less constant. However, since transports drop dramatically with decreasing velocity amplitude (because of the highly nonlinear transport) many inlets in practice will be stuck with rather similar velocity amplitudes.

In the case there is also a significant longshore transport, it is reasonable to assume, as in [Kraus (1998)], that the longshore transport enters the gorge directly, but the wave-driven transport on the downdrift side takes place from the ebb delta to the downstream beach; see panels (e) and (f).We can approximate this mean gross transport by considering a simple transport formula like Engelund and Hansen or Meyer-Peter and Muller:

$$Meyer-Peter-Muller: \langle|S|\rangle = \frac{8C_f^{3/2}}{g\Delta}\left\langle|V|^3\right\rangle \approx \frac{8C_f^{3/2}}{g\Delta}\left(0.42 v_{\max}^3\right)$$
(5.45)
$$Engelund-Hansen: \quad \langle|S|\rangle = \frac{0.05 C_f^{3/2}}{g^2\Delta^2 D_{50}}\left\langle|V|^5\right\rangle \approx \frac{0.05 C_f^{3/2}}{g^2\Delta^2 D_{50}}\left(0.34 v_{\max}^5\right)$$

where the factor 0.42 resp. 0.34 appears by considering the even velocity moments of a sinusoidal motion, which according to [Guza and Thornton (1985)] are:

$$\left\langle|V|^3\right\rangle = 1.20\left\langle V^2\right\rangle^{3/2} = 0.42\hat{V}^3$$
$$\left\langle|V|^5\right\rangle = 1.92\left\langle V^2\right\rangle^{5/2} = 0.34\hat{V}^5$$
(5.46)

In Figure 5.22 we plot the gross transport per year, multiplied by a factor of 0.1, as a crude indication of the loss of sediment from the inlet gorge, as a function of the cross-sectional area; we use the Meyer-Peter-Muller (MPM) formula, as in [Kraus (1998)]. We can interpret this as follows: suppose we start at a cross-sectional area of 3500 m2 after a breach; the initial velocity amplitude is then approx. 5 m/s. The inlet will quickly be scoured out and the velocity amplitude will move along the Escoffier curve shown in the bottom panel, while the gross transport out of the gorge area reduces rapidly. If there is no source of sediment in the inlet, the process will continue but slow down considerably; going from 5 m/s (the red point) to 1 m/s (the blue point), the transport decreases by more than an order of magnitude, even with the MPM formula, which uses the relatively low power of 3. Still, the process will continue as it is still nowhere near the critical velocity for sediment transport.

For the case with a longshore transport feeding the inlet (in this example 500,000 m3/yr) a quasi-equilibrium is found at the green dot: the total transport out of the gorge (approximated by 0.1 times the gross transport) is matched by the longshore transport into the gorge. We call this quasi-equilibrium, because the ebb and flood deltas will still be expanding, which in the long run increases the flow resistance in the gorge, which in turn will modify the equilibrium.

To come back to the P/A relation, we can plot this evolution in the P/A graph of Figure 5.14. We see that we start from outside the point cloud (the data points are all related to well-established inlets) and move to inside the point cloud. The tidal prism increases slightly because of the increase of the cross-section, until it reaches its maximum. The green dot indicates where the process stops in case of feeding by longshore currents; the blue dot a (still transitional) point in case there is no sediment input into the gorge.

[Van der Wegen et al. (2010b)] come to similar conclusions based on numerical simulations of a schematic tidal inlet, where they find a continuous, if slowly reducing, erosion of the inlet and increase of the cross-sectional area, long after the tidal prism has become independent of the cross-sectional area. Only when feeding the inlet with a source of sediment they find a quasi-steady solution where the amount of sediment transported out of the gorge matches the input of sediment.

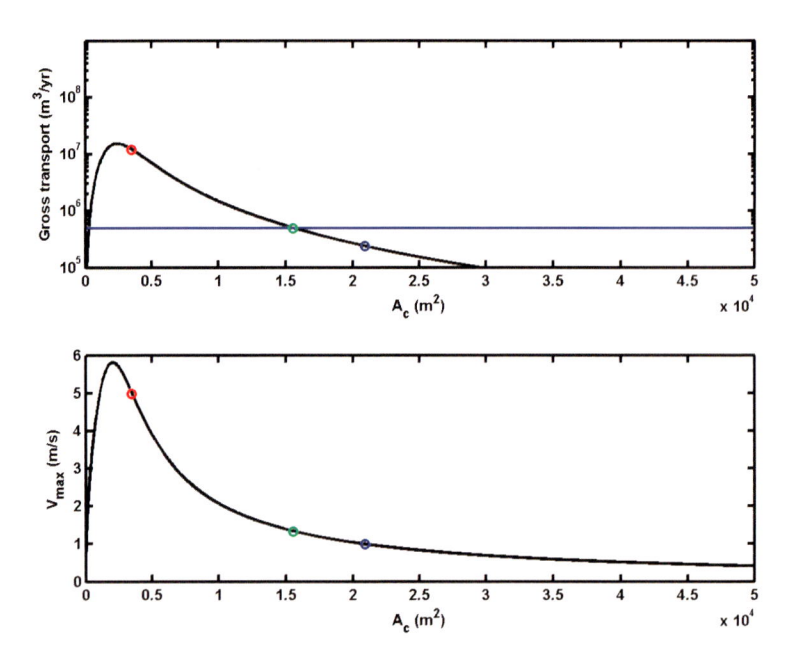

Fig. 5.22 Gross transport (top panel) and velocity amplitude (bottom panel) as a function of the cross-sectional area

The early research on the P/A relationship had a strong focus on the inlet mouth. However, the relationship appears to be valid along tidal channels as well. [De Jong and Gerritsen (1984)] and [Eysink (1990)] observed a constant P/A relationship along the Western Scheldt estuary; [Friedrichs (1995)] summarizes observations for other estuaries and sheltered embayments. This suggests that the P/A relationship is valid for larger spatial scales and a larger range of tidal environments. [Van der Wegen et al. (2010b)] again find based on numerical simulations of a schematic elongated rectangular basin, that the different cross-sections along the channel line up on a P/A curve during all stages of development, but that the lines themselves shift towards relatively larger cross-sections.

The explanation for this is that because of the high sensitivity of transport amplitude to the tidal velocity amplitude, even relatively small differences in velocity amplitude between neighboring cross-sections lead to large differences in transport amplitude, which leads to a gradient-type exchange of sediment from the section with higher transports to the section with lower transport; this evens out the differences between the velocity amplitudes of different cross-sections, and thereby lines up the cross-sections on the P/A curve. If however there is a net import or export of sediment into or from the estuary as a whole, the whole curve will slowly move up or down.

5.4.2.2 Channel volume vs. tidal prism

By integrating the P/A relationship along the channel(s) in a tidal basin, [Eysink (1990)] derived a relationship between the tidal prism and the total volume of channels below MSL, V_c:

$$V_c = CP^{3/2} \qquad (5.47)$$

Such a relationship is useful to estimate the infilling of tidal channels after reduction of the tidal prism, for instance by closure of part of the tidal basin. However, the coefficient C depends on the shape of the basin, so it has to be derived from measurements for each basin.

5.4.2.3 Relative flat area vs. basin area

The inter-tidal flat area is defined as the area between MLW and MHW. In literature there are various suggestions for flat areas in equilibrium conditions. [De Vriend et al. (1989)] showed a general relation between the flat area and the total area of the basin. [Renger and Partenscky (1974)] worked on the same kind of relation for inlets in the German Bight. [Eysink (1990)] used the same idea (A_f/A_b as a function of A_b) to analyze the available data in tidal inlets and estuaries in The Netherlands. There is a general tendency of the relative flat area to decrease with increasing basin size, possibly related to the importance of waves but also quite possibly the result of the fact that it takes much longer to fill up a large basin than a small basin.

5.4.2.4 Height of flats

[Eysink (1990)] claims that one of the first parameters that aims for equilibrium in a relatively short time is the height of flats, which is somewhat correlated to the tidal amplitude.

5.4.2.5 Ebb and Flood Dominance

[Speer and Aubrey (1985)] used a 1D numerical model to study the influence of geometry and bathymetry on tidal propagation of short, friction-dominated and well-mixed estuaries. They suggested that two non-dimensional parameters can be used to characterize the tidal basins into ebb or flood dominant ones. The first one is a/h, the ratio of the tidal amplitude and the depth of the channel with respect to MSL, which shows the relative shallowness of the estuary. The second parameter is the ratio of the volume of inter-tidal storage and channel volume (V_S/V_C). Larger values of a/h (shallower basin) means longer ebb duration (due to larger effect of friction and different wave propagation velocity), while increase in inter-tidal storage will decrease the flood propagation and duration. Later, [Friedrichs and Aubrey (1988)] confirmed the Speer model against measured data along the Atlantic coast of the United States.

5.4.2.6 *Ebb delta volume vs. tidal prism*

The volume of the ebb delta is also closely related to the tidal prism. If the ebb delta volume is defined as the volume of sand relative to a hypothetical undisturbed sea bottom, [Walton and Adams (1976)] found that it correlates with the tidal prism to the power of 1.23.

5.4.3 *Discussion*

Empirical relationships such as the ones discussed above can be used directly in trying to estimate the effects of interfering in the natural system (closure of tidal inlets, land reclamations) or the effects of sea level rise. [Eysink (1990)] gives some examples of how to use these relationships, together with estimated timescales of adaptation, to predict to which new equilibrium situation a certain morphological unit may tend, and at which speed. [Van de Kreeke (1992)] used Escoffier's curve with an equilibrium inlet cross sectional area-tidal prism relationship to examine the (in)stability of an inlet.

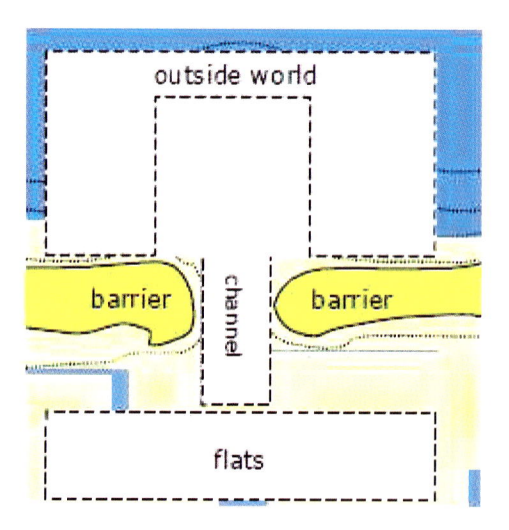

Fig. 5.23 Schematization of tidal inlet elements according to ASMITA

However, in a system with elements (ebb delta, flat, channels) competing for sediment, such a direct application is more difficult; in such cases an approach like ESTMORPH ([Wang *et al.* (1998)]) or ASMITA ([Stive and Wang (2003)]) is more appropriate. In such models a tidal system is divided into large elements (see Figure 5.23), which are described by one variable representing their morphological state, e.g.:

- Ebb-tidal delta: Integral sediment volume above a fictitious sea bottom, which would exist if the inlet were absent;
- Flats: Integral sediment volume of the tidal flats above MLW;
- Channels: Integral channel volume below MLW.

Each element strives toward its own equilibrium, but the speed at which this occurs depends on the transports between the elements, which in turn depends on how far each element is from equilibrium. This allows for much more complex behaviour, where elements do not monotonously tend towards their equilibrium but may even temporarily move away from it. Models like this have been successfully applied to evaluate the response of whole tidal systems on changes such as due to increased sea level rise ([Van Goor *et al.* (2001)]).

Finally, empirical relationships are very useful to evaluate the long-term behaviour of process-based morphological models, as we will see in the case study of long-term modeling of tidal inlets.

Chapter 6

Modeling Approaches

6.1 Coastal profile, coastline and area models

Traditionally, three types of models have been developed: *coastal profile* models, where the focus is on cross-shore processes and the longshore variability is neglected (for reviews see [Roelvink and Broker (1993)]; [Schoonees and Theron (1995)]), *coastline models*, where the cross-shore profiles are assumed to retain their shape even when the coast advances or retreats (e.g. [Szmytkiewicz et al. (2000)]) and *coastal area models*, where variations in both horizontal dimensions are resolved (e.g. [De Vriend et al. (1993)], [Nicholson et al. (1997)]. These coastal area models are further subdivided into two-dimensional horizontal (2DH) models, which use depth-averaged equations, and three-dimensional (3D) models (e.g. [Lesser et al. (2004)]), which resolve the vertical variations in flow and transport.

The general setup if these models is as follows (see Figure 6.1 below):

(1) We start from an initial profile, coastline or 2D bathymetry;
(2) Given boundary conditions for waves and flow, coupled wave and flow models are run for a certain period, during which the bathymetry is kept constant;
(3) The sediment transport field is computed, based on the flow field, the wave field, the bathymetry and sediment characteristics;
(4) The bathymetry and (when applicable) the bed sediment composition are updated based on the sediment transport gradients;
(5) Back to 3 for an intermediate update of transports, or to 2 for a full update of flow, waves and transport.

Large differences exist in the type of flow and wave models applied (stationary, instationary), the sediment transport models (sand / mud / multi fraction; bed load and suspended load or total transport), the frequency of updating (once per tidal cycle, every timestep), numerical bed updating schemes and morphological acceleration techniques. We will discuss these aspects for each of the model types we consider.

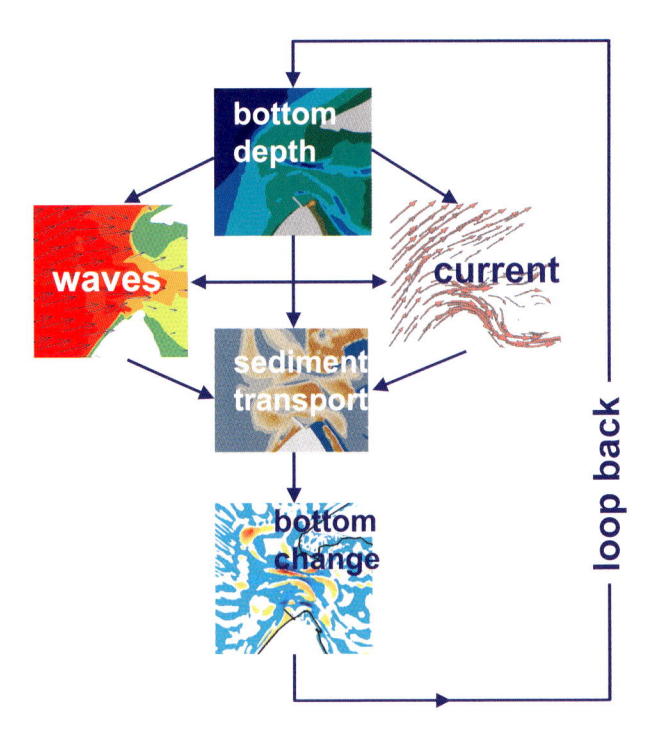

Fig. 6.1 Morphodynamic model loop

6.2 Scales of application

6.2.1 *Coastal profile models*

Coastal profile models are typically applied for two kinds of application: evaluation of storm impacts on a coastal profile, with or without hard structures, and evaluation of longer-term behaviour of sandbars and nourishment schemes, both on the beach and on the shoreface.

The first type is the oldest, and in a way the simplest, since we try to model a situation with a very dominant seaward directed transport, where the main difficulties lie in predicting the right speed of erosion and in achieving the correct profile shape after the storm. Most of the profile changes take place in a few hundreds of metres seaward of the dune front or sea wall, and the duration of such events is at most a couple of days.

The second type involves the whole active profile, typically in the order of a kilometer cross-shore, down to a depth of some 6-15 m depending on the wave climate and on the simulation duration. Changes in the pattern of bars and troughs take place over years, where in some systems a single bar moves back and forth, and in others bars are created near the shoreline, move seaward while growing until they damp out, allowing a new cycle to begin. This process may take from a couple of years to more than a decade. Such behaviour is the result of a subtle balance between a various counteracting processes, none of which are modelled very accurately. [Roelvink et al. (1995)] obtained somewhat realistic, almost periodic bar behaviour on a timescale of approx. a decade; [Ruessink et al. (2007)] after rigorous model sensitivity study and extensive calibration reproduced bar behaviour on a timescale of months for three different sites. Systems like this are affected by beach or foreshore nourishments that are of similar scale and clearly interact with the bar pattern. This makes modeling the behaviour and lifetime of nourishments very difficult: only when the natural bar behaviour is modelled somewhat realistically can we have some trust in the model's capability to predict the relative impact of the nourishment. Similar arguments are valid for hard structures, which also interact with the profile on the same scale as the longshore bars.

6.2.2 *Coastline models*

Coastline models assume gradually varying flow conditions and approximately parallel depth contours. Under such circumstances the longshore transport can locally be treated as if it is fully adapted to the local incident wave conditions relative to the coast orientation. Since it typically takes hundreds of metres for the longshore current to spin up, coastline models should in principle be applied for large-scale applications, over alongshore distances of many kilometres. Good examples are the accretion of beaches updrift of a harbour [Szmytkiewicz et al. (2000)]), the large-scale evolution of river deltas, longshore spreading of large beach nourishments (e.g. [Dean (1992)]), evolution of headland embayments and realignment of coastlines due to changes in wave climate (e.g. [Buijsman et al. (2001)]).

However, with the help of various additional techniques coastline models are also frequently applied at relatively small scales, such as behind emerged or submerged offshore breakwaters, around T-groynes and near tidal inlets. This is achieved by combining a coastline model with a 2D wave model to predict nearshore wave climates, by extra terms in the longshore transport equations to account for set up differences, and sometimes by schematising the coast into two interconnected coastlines. Such applications must be treated with much caution, since a lot of the processes acting on such small scales are not represented in this approach; even when a successful calibration is shown, its value may be limited if the wrong processes were 'tuned'. Sometimes small-scale structures are somehow represented in a coastline model that is aimed at predicting large-scale trends. This in our opinion is fine, as long as the details of the evolution in the vicinity of the structure are not taken too seriously.

6.2.3 *Coastal area models*

Coastal area models are applied where a separation between longshore and crossshore scales is not possible, for instance in the vicinity of a tidal inlet, with a complex bathymetry with channels and shoals at varying angles with respect to the undisturbed coast orientation. Area models are applicable at a range of scales, from small-scale coastal engineering problems to macro-scale evolution of tidal basins.

At the smallest scales, we are talking about the details of the waves and wave-driven flow patterns and resulting morphology changes around small structures such as groynes and detached breakwaters, or the erosion of dunes where the coast is not alongshore uniform. The small grid cell sizes required to resolve the main processes and associated small time steps make such problems computationally intensive, especially when predictions over longer time than some storms are required. An important choice is whether to treat the problem as two-dimensional, where the focus is on the effect of the horizontal circulation patterns, or as (quasi-) three-dimensional, where vertical circulations (undertow) are accounted for as well as additional wave-related effects providing an onshore component (e.g. [Lesser et al. (2004)]). In the first case, beach profiles may evolve that are far from an equilibrium shape; in the second case, there will be a tendency towards restoring a certain profile shape, but much care must be given to tuning the cross-shore processes in such a way that reasonable profiles develop or are maintained. At these scales, sand bars may also be resolved, but as we discussed under profile models, obtaining the right bar behaviour is a tight balancing act. An especially difficult challenge is to model the natural evolution of complex three-dimensional beach topography, where it is key to get the transitions between different beach states right (e.g. [Reniers et al. (2004a)], [Smit et al. (2005)]).

At a larger scale, such as for instance in studies of harbour extension, inlet stabilization or large-scale nourishments or land reclamations, details of processes within the surfzone usually cannot be resolved, but the surfzone must be resolved at least enough to generate a decent longshore current and transport; this means a minimum of 5-10 grid cells across the surfzone. As these studies are often aimed at predicting the evolution over months to years, it is important that some cross-shore transport processes are accounted for, so cross-shore profiles stay in a reasonable shape even though bars cannot be represented and must perhaps even be prevented from occurring. Part of the evolution of coastal profiles depends on the interaction with the dunes, in terms of dune erosion and regeneration by wind-blown transport. Such processes are typically ill-represented and some heuristic approaches to exchange sand between the wet model domain and the dry beach and dunes are called for. Development of current-induced scour holes and sedimentation of navigation channels generally can be predicted quite well, although this cannot be said of channels in estuaries, where variable sediment behaviour and three-dimensional effects are very important; in these cases, waves typically are less of a problem.

An interesting application on a very large scale is the study of the long-term evolution of tidal basins and estuaries. As long as waves are neglected (somewhat reasonable in large inlets and when the focus is not on the ebb delta) the scales to resolve are those of the main channels and shoals, which typically are hundreds

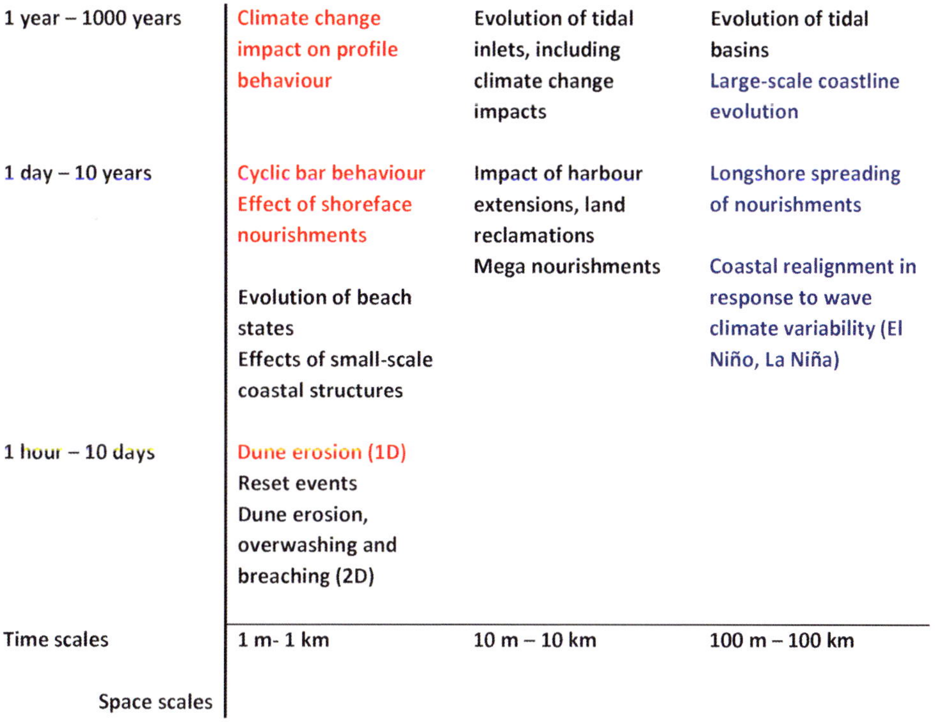

Time scales			
1 year – 1000 years	Climate change impact on profile behaviour	Evolution of tidal inlets, including climate change impacts	Evolution of tidal basins Large-scale coastline evolution
1 day – 10 years	Cyclic bar behaviour Effect of shoreface nourishments Evolution of beach states Effects of small-scale coastal structures	Impact of harbour extensions, land reclamations Mega nourishments	Longshore spreading of nourishments Coastal realignment in response to wave climate variability (El Niño, La Niña)
1 hour – 10 days	Dune erosion (1D) Reset events Dune erosion, overwashing and breaching (2D)		
Space scales	1 m – 1 km	10 m – 10 km	100 m – 100 km

Fig. 6.2 Overview of scales of typical applications and applicable model types; profile models (red), coastline models (blue) and area models (black)

of metres to kilometres wide, so even relatively coarse models with grid resolutions in the order of 100 m can perform reasonably well. Because of the larger space- and time steps, and with the help of new acceleration techniques, simulations over time spans of centuries are possible, allowing to study the evolution of tidal basins to equilibrium and to simulate the effects of sea level rise on such basins (e.g. [Wang et al. (1995)], [Hibma et al. (2003)], [Van der Wegen and Roelvink (2008)], [Dastgheib et al. (2008)] and [Dissanayake et al. (2009)]). Surprisingly realistic results are obtained using relatively simple physics, though a general tendency in such simulations is to generate too deep and narrow channels. Recent research however shows that much better channel profiles are obtained when taking into account the spatial variation in grain sizes and especially the tendency of sediment to coarsen with increasing mean shear stress.

When the adjacent coasts and ebb deltas are to be properly represented wave effects must be included. This poses severe restrictions on the nearshore grid resolution and drastically increases the computational cost; typically, by at least an order of magnitude. Besides, as we will discuss in the next section, we will have to schematize the highly variable wave climate in addition to the schematization of wind, tide and, where relevant, river discharges. The applications and scales discussed in the previous paragraphs are summarized in Figure 6.2.

6.3 Input schematisation

Once we have determined the appropriate model for the problem we need to tackle, we can start to run simulations over certain time periods, using time series of various input conditions. This is relatively straightforward if we want to model a not-too-long measurement campaign where the relevant input conditions have been measured continuously. In that case we know the input conditions as a function of time and we can afford to run the whole time series as it only covers a limited time period.

In the course of calibrating and validating a model, we often want to run so-called 'hindcast' runs over longer periods between bathymetric surveys. Often, time series of input conditions will be available, but it would take too long to run the model through hourly changing conditions over a, say, 5-year period. In that case, we have to apply some way of input schematisation in order to reduce the number of conditions that we actually have to simulate, in combination with some morphodynamic updating technique.

For predictive runs, we do not know the sequence of input conditions at all *and* we usually have to limit the number of conditions that we simulate in order to arrive at sufficiently long simulation periods.

6.3.1 *Input parameters*

The first thing to do in creating representative time series or climate of conditions is to select the basic input parameters. A first list of these parameters in a typical coastal model would be:

(1) tidal amplitude or phase within the spring-neap cycle;
(2) offshore wave height, period, direction and spectral shape at some representative location;
(3) wind speed and direction;
(4) river discharge;
(5) surge level;

It is important to realise that some of these parameters are correlated while others are not:

(1) The tidal boundary conditions are not related to the other parameters, although the tidal motion in a coastal model is modified by the surge, waves and wind;
(2) The surge level is often correlated with the wind, though with considerable scatter; it is also correlated with the wave height for a given directional sector.
(3) The wind wave direction is correlated with the wind direction, especially for strong wind conditions; swell wave direction generally is not correlated much with the local wind;
(4) The wind wave height for a given direction is related to the wind speed;
(5) The wind wave period is correlated with the wind wave height;

Before we can make sensible simplifications to the input conditions, we have to study these relations based on the available data and an analysis of the situation.

As the tidal boundary conditions are independent of the wave and wind input conditions we treat the tidal schematisation separately from the schematisation of the wind/wave climate.

6.3.2 General principle of schematisation

As we have to do the schematisation *before* we can run the full morphological model, it must be based on an analysis of initial transports and bottom changes. This means that in selecting representative conditions, the chronology does not play a role. Later, as we will be assembling simulation scenarios, we will get back to this aspect.

The general principle of the schematisation is now, that we select some output parameter or criterion C and a set of input conditions gathered in vector v. We then evaluate the time-average of this criterion and select a combination of representative conditions and accompanying weight factors for which the same average value of C is found. In equation (6.1) the following steps are outlined:

(1) The time-average is replaced by an average over individual input combinations in a time-series; if the input time series is detailed enough this does not lead to errors.
(2) Instead of computing for each measured combination of input parameters, we can first group them into classes, perform the evaluation of C for the class averages of all input parameters and then multiply the outcome by the probability of each class. As long as the classes are chosen small enough, this step does not lead to significant errors. In fact, C based on this evaluation is often used as the target value as it is close enough to the real solution.
(3) This step is just a reordering of the results of the previous step into larger classes k of input conditions. Here we make the actual selection of situations that we want to represent.

Now we replace all the individual combinations of input conditions l within a larger class k by a single representative condition and probability.

$$\frac{1}{t_{end}-t_{start}} \int_{t_{start}}^{t_{end}} C(t)dt \stackrel{(1)}{\approx} \frac{1}{N}\sum_{i=1}^{N} C(v_i)$$
$$\stackrel{(2)}{\approx} \sum_{j=1}^{N_{classes}} C(v_j)p_j$$
$$\stackrel{(3)}{=} \sum_{k=1}^{N_{cond}} \sum_{l=1}^{N_k} C(v_l)p_l$$
$$\stackrel{(4)}{\approx} \sum_{k=1}^{N_{cond}} C(v_{rep,k})p_k$$

(6.1)

Looking at step 4 in (6.1), we see that for each larger class k we must demand:

$$\sum_{l=1}^{N_k} C(_l)p_l \approx C(_{rep,k})p_k \Rightarrow C(_{rep,k}) \approx \frac{\sum_{l=1}^{N_k} C(_l)p_l}{p_k} \qquad (6.2)$$

where we still have the choice to interpret the probability of the class k as just the sum of the probabilities of the sub-classes it is representing, or as a weighting factor used to adjust the resulting transports to better fit the time-averaged criterion. This concept was introduced by [Latteux (1995)] who proposed to aim at representing the whole sedimentation/erosion pattern of a model. He found that the overall *pattern* for a spring-neap cycle is often best represented by a relatively high tide, but the transport *magnitudes* for such a tide would be too high; therefore he used p_k as a weight factor to compensate for this.

This is a useful concept when there is just one input variable, such as the tidal amplitude. However, its application for more than one input variable is not straightforward; in such cases we advise to use the sum of the probabilities of the sub-classes p_l for the probability p_k.

For one criterion, we can just pick out the condition for which equation (6.2) is best met, assuming that we picked our sub-classes fine enough. If we want to meet more criteria at the same time, we will have to choose a compromise for which all criteria are reasonably well satisfied.

6.3.3 *Tidal schematisation*

Tidal schematization is mainly important for morphodynamic area models, where instead of running a model through 700 tidal cycles per year, each of which produces rather similar patterns of (very small) bottom change, we run the model through some carefully selected representative cycles and use the results to extrapolate morphological changes over a longer period; the methods for this extrapolation or acceleration will be discussed in detail in Chapter 9.

Some important concepts must be discussed before we go into the details of the schematization methods.

6.3.3.1 *Preserving tidal asymmetries*

The tidal motion in coastal areas, estuaries and inlets is the result of the combined action of a large number of tidal components. However, when averaging tidal transports over longer periods, motions due to different components that are uncorrelated filter out, and only some specific combinations leave a net contribution:

- Interaction between the M2 component and its overtides M4, M6 etc.
- Interaction between diurnal components O1 and K1 and the semidiurnal component M2

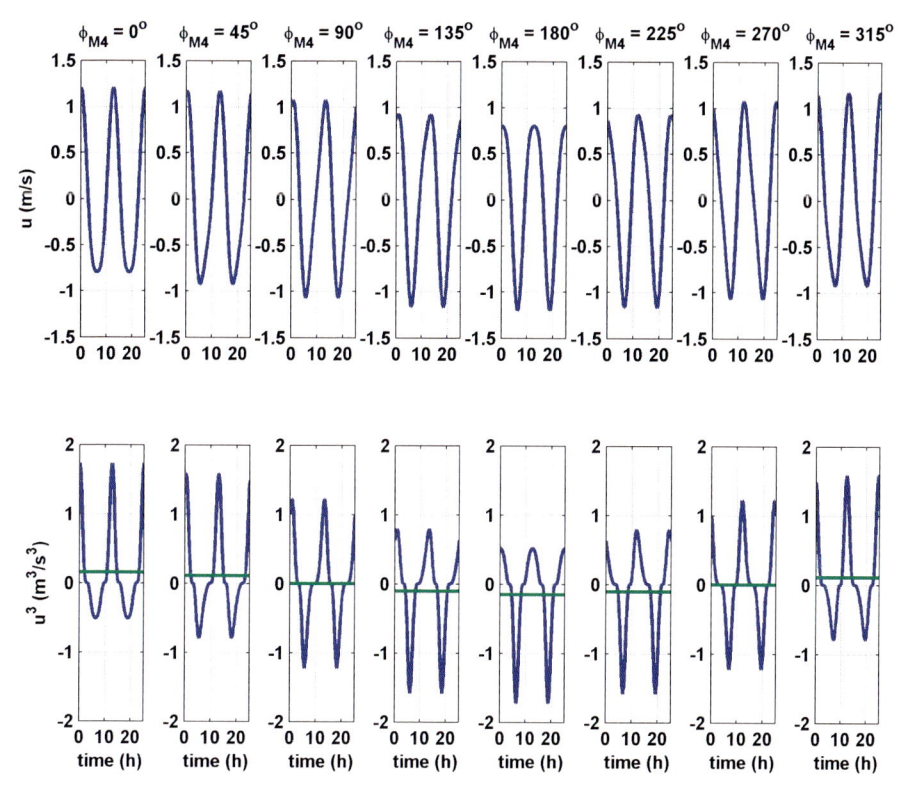

Fig. 6.3 Time series of velocity (upper panels), third power of velocity (bottom panels, blue lines) and mean of third power of velocity (green lines) for M2=1 m/s, M4=0.2 m/s and different relative phases of M4.

- Interaction between the mean current and all tidal components.

The importance of the interaction between the semi-diurnal M2 component and its overtides M4, M6 etc. has been pointed out, among others, by [Van de Kreeke and Robaczewska (1993)]. Limiting ourselves to the interaction between M2 and M4, this effect can be demonstrated by considering as in the previous section, that the transport is proportional to the third power of the velocity due to these components:

$$u = M_2 \cos(\omega_{M2} t - \varphi_{M2}) + M_4 \cos(\omega_{M4} t - \varphi_{M4}) \tag{6.3}$$

The mean of the velocity to the third power is then given by:

$$<u^3> = \frac{3}{4} M_2^2 M_4 \cos(2\varphi_{M2} - \varphi_{M4}) \tag{6.4}$$

This effect is illustrated in Figure 6.3, which shows how the effect of the tidal asymmetry depends on the phase of the M4 component relative to the M2 component.

A more recently found combination is that between O1, K1 and M2, which only plays a role in areas with important diurnal components. This can be explained by the fact that the O1 and K1 components have frequencies that add up to exactly that of the M2 component. As was shown in [Hoitink et al. (2003)], for the case where transport is proportional to the third power of the velocity and this velocity is given by:

$$u = O_1 \cos(\omega_{O1} t - \varphi_{O1}) + K_1 \cos(\omega_{K1} t - \varphi_{K1}) + M_2 \cos(\omega_{M2} t - \varphi_{M2}) \quad (6.5)$$

and the net effect of the interaction between O1, K1 and M2 is:

$$<u^3_{O1K1M2}> = \frac{3}{2} O_1 K_1 M_2 \cos(\varphi_{M2} - (\varphi_{O1} + \varphi_{K1})) \quad (6.6)$$

As was shown by [Lesser (2009)], the same net effect can be achieved by replacing O1 and K1 by an artificial diurnal component with a frequency of exactly half that of M2, and the following amplitude and phase:

$$C_1 = \sqrt{2 O_1 K_1} \ , \quad \varphi_{C1} = \frac{\varphi_{O1} + \varphi_{K1}}{2} \quad (6.7)$$

The effect of the interaction of the different components with the mean current is given by eq. (6.8). We see that the net effect of the components is given by a quadratic addition of the velocity amplitudes of the components:

$$u_0(u_0^2 + \frac{3}{2}\sum_{i=1}^{n}\widehat{u}_i^2) = u_0(u_0^2 + \frac{3}{2}\widehat{u}_{rep}^2) \Rightarrow r = \frac{\widehat{u}_{rep}}{\widehat{u}_{M2}} = \frac{\sqrt{\sum_{i=1}^{n}\widehat{u}_i^2}}{\widehat{u}_{M2}} \quad (6.8)$$

where the enhancement factor r is thus the ratio between the rms value of the velocity components and the M2 velocity amplitude. To directly relate the enhancement factor to the water level amplitudes we have to take into account that the velocity amplitude scales with the period of each component, so:

$$r = \frac{\sqrt{\sum_{i=1}^{n}\widehat{u}_i^2}}{\widehat{u}_{M2}} = \frac{\sqrt{\sum_{i=1}^{n}\left(f_i\widehat{\eta}_i\right)^2}}{f_{M2}\widehat{\eta}_{M2}} \quad (6.9)$$

From these considerations it follows that, to a good approximation, the morphological effects of the full tide can be represented by a periodic tide consisting of the artificial diurnal component C1, M2 and the overtides M4, M6 etc.

In the following we will first discuss how to create representative boundary conditions for open-ocean conditions, where the tide is driven by just a few tidal components; after this, we will deal with schematizing boundary conditions for more complex shallow environments.

6.3.3.2 Open ocean coasts

As an example we will use the tidal motion around Grays Harbor, Wa. on the US west coast. The main offshore tidal components of the water level at this location are:

As we see, the tide at this location has important diurnal components and semi-diurnal components S2 and N2 of similar magnitude. These features lead to a rather

Table 6.1 Tidal amplitude and phase for Grays Harbor, WA, USA.

Component	Frequency (1/hr)	Amplitude (m)	Phase (deg.)
M2	.0805114006	0.941	229
S2	.0833333333	0.210	247
N2	.0789992487	0.176	209
K1	.0417807462	0.500	230
O1	.0387306544	0.274	221

variable tidal motion from one spring-neap cycle to the next. We will analyse the effects of this variation on the sediment transport through the gorge. We take the velocity to the third power as a proxy for sediment transport and separately look at three time-averaged parameters:

(1) The cumulative transport, or time-integral of U^3, a measure of the net in- or export;
(2) The cumulative positive transport, a measure of the behaviour of flood channels;
(3) The cumulative negative transport, a measure of the behaviour of ebb channels.

The gross cumulative transports are also an indicator of how constant or variable in time erosion around a fixed structure would be, as such erosion can be expected to take place both during ebb and during flood flow.

The Grays Harbor tidal basin has a mean area of approx. 236 km^2. The gorge has a typical width of approx. 2.1 km and a cross-sectional area of approx. 24,000 m^2. Using these data we can make a very simple estimate of the flow velocity through the gorge over a whole year, say 2003, with the help of eq. (3.77) and neglecting friction.

In Figure 6.4 we have plotted the result, gradually adding complexity by including more tidal components. The upper panel shows the result for just the M2 component, with a zero net cumulative transport and equal gross positive and negative transports. Adding S2 leads to a sinusoidally varying amplitude and somewhat higher gross transports. Adding N2 to this leads to much more variable instantaneous velocities and transports, but only slightly different cumulative transports. The individual contribution of each tide and even each spring-neap cycle to the cumulative transports is very small.

The effect of adding the diurnal components O1 and K1 leads to a small net contribution to transport, which is modulated through the year. In a situation dominated by the semi-diurnal component, the interaction between diurnal and semi-diurnal components is negligible.

Here, it is still rather small but not negligible. We see that for all the variation in the instantaneous velocity and transport, the cumulative transport seen over a period of one year is almost linear. We see that this cumulative transport, which is the filtered result that ends up in the morphological changes, can easily be mimicked by a single enhanced M2 component, plus the artificial diurnal component C1, plus the mean component.

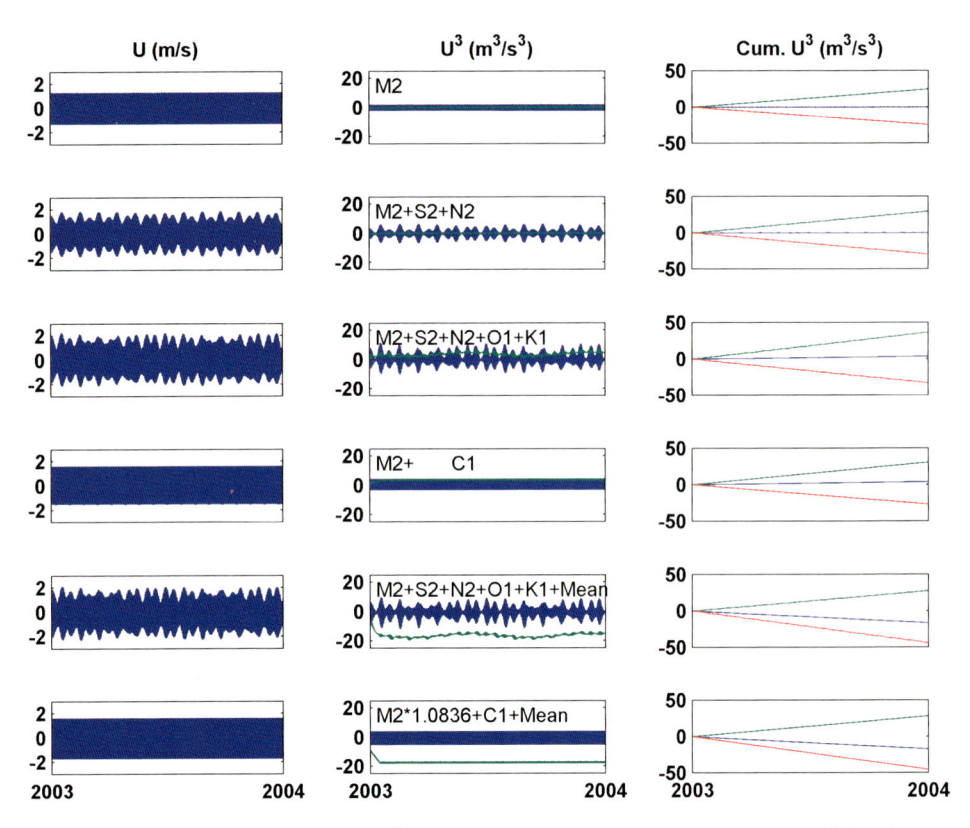

Fig. 6.4 Tidal velocity (left panels), U^3 (middle panel, blue line), monthly average U^3 *50 (middle panel, green line); cumulative U^3, net and gross over one year, for various combinations of tidal components; schematic case Grays Harbor.

For this particular case applying eq. (6.9) leads to an enhancement factor for M2 of 1.0836. We see in the lower panel of Figure 6.4 that this approximation leads to very similar gross and net cumulative transport rates through the year.

In Figure 6.4 the results are enlarged for the last two scenarios, where we see that the simple combination of and enhanced M2, an artificial diurnal component C1 and the mean component give a very accurate representation of the gross and net cumulative transport rates. Also, it shows that we do not have to worry too much about the variation in spring-neap cycles through the year, as this gets filtered out in the cumulative transports.

[Lesser (2009)] shows that using such a simplified tide, which is exactly periodic with a period of approx. 24 h 50 min gives a very accurate representation of the tidal effects in a complex model of Willapa Bay, just south of Grays Harbor.

The tide in these cases is relatively easy to schematize since, on the open ocean coast, only a few tidal constituents suffice to run a tidal model and typical shallow-water components such as M4 are negligible at the sea boundary (though they will be generated inside the model). Additionally, as we discussed in Section 3.5, since

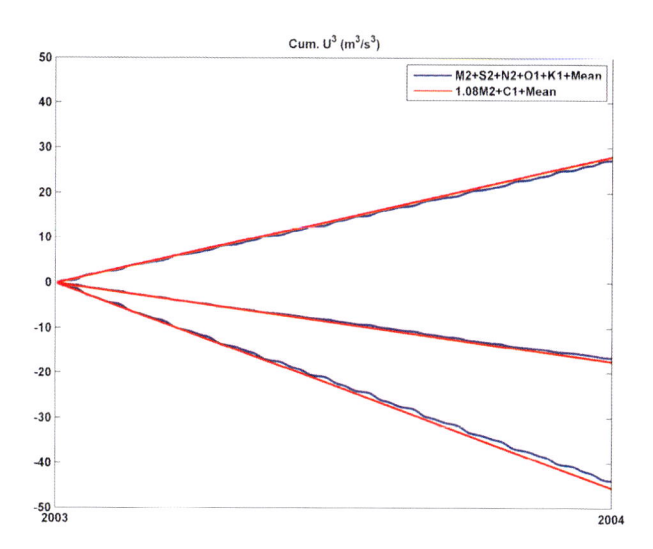

Fig. 6.5 Cumulative U^3, gross and net, for schematized Grays Harbor model; blue lines: full year simulation with all components; red lines: results for run forced by representative tide = 1.08 * M2 + C1

the tide propagates alongshore at a very high celerity, phase differences alongshore are negligible, and the offshore boundary of tidal models on such coasts can be treated as uniformly moving up and down.

6.3.3.3 Selecting morphological tide in shallow seas

In shallower environments such as the North Sea, the Adriatic or the Gulf of Tonkin, a much larger set of tidal components must be taken into account in driving a model, and alongshore variations in phase and amplitude must be imposed. Usually this means that our morphological model must be 'nested' in a larger-scale regional model. For deriving a morphological tide, a simple option is then to use a similar schematization procedure for the regional model, i.e. run it with just the C1 and enhanced M2 components at its ocean boundaries, and to impose the resulting water levels and water level gradients or velocities on the boundaries of our morphological model. The M4 and higher overtides of M2 are automatically generated in the regional model and their amplitudes and phases can be checked against tidal stations in our area of interest.

However, there are some uncertainties related to this method. First, the propagation and generation of the relevant components through the regional model may be affected by the presence of the other components. Second, the mean currents generated inside the local morphological model may be different for different phases of the spring-neap cycle. therefore, an alternative method is usually applied, similar to the one proposed by [Latteux (1995)]. The method can be summarized as follows:

- A simulation of flow, sediment transport and bottom changes over a full spring-neap cycle is carried out as a reference simulation
- The pattern of sedimentation and erosion (or sediment transport rates) over the full simulation is compared with the patterns over smaller time periods, usually taken as 24 hours and 50 minutes to capture diurnal and semidiurnal components.
- For each consecutive time period the following parameters are evaluated:

(1) the correlation between tide-averaged transport rates in all grid points for the selected period and that for the spring-neap period. This parameter indicates if the overall pattern is represented correctly;
(2) the slope of the linear regression between the transport rates over the spring-neap period and those over the selected double tide. This slope can be seen as a time-scale factor. The computed transport rates using a certain representative tide should be multiplied by this factor to obtain the actual transport rates. However, in general we prefer to apply a representative tide for which this factor is close to 1.

- For the selected time period we carry out a harmonic analysis of the time series of boundary conditions, in order to create a perfectly periodic boundary condition, with a base frequency of half the M2 frequency.

As an example we show how this procedure was applied for a morphological study of the Humber Estuary. The bathymetry is shown in Figure 6.6. The water level variation over a half spring-neap cycle is shown in Figure 6.7, and the 7 double tides that were investigated are indicated in the table.

Figure 6.8 shows the correlation plots for the transport rates in the mean flow. The dots represent the averaged transport rates in each grid cell for each of the 7 cycles. On the horizontal axes, the average transport rates for the full spring-neap cycle are specified and on the vertical axes, the transport rates of the particular double tide. Table 6.2 gives the results of the comparison.

Clearly, tide no.5, gives the best fit. It gives a good representation of the transport rates and is therefore selected as the representative, or morphological tide. However, this tide does overestimate the average transport rates by a factor of $1/0.71$. In effect, when this morphological tide is used, the computational times should be reduced by a factor of $1/0.71$ (=1.4). As an alternative, especially when the morphological tide has to be combined with other effects (waves, wind, etc.) the second-best tide in terms of correlation (no. 4) can be chosen, which has the advantage that no correction factor has to be applied.

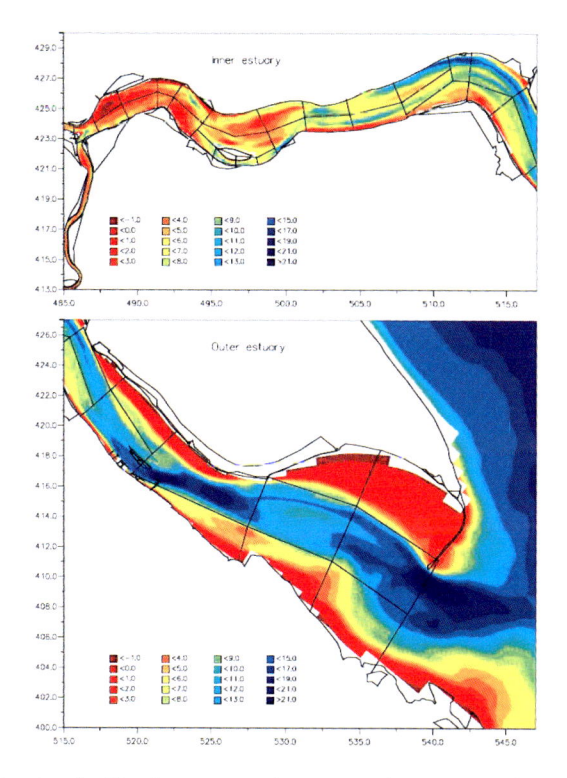

Fig. 6.6 Bathymetry for the Humber estuary (upper panel connects to lower panel at the upper left corner).

Table 6.2   Derivation of morphological tide

Tide no.	Start time	Corr. coeff. mean flow direction	Slope mean flow direction	Corr. coeff. perp. flow direction	Slope perp. flow direction
1	02 10 1997 01 00	0.6924	3.2969	0.6961	3.9413
2	03 10 1997 02 00	0.8207	2.6730	0.8510	3.1553
3	04 10 1997 02 50	0.9204	1.6685	0.9435	1.8616
4	05 10 1997 03 40	0.9793	1.0355	0.9850	1.0459
5	06 10 1997 04 30	0.9974	0.7123	0.9977	0.7036
6	07 10 1997 05 20	0.9910	0.5564	0.9947	0.5493
7	08 10 1997 06 10	0.9783	0.4768	0.9864	0.4804

6.3.4 *Schematisation of wind/wave climate*

As an example we may take the case where we want to study a part of the Dutch coast, situated on the North Sea, near Hoek van Holland (see map in Figure 6.9) with a coast orientation of approx. 30 degrees relative to North. We can obtain wave data from the permanent network of monitoring stations operated by Rijkswaterstaat (available at www.golfklimaat.nl).

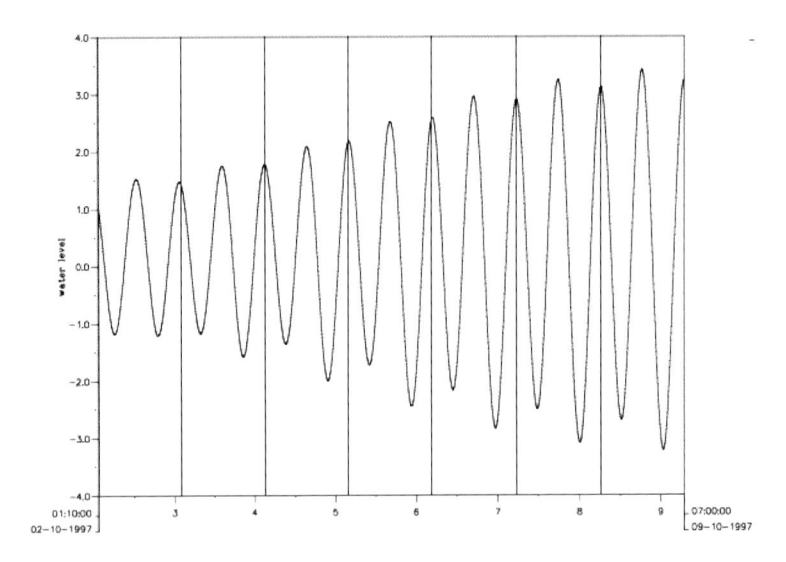

Fig. 6.7 Water level variation over a half neap-spring cycle at the Humber estuary

We retrieved 23 years of data for the location Euro platform, which has directional wave information. Figure 6.10 shows the percentage of occurrence of significant wave height and wave direction classes. From this it is evident that there are two main wave directions, viz. from the southwest, 210-240 degrees, and from northwest to north, 330-360 degrees. This is explained by the dominant wind direction from the southwest and the largest fetch from southwest and the north.

In Figure 6.10 we show some typical relations between the parameters mentioned above, for the directional sector 330-360. The peak period is strongly correlated with the wave height, as is the wind speed and (for this sector) the surge, the residual water level set-up after filtering off the tidal variation. In the same figure we show the correlation between wave direction and wind direction. In Figure 6.11 the mean value of these parameters per wave direction / wave height class are shown, which reveals some clear trends, such as a strong correlation of the surge level with the wave height for north-westerly conditions.

In this case we can use these correlations to reduce the number of independent input parameters that we have to consider, and thus reduce the dimension of the schematisation problem. We can consider wave direction and wave height as the most important parameters governing the sediment transport and schematise these to a number of representative conditions. For each of these conditions we may then assign the mean value of peak period, surge level, wind speed and wind direction based on the relations shown above.

The next step is to define criteria by which to select a reduced number of input conditions. Such criteria may be:

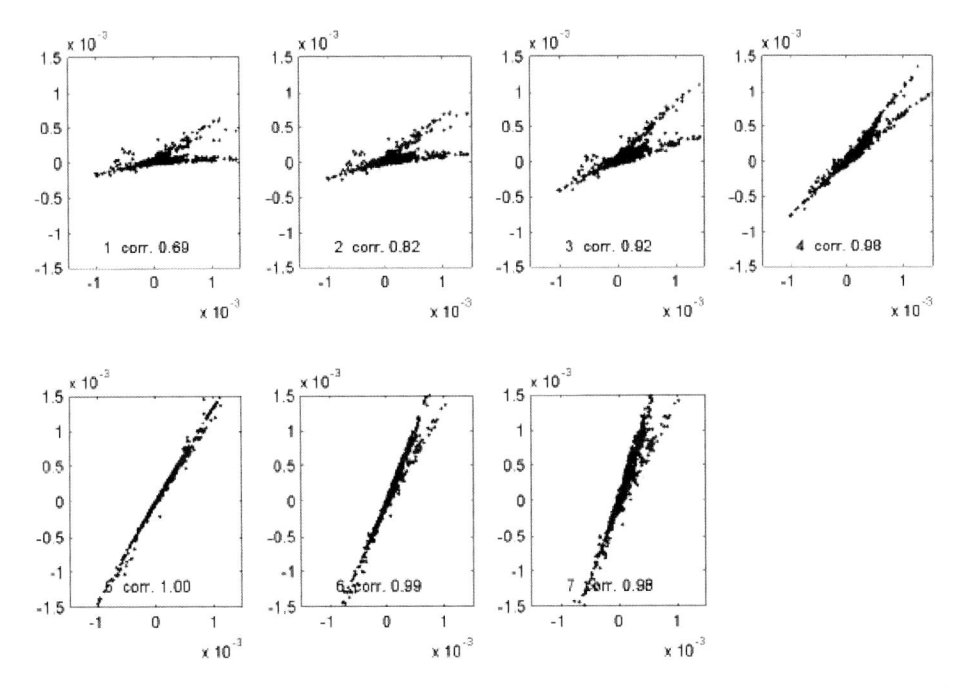

Fig. 6.8 Correlation for transport rates obtained from a single tide versus the transport obtained from the neap-spring tidal cycle.

(1) Representing the average longshore transport;
(2) Representing the cumulative transports through important control sections;
(3) Representing the overall sedimentation/erosion pattern;
(4) Representing the overall sediment transport vector field.

We can try to evaluate these criteria by using

(1) simple relations that capture important effects
(2) a simplified model, e.g. a profile model, to evaluate the distribution of transport across coastal profiles
(3) the full model, but run for short periods for all conditions, so total computation time is still manageable.

6.3.4.1 Using simple relations

Let us take the CERC formula as an example, and simplify this even more to the simple expression:

$$S_{long} = AH_{m0}^{2.5} \sin\left(2(\varphi_w - \varphi_c)\right) \qquad (6.10)$$

This can be used easily to select a single representative wave height for a given direction class. We select as a representative direction the class middle, and then

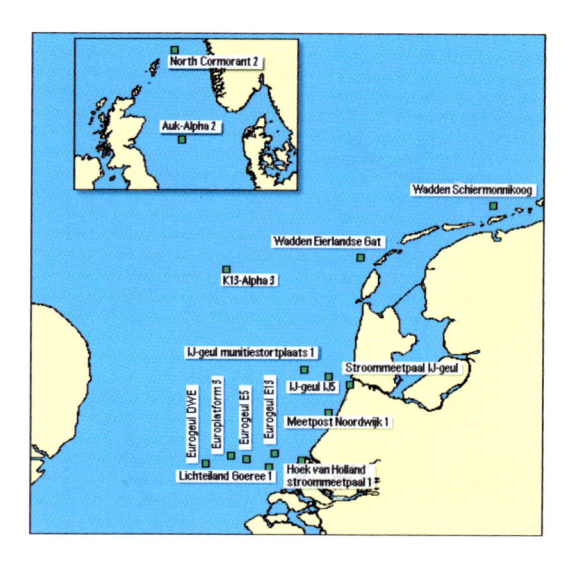

Fig. 6.9 North Sea monitoring stations Rijkswaterstaat

compute the representative wave height from:

$$\sum_i \left[p_i H_{s,i}^{2.5} \sin\left(2\left(\varphi_{w,i} - \varphi_c\right)\right) \right] = \sum_i (p_i) H_{s,rep}^{2.5} \sin\left(2\left(\varphi_{w,rep} - \varphi_c\right)\right) \Rightarrow \quad (6.11)$$

$$H_{s,rep} = \left(\frac{\sum_i \left[p_i H_{s,i}^{2.5} \sin(2(\varphi_{w,i}-\varphi_c)) \right]}{\left[\sum_i (p_i)\right] \sin(2(\varphi_{w,rep}-\varphi_c))} \right)^{1/2.5}$$

Or, for a narrow range of wave angles, just:

$$H_{s,rep} = \left(\frac{\sum_i \left[p_i H_{s,i}^{2.5} \right]}{\sum_i (p_i)} \right)^{1/2.5} \quad (6.12)$$

The number of wave directions chosen should be 2 at the very minimum, since it is not only important to obtain reasonable estimates of the net longshore transport, but also the gross transports in both direction should be represented. It must be stressed that this simple method is not very accurate and results may deviate because the transport relation in the full model is more complicated than the simple power of 2.5 relationship.

6.3.4.2 Using a profile model as a proxy

With the previous method, we can more or less guarantee the longshore transport rates in both direction, but the cross-shore distribution can still be completely

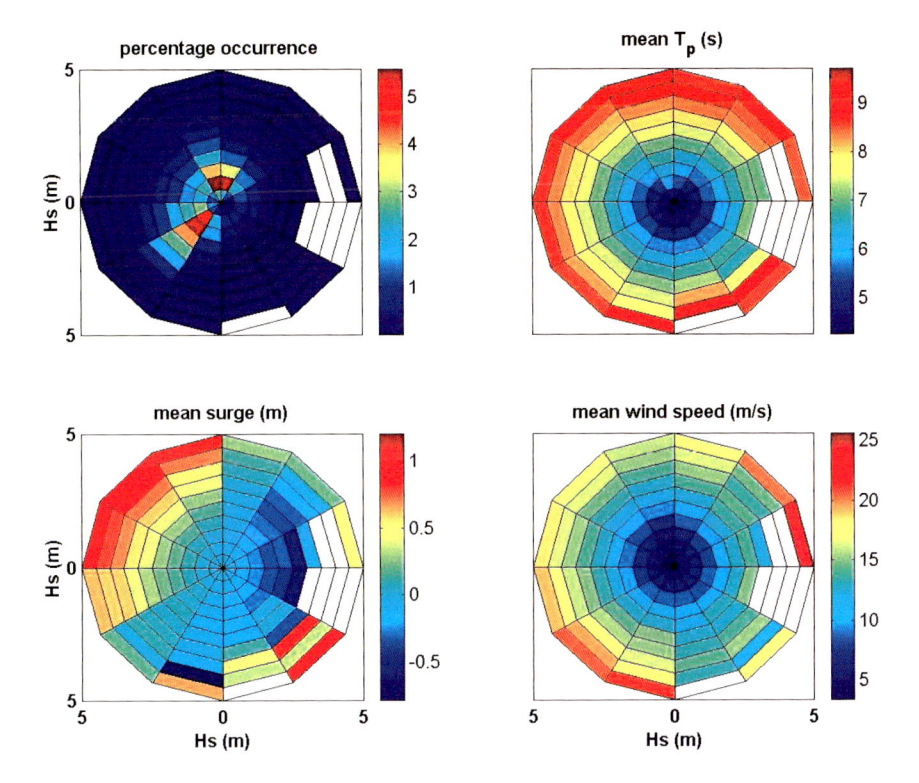

Fig. 6.10 Percentage of occurrence of 0.5 m H_s classes and 30 deg. wave direction classes (top left); mean T_p (top right), mean surge level (bottom left) and mean wind speed (bottom right) per direction/wave height class.

wrong; for instance, when looking outside the surf zone for the given representative wave height, the local transport in the schematized wave climate will be zero, even though there obviously are higher wave conditions where the surf zone extends further and transports can be significant. An improvement in this representation is therefore to consider more wave height classes and to apply the criterion above per group of wave conditions. We can do better than this, however, and use a profile model as a good proxy of the full model, in that at least it will give us an idea of how the transport is distributed across the surfzone.

We use as an example the situation that we want to find representative wave conditions for a given wave direction. We can run a profile model that gives us the cross-shore distribution of transport as a function of the wave height (in 0.5 m classes); by multiplying the transport profile per wave class with its probability of occurrence we get the weighted average distribution of the transport. We see that the average distribution is quite different from the one with the schematization based on $H_{m0}^{2.5}$: high waves with a low frequency of occurrence lead to a long tail that is totally missed in the simple schematization. A better method is to make use

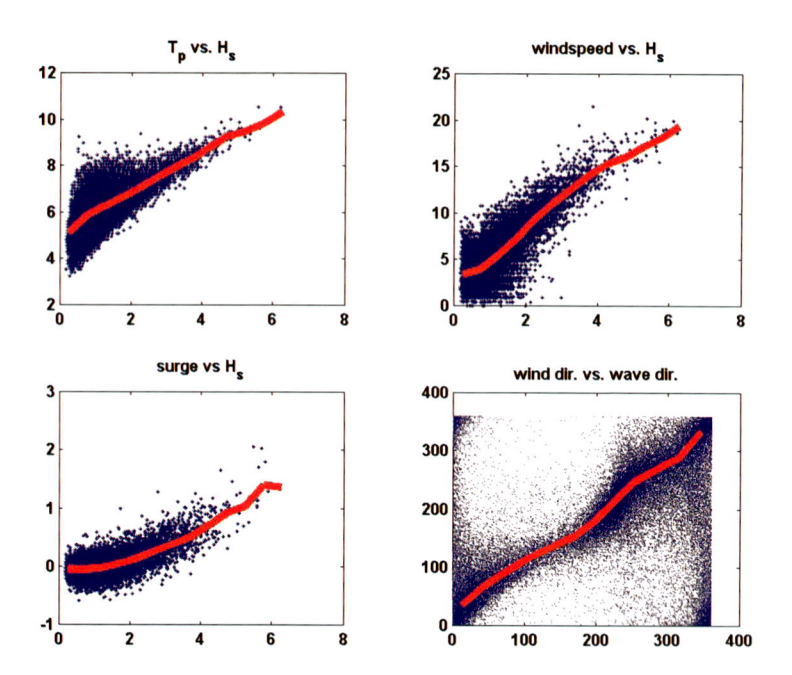

Fig. 6.11 Example scatter plots and trends of T_p vs H_s, windspeed vs. H_s and surge level vs. H_s, for sector 330-360 deg. N; scatter plot of wind direction vs. wave dir. for all directional sectors; data Euro platform, North Sea, 1979-2001, courtesy Rijkswaterstaat, the Netherlands.

of actually computed transports. We select a number of wave classes that we want to group together, and for these wave conditions we require:

$$\sum_i p_i S_i = \left(\sum_i p_i\right) S_{rep} \Rightarrow S_{rep} = \frac{\sum_i p_i S_i}{\sum_i p_i} \qquad (6.13)$$

where S is the total transport, computed by the profile model. We then compare this representative transport with the total transports computed as function of the wave height. By interpolation we can then obtain the representative wave heights for the group of wave classes selected. By combining the resulting transports for these wave heights and multiplying them by the probability of the group of wave classes we get an approximation of the weighted average transport that we can check against the full result. This is illustrated in Figure 6.12 for two groups of wave height classes. The selection of where to divide the waves into low waves and high waves is still arbitrary, and can be varied until a good overall fit is obtained. In Figure 6.12 we compare the resulting distributions based on one representative condition and on two conditions. In both cases the total transport is equal to that of the full climate, which is an improvement on the case with the a priori estimate of the representative wave height. The schematization to two conditions is a marked improvement on the distribution based on one condition and is for all practical purposes sufficiently accurate.

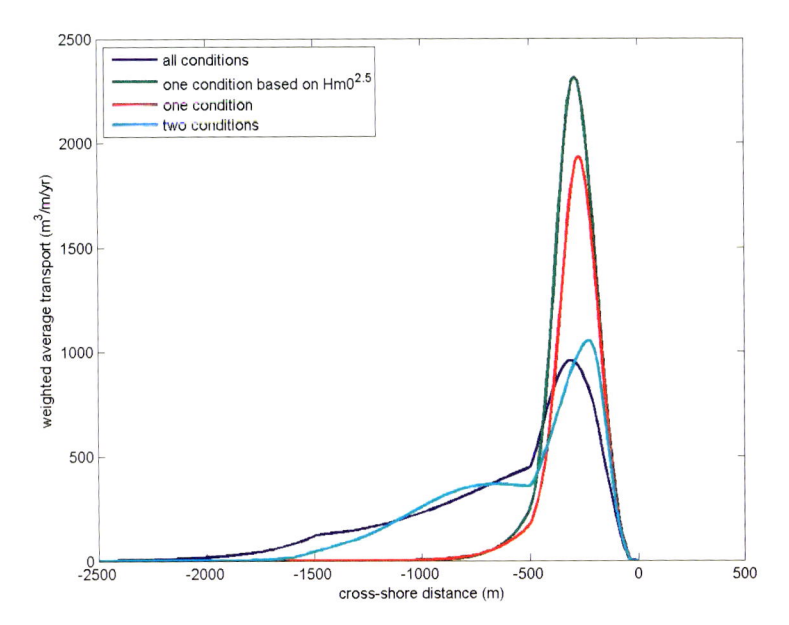

Fig. 6.12 Cross-shore distribution of longshore sediment transport for a given direction, and full wave height distribution in 0.5 m classes, compared with schematizations based on one or two representative wave heights

6.3.4.3 *Optimization procedure 'OPTI'*

The relatively simple methods shown above are not easily applicable in complex areas with multiple coast orientations and different areas with different dominant processes. In such cases we can use a more sophisticated technique, where we try to reproduce a complete pattern of sedimentation/erosion or sediment transport. The outline of this method, called 'OPTI', is as follows:

- Set up a (coarse-grid if necessary) flow, wave and morphology model that runs over one tidal cycle, computing the tide-averaged sedimentation/erosion pattern
- Run this model for each of the wave/wind/tide conditions
- Compute the 'target' sedimentation/erosion pattern as the weighted average over all sedimentation-erosion patterns, taking into account the probability of occurrence or 'weight' of each condition
- Find a reduced set of conditions and weight factors that produce the same sedimentation-erosion pattern as the 'target' one.

It does require that we first run the complete model through all possible conditions, though only for a short time. This is not a bad idea anyway, since it will allow us to check if the model behaves reasonably for all input conditions. The optimization procedure we developed is as follows:

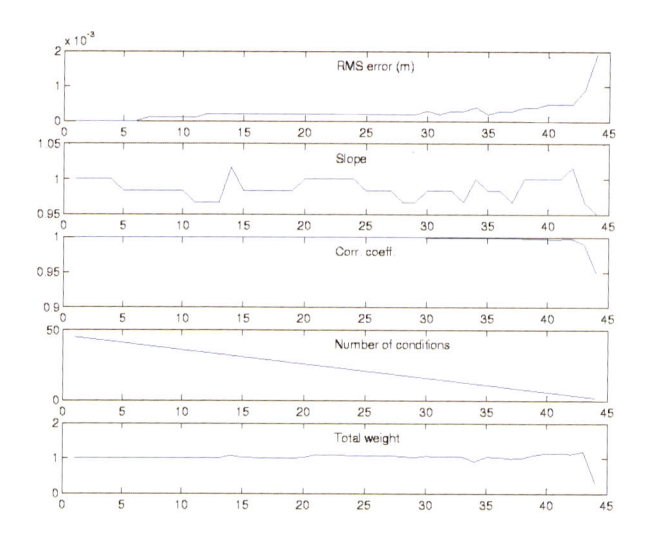

Fig. 6.13 Evolution of various error parameters through an OPTI procedure. We start with 45 conditions; after iteration 44 only one condition is left. The result for 4 remaining conditions (after iteration 41) was used in the further morphological study.

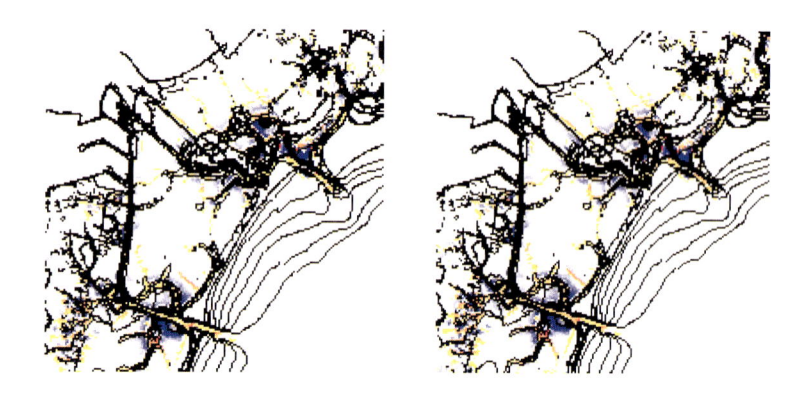

Fig. 6.14 Comparison between weighted average sedimentation/erosion patterns for all 45 conditions (right panel) and 4 optimized conditions (left panel)

(1) We start with the given set of weight factors. If we add up all the individual patterns multiplied by these weight factors we get the target pattern. By the way, this adding up does not cost any significant computation time.
(2) Create a large number of 'mutations', typically in the order of 1000, where we vary the weights randomly within a certain range. For each of these mutations we compute the weighted average pattern and compute some error statistics in this pattern compared to the target pattern.
(3) Select the mutation with the smallest error and keep track of error parameters

(4) Remove the condition with the smallest contribution to the average pattern
(5) Return to 2

What happens in this process is that weights gradually grow or reduce until they become either dominant or extinct, and that the overall pattern remains almost unchanged until we're down to less than 10 or even 5 conditions. When we plot the error statistics against the number of conditions remaining we usually see a point where the error increases sharply; obviously we want to stay away from this point.

As an example we use a study on the long-term morphological development of the lagoon of Venice. The wave climate inside the lagoon is dominated by locally generated waves; the wind climate is very variable and the morphological changes are due to a combination of tidal currents and waves.

From both the error statistics in Figure 6.13 and the visual comparison of the sedimentation/erosion patterns in Figure 6.14 we can see that there is a very good match between the two patterns, and that reducing the wind climate to just 4 conditions does not lead to any degeneration of model skill, particularly when we compare this kind of agreement with the match between model and observations, which of course never comes close to such agreement.

Chapter 7

Coastal Profile Models

7.1 Introduction

7.1.1 *Principles and Approach*

In "process-based" or "deterministic" profile models the different processes, which contribute to profile development are explicitly taken into account ([Roelvink and Broker (1993)]. And although process-based profile models ignore alongshore variation, they capture an important part of the processes that shape our coasts and so provide insight and guidance for the more complex area models disussed in Chapter 9. The earlier process-based profile models ([Dally and Dean (1984)], [Stive (1984)] and [Steetzel (1987)], [Steetzel (1990)]) mainly consider transport caused by suspended sediment carded offshore by the return flow, which works well for the case of dune erosion during storm conditions (see Sections 5.3.3 and 10.3). More moderate wave conditions, which rebuild the beach, require additional transport mechanisms. [Watanabe and Dibajnia (1988)], use various empirical formulations for on- and offshore transport as a function of the bottom shear stress. [Stive (1986)] apply the energetics model by [Bailard (1981)] and consider contributions from wave asymmetry terms. [Roelvink and Stive (1989)] add spatial lag effects in the return flow description, additional stirring due to breaker induced turbulence and a first attempt at accounting for long wave effects. The latter are treated in a more sophisticated way by [Sato and Mitsunubo (1991)]. [Nairn *et al.* (1990)] discuss the transition zone between initiation of breaking and the onset of wave setup and return flow, explained by the effects of rollers. The vertical variation of the mean flow within the surfzone can be strong and clearly affects the pick-up and transport of sediment in the watercolumn and near the bed. [Broker-Hedegaard *et al.* (1991)] use a boundary layer model to describe the variations in the vertical and in time within the wave period, and use tabulated results during a morphodynamic run. A comparison of several morphodynamic models is given in [Broker-Hedegaard *et al.* (1992)]. A comprehensive assessment of various profile transport models is given by [Schoonees and Theron (1995)].

Over the last decade, much experience has been built up about the model components and about their interactions and resulting capabilities. These include additional sediment transport contributions by boundary layer streaming, intra-wave pressure gradients (acceleration), Stokes drift, variable grain size and improved descriptions of the sediment transport processes (see Chapter 4 and references therein). We are at a point that these models have shown considerable skill in predicting the profile evolution ([Thornton et al. (1996)], [Gallagher et al. (1996)], [Hoefel and Elgar (2003)], [Henderson et al. (2004)], [Ruessink et al. (2007)]) over times scales of days to even years [Walstra and Ruessink (2009)] during both moderate and storm conditions. The strong point in these models is that their applicability is governed by processes rather than by geography. Surprisingly, much in the processes that drive the cross-shore changes in the profile is still poorly understood, and in general they still require a site-specific calibration to perform well over longer time scales.

7.1.2 *Profile modeling*

An important aspect of coastal profile modeling is the ability to generate a barred profile consistent with observations. As described in section 5.3.1., a key element in this is predicting the correct phase shift between the bed elevation and the concurrent sediment transport. In the following we use a profile model to compute the sediment transport over the barred profile given in section 5.3.1 and examine the contribution of the individual sediment transport processes to the cross-shore distribution of the total sediment transport and its ability to generate and propagate bars ((Matlab-code to calculate the wave transformation, flow velocities and sediment transports has been included as **profilemodel.m**).

The initial profile has a clear bar-trough signature with the crest located approximately 120 m from the shore line (panel A of Figure 7.1).
To model profile evolution we incorporate the following sediment transport processes (see Sections 4.3 and 4.4 for a description of the individual processes):

$$S_x = a_b S_{b,x} + a_s S_{s,x} + a_{sk} S_{sk,x} + a_{as} S_{as,x} + a_{lw} S_{lw,x} + a_{sl} S_{sl,x} \qquad (7.1)$$

where $S_{b,x}$ represents the bed load transport, $S_{s,x}$ the suspended transport, $S_{sk,x}$ the skewness related transport, $S_{as,x}$ the transport related to wave asymmetry, $S_{lw,x}$ the long wave related transport and the bed-slope transport is given by $S_{sl,x}$ with corresponding expressions in Sections 4.3 and 4.4.

Each of these transport mechanisms has a corresponding calibration factor, a_i, which represents the incomplete knowledge in the understanding of these processes which requires a site-specific morphodynamic calibration . Once properly calibrated, a comprehensive cross-shore profile model may predict the bar-dynamics on the time scale of days to weeks ([Hoefel and Elgar (2003)], [Ruessink et al. (2007)]) or even years ([Roelvink et al. (1995)], [Walstra and Ruessink (2009)]). It must be noted that this calibration is non-trivial as a large number of model coefficients is involved typically requiring a large number of computations and optimizing strategies ([Ruessink et al. (2003)]). The calibration coefficients and model bound-

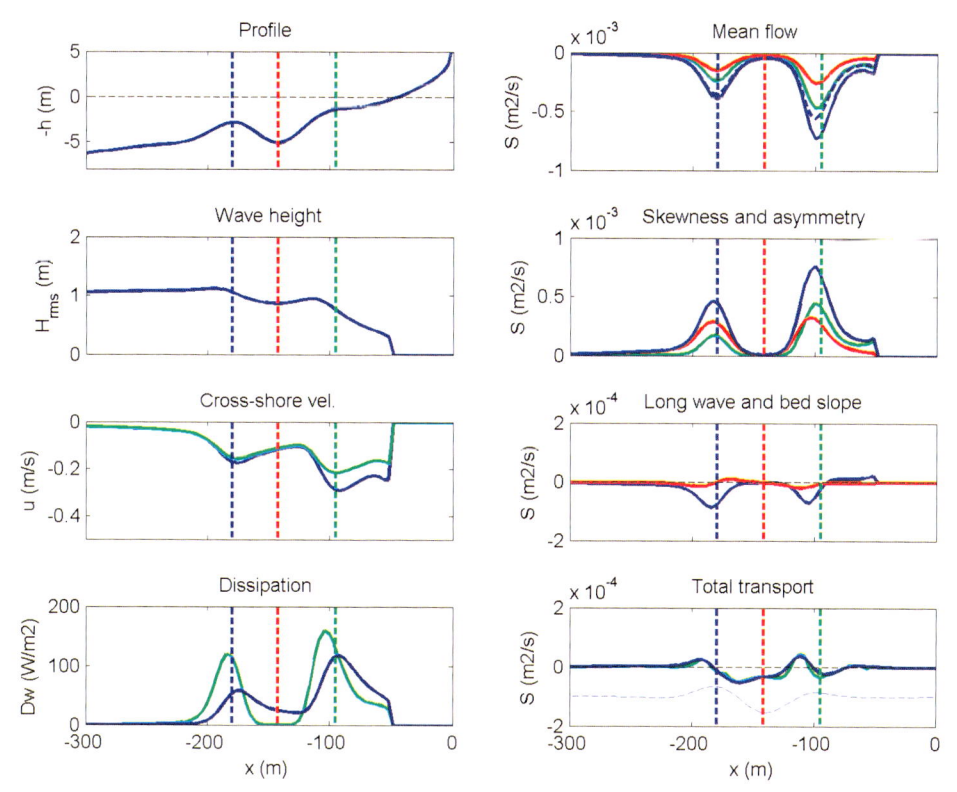

Fig. 7.1 Profile model prediction of wave transformation, hydrodynamics and sediment transport across the surfzone. Panels labeled A-H in counter clockwise direction starting in the upper left corner. Panel A: Bottom profile with position of outer bar crest (blue dashed line, trough (red dashed line) and inner bar crets (green dashed line). Panel B: Cross-shore wave heigth distribution. Panel C: Return flow with (blue) and without (green) roller. Panel D: Wave energy dissipation (green) and roller dissipation (blue). Panel E: Suspended (red), bed-load (green) and total (blue) sediment transport due to return flow and corresponding total transport without roller (blue dashed). Panel F: Skewness (red), asymmetry (green) and combined (blue) transport. Panel G: Long wave (blue) and bed slope (red) transport. Panel H: Total sediment transport with (green) and without bed-slope effect (blue). Bed-anomaly given as a reference (dashed line).

ary conditions used in the profile model simulations discussed below are given in Table 7.1.

Table 7.1 Model parameter settings

A_b	A_s	A_{sk}	A_{as}	A_{lw}	A_{sl}	H_{rms}	T_{m01}	θ
1	1	0.2	0.2	0.1	0.1	1.0 m	8 s	0°

Using a profile model to compute the cross-shore wave height, return flow and just the mean flow-related bed load and suspended load sediment transport distributions typically results in a negative phase lag between sediment transport and underlying bottom profile. As discussed earlier, such a transport pattern will lead to the undesired flattening of the bar while propagating offshore. However, a number of processes contribute to an onshore shift of the sediment transport pattern. The roller energy, resulting from the transition of organized wave energy into a turbulent bore at wave breaking, creates an onshore shift in the radiation stresses and mass flux (e.g., [Nairn et al. (1990)]). As a result the return flow and longshore current, which are important agents for the sediment transport in the surf zone, are shifted in the onshore direction (e.g. [Reniers et al. (2004b)], [Reniers and Battjes (1997)]). This transition effect is modeled as a simple filter as outlined by [Dronen and Deigaard (2007)]:

$$\frac{\partial}{\partial x}\left(\lambda_r \frac{D_r}{c}\right) = \frac{D_w}{c} - \frac{D_r}{c} \qquad (7.2)$$

where λ_r is the spatial lag set to ten times the local water depth. The roller dissipation distribution is shifted onshore consistent with the presence of continued wave breaking in the trough. From the roller energy dissipation, D_r, the corresponding roller energy can be inferred:

$$E_r = \frac{D_r c}{2g \sin \beta} \qquad (7.3)$$

from which in turn the roller related mass flux is obtained:

$$M_r = \frac{2E_r}{c} \qquad (7.4)$$

The depth-averaged return flow compensating for the wave-and roller related mass flux is now given by:

$$\bar{u} = \frac{E_w}{\rho h c} + \frac{2E_r}{\rho h c} \qquad (7.5)$$

Next the mean flow velocity is evaluated at 10 cm above the bed using the vertical profile model discussed in Section 3.7.5, using the total mass flux, composed of the wave- and roller-related contributions, as a constraint. Comparing the near-bed velocities with and without the roller effect it is seen that the spatial lag results in an onshore shift of the maximum near-bed velocity (compare green and blue lines in panel C of Figure 7.1). As a result there is also a small onshore shift in the return flow related sediment transports (compare dashed and solid blue lines in panel E of Figure 7.1). The inclusion of the roller-related mass flux also leads to an overall increase of the sediment transport rate.

An additional effect is the breaker delay, accounting for the fact that the waves take some time to react to the local bed changes due to the presence of a bar [Roelvink et al. (1995)]). This effect is most apparent in the case of narrow-banded swell, where the plunging waves can occur inward of the bar crest location. This effect is partly represented by the spatial lag in the roller dissipation equation used here.

The onshore shift in the offshore directed sediment transport related to the return flow is important for the offshore motion and concomitant growth of the bar during more energetic conditions when waves are breaking on the bar (see Section 4.5) . The frequently observed onshore motion of the bar during quiescent conditions also requires an onshore shift in the onshore directed sediment transport. This can be achieved by accounting for the wave asymmetry-related sediment transport ([Elgar et al. (1997)], [Hoefel and Elgar (2003)]), which persists into the trough, in contrast to wave skewness related transport which typically has its maximum at the seaward side of the bar crest (see panel F in Figure 7.1 and the description in Section 4.4.1).

The bound long wave-related sediment transport maxima occur at the seaward side of the bar thus contributing to the flattening of the barred profile (panel G of Figure 7.1). This observation is consistent with other model studies that include long-wave related sediment transport ([Roelvink (1993)], [Reniers et al. (2004a)], [Smit (2010)]).

Bed slope-related transport is included according to eq. increasing (decreasing) the down- slope (up-slope) sediment transport and is at least an order of magnitude smaller than the return flow or short-wave related transports (compare panels E,F with G of Figure 7.1).

The sediment transport descriptions in the profile model do not account for the fact that in reality time is required for the sediment to be either picked up from the bed or to settle on the bed. This effect is accounted for by introducing a diffusion operator ([Dronen and Deigaard (2007)]):

$$\frac{\partial}{\partial x}\left(D_h \frac{\partial S_x}{\partial x}\right) = S_x - S_{x,0} \qquad (7.6)$$

where $S_{x,0}$ corresponds to the total instantaneous sediment transport calculated from the individual transport descriptions and Sx the diffused sediment transport rate used to update the bed level. Following [Dronen and Deigaard (2007)] the diffusion coefficient is set at $(10h)^2$ m^2. The differences between the instantaneous sediment transport distribution and the diffused transport are relatively small (compare green and blue lines in panel H in Figure 7.1). The total transport, i.e., including all transport contributions described above, shows an onshore displacement with the underlying bathymetry (bathymetric variability shown by the dashed blue line in panel G of Figure 7.1), and is therefore expected to lead to a growing bar amplitude. This is evaluated next.

7.2 Short-term event modeling

Calculating the bed evolution for a period of two days with constant offshore boundary conditions confirms the offshore propagation of the offshore bar and concurrent growth of the bar-trough amplitude (panel A of Figure 7.2). In addition the inshore bar has become more pronounced and shore line erosion is mostly absent. Next the individual transport contributions are turned off one by one to examine the effect on the profile evolution. Note that the effects are in a significant part determined by the calibration coefficient settings that have been kept constant.

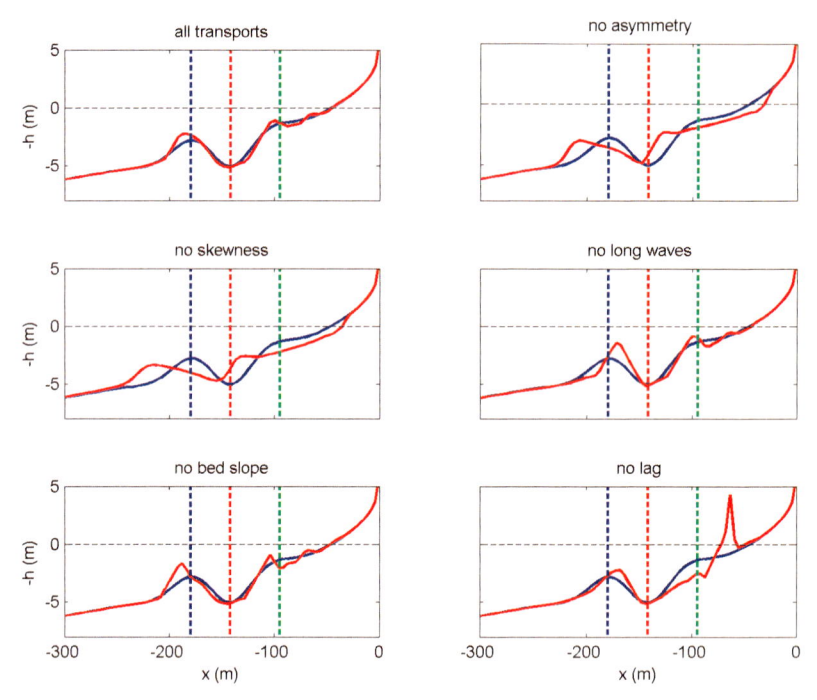

Fig. 7.2 Short-term profile response with initial profile (blue) and final profile (red). Panels labeled A-F in sequential order starting in the upper left corner. Panel A: All processes included. Panel B: No asymmetry. Panel C: No skewness. Panel D: no long waves. Panel E: no bed slope transport. Panel F: no spatial lag.

Excluding the transport associated with the wave asymmetry, S_{asx}, reduces the onshore sediment transport (panel F Figure 7.2), leading to predominantly offshore directed transport that erodes the beach and propagates the bars offshore. The bar shape becomes more 'shock-like', i.e., with a steep seaward front and mild shoreward back (see also Figure 5.1 in Section 5.2.1). A similar result is obtained with the omission of the wave-skewness related sediment transport although there are differences in the bar-shape related to the differences in the sediment transport distribution of the skewness and asymmetry related contributions (panel F of Figure 7.2).

Without long waves the bars become significantly more pronounced and move onshore. The former is consistent with the observed negative phase lag between the long-wave related transport and the underlying bed profile (panel G of Figure 7.2).

Without bed slope effects, the bars become again more pronounced, not unlike what happened by excluding the long wave transport. However, their propagation direction is still offshore.

Omitting the spatial lag and concurrent roller contribution yields unrealistic accretion at the beach, and a shock-like onshore propagation of the offshore bar, consistent with a predominantly onshore directed sediment transport.

As mentioned earlier, after carefull site-specific calibration of the individual transport process parameters cross-shore profile models have shown considerable skill in predicting profile changes during storms and more moderate wave conditions, thus covering both offshore and onshore bar migrations ([Hoefel and Elgar (2003)], [Ruessink *et al.* (2007)]).

7.3 Long-term evolution of barred profiles

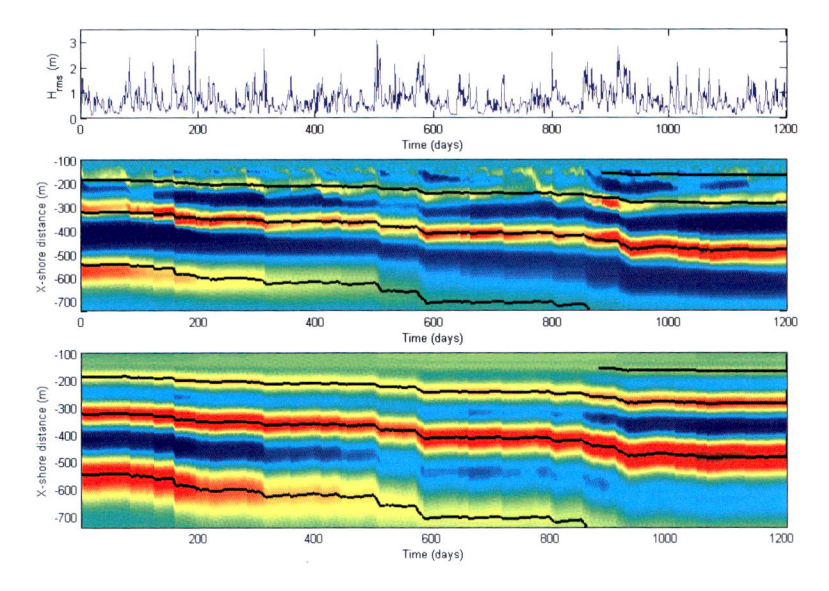

Fig. 7.3 Upper panel: Observed offshore root mean square wave height at Noordwijk. Middle panel: Daily bed-anomaly showing the bar crests and troughs as they respond to the offshore wave forcing. Lower panel: CEOF filtered bed anomaly reponse showing the cyclic bar response at Noordwijk. (Figure courtesy of Dirk-Jan Walstra)

[Walstra and Ruessink (2009)] have shown that process-based profile models can be used to simulate the multi-year cross-shore bar-cycling with significant skill. Comparisons of bar position, amplitude and bar-separation showed good correspondence with approximately 40 years of annual JARKUS profile measurements at Noordwijk, the Netherlands. Using the measured offshore wave conditions the daily changes in the cross-shore profile from 1984-1988 have been simulated (see

Figure 7.3). Both the onshore motion of the bar during moderate wave conditions and the offshore motion of the bars during storm conditions are apparent. By filtering the model predicted profile response with CEOF ([Ruessink et al. (2001)]) the longterm behaviour is revealed showing the typical 4 year bar cycle observed at Noordwijk (compare with Figure 5.5).

The good quantitative comparison between model predictions and observations allows for analyses of the processes contributing to the longterm bar behavior thereby providing more insight into the underlying mechanisms and potentially resulting in more generally applicable profile models, i.e. without the need of a site specific calibration.

7.3.1 Including longshore transport gradients

In many cases the longterm shore line is not in equilibrium, but subject to erosion or accretion. This in itself does not prevent the use of a process based profile model to simulate the (barred) profile changes over time, provided an alongshore sediment transport gradient is added to the bed-updating scheme. This can be readily achieved by prescribing an alongshore transport gradient as a fraction of the calculated alongshore sediment transport rates:

$$\frac{\partial S_y(x)}{\partial y} = \frac{S_y(x)}{L_y} \qquad (7.7)$$

where L_y is some appropriate length scale which can be estimated from the calculated yearly-averaged cross-shore integrated alongshore sediment transport and the observed erosion/accretion rate.

7.4 Nourishments

Once a cross-shore profile model is properly calibrated it can be used in assessing different nourishment scenarios. [Roelvink et al. (1995)] describe the process of morphodynamic calibration, wave climate reduction (see also section 6.3.4) and subsequent morphological prediction of the effectiveness of a shore face nourishment positioned between the inner and outer bar crest at one of the Dutch barrier islands over a ten year period. The hydrodynamic calibration focuses on the cross-shore wave transformation, i.e. wave breaking, over the multiple barred profile.

The sediment transport rates are verified with sediment transports inferred from the observed profile changes 150 days after the initial deployment of the nourishment combined with a pre-nourishment period to verify the autonomous behaviour. Satisfactory results were obtained for both periods (see Figure 7.4) with the most important calibration parameters being the breaker delay and the bottom slope effect. The latter plays an important role in the decay of the offshore bars as they enter deeper water and the underlying processes are not well understood at present.

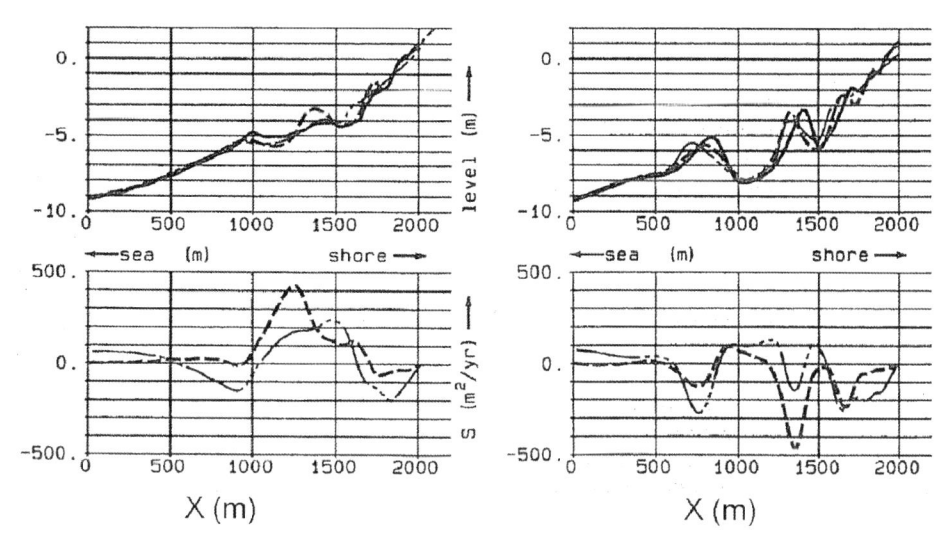

Fig. 7.4 Comparison bewteen observed and predicted profile changes and inferred transport rates in with and without the presence of a shore face nourishment.

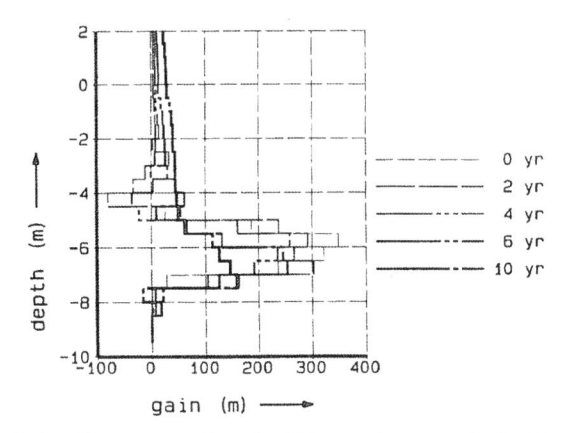

Fig. 7.5 Gain of bed-elevation expressed as the difference between the bed level with and without the presence of the shore face nourishment.

Deposition of the nourishment in between the two breaker bars is predicted to lead to a gradual onshore transport of sediment effectively nourishing the beach for a period over 10 years (see Figure 7.5). Shoreface nourishments have the advantage that they protect the beach from severe wave attack during storms thereby providing beachface stability. In contrast, beach nourishments provide immediate relief for the beach, but are susceptible to significant beach erosion during severe wave attack.

A typical response for a shore face nourishment in a barred profile is the onshore motion of the bars located inshore of the nourishment ([Rijkswaterstaat (1988)]). This effect can be simulated well with a profile model. 2D effects such as the creation of a shadow zone in the case of oblique wave incidence, affecting the shore line downstream of the longshore current drift ([Rijkswaterstaat (1988)]), as well as eddy-like circulations at the lateral ends of the finite length nourishments require more sophisticated modeling (see Section 10.1). As time progresses the nourishment becomes more diffused and two-dimensional effects become increasingly important.

Chapter 8

Coastline Models

8.1 Principles

Coastline models are built on the assumption that cross-shore processes are relatively fast and are able to maintain the shape of the coastal profile over a certain profile height, which runs from the so-called depth of closure to the top of the active profile, usually the first dune top. Short-term fluctuations of the profile are smoothed out and bar behaviour is not considered. The general principles of coastline theory have been outlined in Section 5.3, here we will explain the workings of more generic coastline models.

The typical setup of such models is that the computation is split up into two parts. First, a profile model is run where the longshore transport rate is computed as a function of the coast angle and wave climate and stored in tables. Second, a coastline model where the longshore transports are computed as a function of the coast angle by looking up the transports in these tables, after which the coastline positions can be updated, coast angles can be recomputed, and so on.

Effects of structures such as groins and breakwaters are simulated by blocking the transport rate in proportion to the part of the cross-shore distribution of the longshore transport that they cover, and by selectively shielding certain wave conditions.

8.2 Existing models

A number of coastline modeling packages is available commercially or otherwise. We will briefly discuss their main features here:

GENESIS - a package developed by the US Army Corps of Engineers and freely available (just google 'genesis cerc' and you'll find the history and complete documentation). It is based on the extended CERC formula including the effect of longshore gradients in wave height and has advanced features for describing the effects of coastal structures on the local wave climate. It has been widely used in coastal engineering projects and is a standard design tool in the US. GENESIS can be operated from the SMS user interface.

LITPACK - a package developed by DHI in Denmark that includes the modules STP for calculating non-cohesive sediment transport in a combined waves and current environment; LITDRIFT, which combines STP with a coastal hydrodynamic module to give a deterministic description of littoral drift; and LITLINE, a coastline evolution model that simulates coastal response to gradients in the alongshore sediment transport capacity arising from the effect of natural features and a wide variety of coastal structures. The main difference with GENESIS is in the process-based sediment transport description.

UNIBEST-LT/CL - developed by Delft Hydraulics (now Deltares); this package combines longshore transport (LT) calculations based on a coastal profile model approach with coastline computations on a curvilinear coastline (CL). A range of sediment transport formulations is available, allowing to review the uncertainty inherent in such models. A detailed wave climate can be considered and wave transformation can be simulated using the SWAN spectral wave model.

BEACHPLAN - developed by HR Wallingford, this model is based on the CERC formula including the effect of longshore gradients in wave height. It has special features for wave transmission through structures, bypassing of groynes and breakwaters and for the effect of seawalls on the sediment transport.

Though these models are similar in many respects, their user-friendliness and extent of validation will vary, so we advise you to carefully study the track record of these models for your particular problem and to test their ease of use for yourself. An excellent study to compare three of these models (GENESIS, UNIBEST and LITPACK) was presented in [Szmytkiewicz *et al.* (2000)] and may serve as an example of how to compare such models. In the following we will develop a relatively simple coastline model in Matlab, which should provide more insight into how such models work.

8.3 A simple Matlab version

In the following we will build a simple Matlab version of a coastline model. This is done in two steps:

- A profile model that computes the longshore transport for a large number of representative wave conditions, and for different wave angles. It stores the longshore transport distributions and total longshore transport as a function of wave condition and coast angle.
- A coastline model that uses an initial coastline position, additional information about structures and the longshore transport tables (S-phi curves) to predict the future coastline positions

8.3.1 *Profile model to generate S-phi curves*

The profile model in our case builds on the hydrodynamic model developed in Section 3.5, adds the computation of longshore transports according to the Soulsby-van Rijn formula and puts the computation of wave height, water level, longshore velocity and longshore transport across the profile in a loop over all wave conditions; this loop is again put in an overall loop over the expected range of coastline orientations. To illustrate this we show the rump of the profile model given in the enclosed matlab model **profile_model_v00.m**.

The result is a datafile **S_phi.mat** in which net, gross positive and gross negative longshore transport rates are stored as function of coast angle φ and wave direction θ. These wave directions can be used to selectively shield transports from certain wave directions, as will be explained in the next paragraphs.

8.3.2 *Coastline computation*

Similar to Unibest-CL we use a curved reference line and a curved coordinate system where the *s*-coordinate runs along the reference line and the *n*-direction normal to it, facing the sea. We can build this reference line by reading a number or x, y world coordinates, computing the cumulative distance along them and subdividing the line into *ns-1* grid sections.

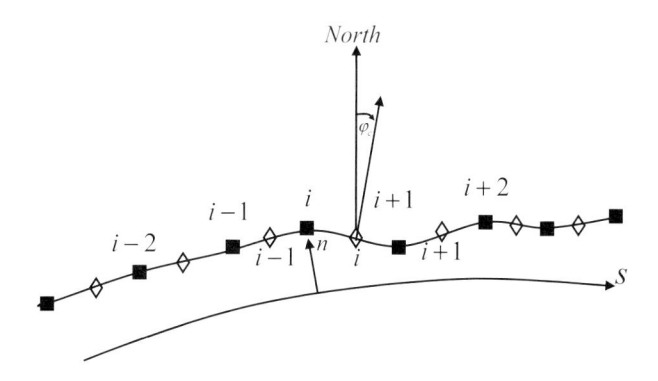

Fig. 8.1 Coastline model grid

The initial coastline position is determined by converting its given x, y positions to s, n coordinates and then interpolating the n values to the s grid (see Figure 8.1).

8.3.3 *The basic version based on S-phi curves*

Next, we read in a table of sediment transports S as a function of coast angle φ_c, as determined by the coastal profile model.

We can now step into a time loop where at every step we do the following actions:

- Compute the actual coast angle at each grid point;
- Compute the longshore transport by interpolating in the $S - \varphi_c$ curve
- Compute the transport gradients
- Update the coastline positions applying a given time step.

A convenient and accurate way to compute sediment transports as a function of coast angle is to put the transport points (open diamonds) in between the coastline grid points (closed squares). In using such a *staggered* grid we can unequivocally compute the coast angle at a given transport point i as:

$$\varphi_{c,i} = 2\pi - \text{atan2}\,(y_{i+1} - y_i, x_{i+1} - x_i) \tag{8.1}$$

where we use the *atan2* function with two arguments to exclude ambiguity in the *atan* function; we apply the normal direction relative to North as indicated in the figure, in line with the nautical convention for wave angles.

The change in coastline position Δn over a time step Δt can now be computed as:

$$\Delta n_i = -\frac{\Delta t}{d}\frac{(S_i - S_{i-1})}{\Delta s} \tag{8.2}$$

We can now update the coastline positions n and from there we can compute the updated x, y positions of the coastline

With this version we can compute the deformation of an open stretch of sandy coastline, for instance after a large-scale nourishment. The assumption is that the same wave climate holds all along the coast. However, this is often not the case, as certain wave conditions may be shielded by headlands, large breakwaters, islands or spits. In that case the $S - \varphi_c$ curve may vary along the coast. A typical example is the coastline development in the vicinity of a large harbour, such as for instance the port of IJmuiden in the Netherlands. The breakwaters were extended to some 2.5 km in the late '60s. This substantially modified the wave climate at the southern side (shielding it from northerly waves) and on the northern side (shielding it from southerly waves). As a result, the equilibrium coast orientation was modified on both sides, leading to very strong accretion on both sides and erosion problems some kilometres away from the harbour.

8.3.4 *Including large-scale variations in wave climate*

We can represent this effect in our coastline model if we store the net longshore transport as function of coast angle, subdivided over the different wave directions in our schematised wave climate, and per alongshore location p only consider wave directions within a specified window (α_1, α_2) as indicated in Figure 8.2 below.

Compared with the basic version this means that before we enter the time-loop we read a table of wave windows at a number of locations; we interpolate the wave windows to all grid points and in each grid point we construct an adapted $S - \varphi_c$

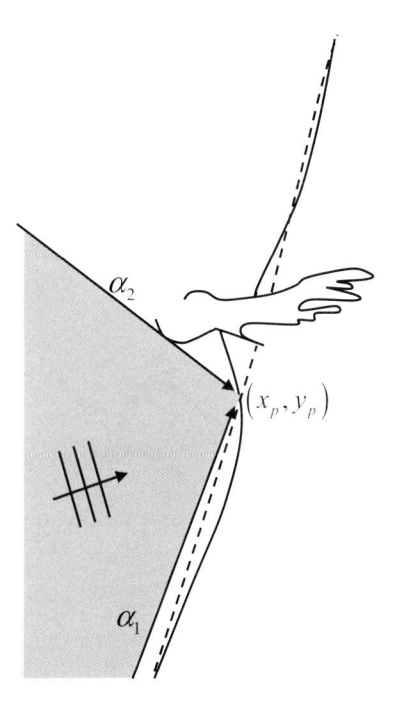

Fig. 8.2 Representation of effect of wave shielding

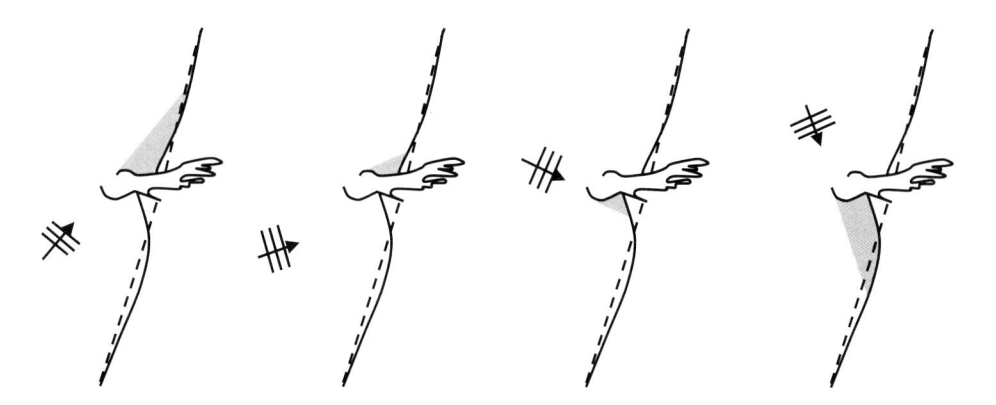

Fig. 8.3 Areas affected by wave shielding for different wave directions, IJmuiden area, the Netherlands

curve. Within the time loop, we use this adapted curve in each point instead of the global one.

The Matlab code for this coastline model is contained in the enclosed **coastline_model_v00.m** and example input files for the case of IJmuiden as called from this script are inluded as well.

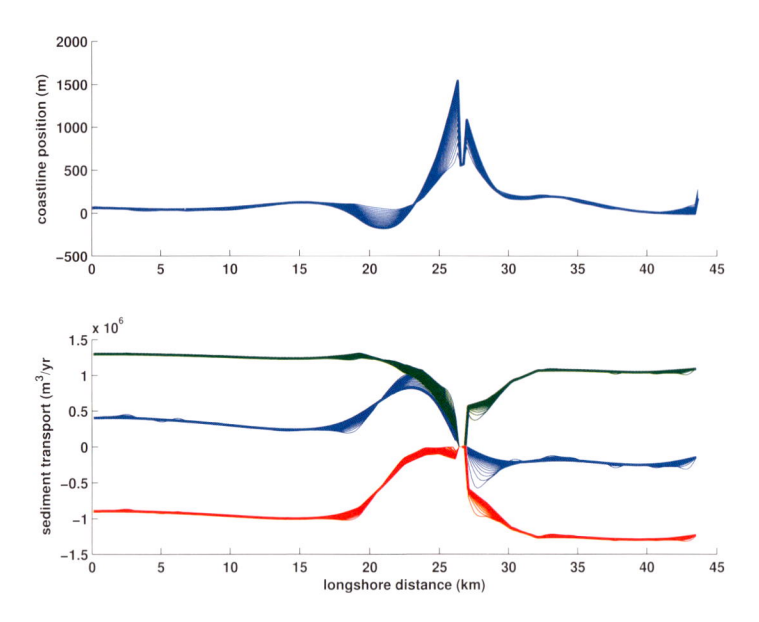

Fig. 8.4 Modeled development of coastline cross-shore distance (top panel) and gross northward (green),gross southward (red) and net longshore transport (blue), IJmuiden area, the Netherlands

8.3.5 Representing small-scale features

Small-scale features such as groynes are too small to influence the wave climate at the scale we are modeling. We can still represent some of their effect by considering that they block part of the longshore transport. In our present model we can indicate the tips and landward ends of a series of groynes. Over the distance covered by each groyne the longshore transport is blocked. We do this based on the shape of the annual mean longshore transport as a function of cross-shore distance, and take into account that this curve shifts in seaward or landward direction as the coastline changes. The result is a reduction factor that is applied to the positive and negative transports at either side of the groyne.

8.4 Case study of IJmuiden, the Netherlands

The harbor moles of the port of IJmuiden, the sea port of Amsterdam, were extended by approx. 2500 m in the period of 1962-1968. A scour hole has developed in front of the entrance, due to the strong tidal current; this has been the subject of 2DH modeling in Roelvink et al. (1998), among others. Closer to shore the coastline on either side has strongly accreted, especially in the South, while further away from the structures considerable erosion has taken place.

Coastline positions at approximately 200 m intervals were extracted from the JARKUS database, for the years 1967 and 2007; see Figure 8.5. A slightly curved

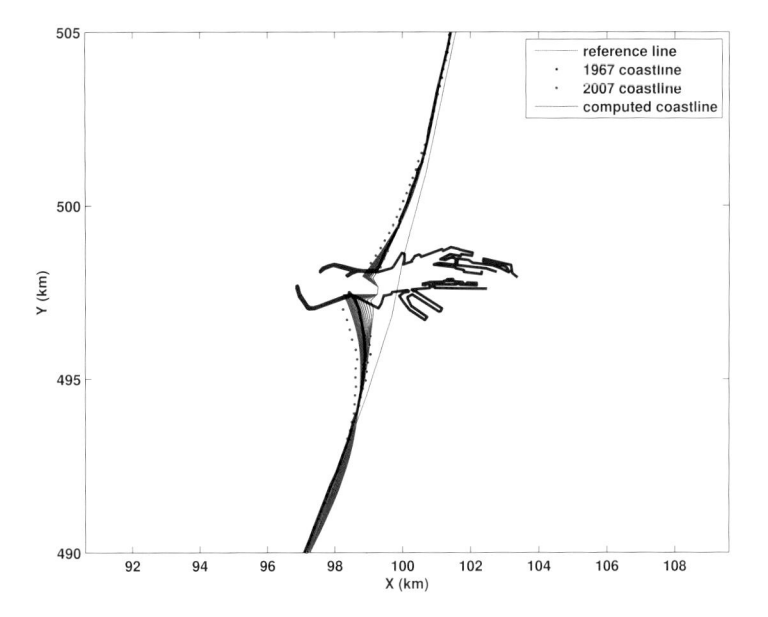

Fig. 8.5 Observed and modeled coastline development 1967-2007, IJmuiden area, the Netherlands

reference line was defined at what could roughly have been an undisturbed coastline. It must be noted that in 1967 the coast had already accreted during the construction of the dams.

We specified the 1967 coastline as the initial condition and specified an equidistant grid of 200 cells over an approximately 40 km long stretch of coastline, which means a longshore grid cell size of approx. 200 m. For wave climate data we selected the Meetpost Noordwijk long-term wave data set, and we divided the wave conditions into bins of 0.5 m H_{m0} wave height and 20 deg. wave direction bins; per bin, the mean T_p period was computed. Based on this wave climate (given in file **climate.txt** and an average coastal profile given by file **profile.txt** we computed the $S - \varphi_c$ curves as described above. The wave shielding angles were specified in a file **wavewindows.txt**. For the specified grid a time step of 0.1 year turned out to be stable and accurate enough. In Figure 8.4 we show the development of the coastline position and the gross and net longshore transport rates in time (lines plotted every two years). We see the coast on both sides of the harbor building out, though more at the southern side. This is due to the *reduction* of the gross southward transport at the southern side, which leads to a strong *increase* in the net northward transport at this end. Likewise, the southward (negative) net transport at the northern side is explained by the reduction of the northward (positive) transport at this side.

The model results are shown as the series of thin green lines in Figure 8.5 and compare favorably with the observed trend. First of all, the model clearly tends towards a plan shape that is very similar to the observed shape; secondly, the rate of change near the breakwaters is quite comparable with the observed changes. More towards the South, the model predicts an eroding coastline, which would surely have happened but for the substantial beach nourishments carried out in this area. When we look at the evolution of the net transport curve, we see that the gradients have been substantially reduced over the 40-year period.

Chapter 9

Coastal Area Models

9.1 Introduction

Coastal morphological area models have been developed since the early '80s, see for instance [De Vriend *et al.* (1993)], [Nicholson *et al.* (1997)] for reviews. Roughly speaking, the first generation of models consisted of existing flow, wave and transport codes patched together with batch files or steering modules and lots of semi-manual interpolation and file transfer between wave, flow and morphology grids. Not surprisingly, these models only went as far as ISE: Initial Sedimentation and Erosion, since it already took a lot of effort to go from all the tidal simulations to running wave models and feeding back the information to the tide models, then passing wave and flow information to the transport model and finally evaluating the transport gradients to come up with an estimate of the rate of change of the bed level. Feeding back the results and repeating the process many times was clearly too much.

In river engineering, much progress was made in morpho*dynamic* or MTM (Medium-Term Morphodynamic) modeling, see e.g. [Struiksma *et al.* (1985)], who clearly got beyond ISE and modelled the time-varying evolution of river bends and bifurcations, including advanced quasi-3D flow description. Of course they lacked the complication of waves, with their separate models and grid requirements.

In the late '80s and early '90s, the large European institutes combined in the EU MaST-G6M and G8M projects carried out major developments and restructuring of their coastal area models, which led to much more robust and flexible codes that are still leading today, such as Delft3D (Delft Hydraulics, now Deltares, Netherlands), Mike21 (Danish Hydraulic Institute) and Telemac (Laboratoire Nationale d'Hydraulique, France). Through a number of model intercomparisons, modellers were challenged to improve their models' capabilities and gained trust as different models and approaches came up with rather similar solutions. These proprietary models have been developed further, are available for research or can be used commercially and are professionally supported.

In the field of coastal morphodynamics, surprisingly few alternatives are available; ECOMSED, which is based on the Princeton Ocean Model (POM), has much capability in (fine) sediment modeling; ADCIRC, a finite-element model much used by/for the Corps of Engineers, is developing morphodynamic capabilities including wave-driven processes. Recent open-source models such as ROMS-SED are catching up fast, and are venturing closer and closer to shore, from a background that is more in coastal ocean modeling, typically at shelf scales. An exciting development of the last few years is XBeach [Roelvink *et al.* (2009)], an open-source model for nearshore morphodynamics, which focuses on extreme events such as hurricanes, with processes like overwashing and breaching included, but can also be applied for small-scale coastal engineering problems.

In the following sections we'll discuss differences in wave drivers, and flow, transport and morphology modeling approaches, before we draw up a summary table.

9.2 Wave drivers

In this section we will briefly discuss the different wave drivers available in various models

9.2.1 *Wave-averaged*

Wave-averaged models consider the wave field averaged over both individual waves and wave groups. In most morphodynamic model applications a spectral model such as HISWA or SWAN is used. HISWA [Holthuijsen *et al.* (1989)] represents the spectrum by a single representative frequency and a discrete directional spectrum, whereas SWAN [Holthuijsen (2007)] models the evolution of the two-dimensional spectrum. SWAN is most used lately, as its numerical scheme is more convenient and it supports curvilinear grids similar to models like Delft3D and ROMS; since recently, even unstructured grid are supported. It can run both in stationary mode and in nonstationary mode; in most morphological applications, which typically cover relatively small areas, stationary mode is applied.

Such spectral models do not easily represent diffraction since the necessary phase information is not available, though some simplified form of diffraction can improve results, for instance behind offshore breakwaters (e.g. [Ilic *et al.* (2007)]).

Models that do include diffraction can be either elliptical or parabolic. For instance in the Mike suite, both options are available. They will give improved wave height and direction patterns behind structures, but need to have a lot of spectral resolution in order to create smooth patterns. This makes their application relatively expensive; especially where there is a significant and variable longshore current, the effect of diffraction on the morphological result is usually not very large.

In XBeach, a simple stationary wave driver is available, which only contains processes of refraction and wave breaking, but is quite fast compared to SWAN and has no disturbance at the lateral boundaries. This wave driver is also available as an option in Delft3D.

Most models also contain some form of a roller model, which generally leads to a better distribution of the longshore current. Delft3D and XBeach apply the roller model formulations described in Chapter 2; other models use a purely local parameterization.

In the days when numerical schemes of wave models were not very accurate and smooth, using wave forces derived directly from radiation stress gradients produced a lot of spurious patterns, as was pointed out by [Dingemans et al. (1987)] . A simple alternative is to base the driving forces on the dissipation rate, as we discussed in Chapter 3. However, nowadays with spectral models such as SWAN, this is no longer necessary and we have found significant differences in flow and transport patterns around tidal inlets, for example, between the dissipation approximation and the full radiation stresses. Therefore we now recommend to use the full radiation stress gradients as driving forces.

9.2.2 *Short wave averaged*

Short wave (but not wave group) averaged models resolve variations at wave group time scale. They need a slowly-varying wave driver, which in practice uses the time-dependent version of the wave action balance. In Delft3D, an option exists whereby the mean wave direction from a SWAN run are used to solve the wave action balance for a given time- and space-varying short wave energy at the offshore boundary. In XBeach, the time-varying wave model resolves the directional spectrum and thereby can solve the refraction of the short waves, so it does not need an external model to provide the wave directions. Both models also include a time-varying roller model.

Originally the use of wave group resolving models was aimed mainly at understanding the effects of infragravity waves on the nearshore morphology, particularly bar patterns, and on efficiently modeling long wave motions in harbours. More recently we realized that infragravity waves provide most of the swash motions during storms, and are therefore very relevant to modeling of dune erosion and overwashing; see [Roelvink et al. (2009)], [McCall et al. (2010)].

9.2.3 *Short wave resolving*

The main problem with short-wave averaged and wave-averaged models is that they cannot predict skewness and asymmetry directly and have to rely on local approximations of those terms. There are only a few area models reported in literature that use a phase-resolving wave driver ([Rakha (1998)],[Van Dongeren et al. (2006)],[Long et al. (2006)]); most of the test cases produced are actually 2DV cases. At this moment, the improvements in predicting cross-shore transport and morphology do not seem to justify the very high computational costs of such approaches.

9.3 2DH, Q3D and 3D

When we are talking about the difference between 2DH, Q3D and 3D, we are talking about the flow model, the sediment transport model and, recently, the bed update model; the wave models that are presently in use are all essentially 2DH, and in the vast majority of cases based on the spectral wave action balance;

9.3.1 *Flow model*

The flow model, when run in 2DH mode, is based on the depth-averaged shallow water equations. In most cases this means that the sediment transport direction is the same as the depth-averaged flow direction, although sometimes the mean return flow is taken into account; we will treat that as the simplest quasi-3D concept rather than real 2DH. In such a model setup, the transport generally follows the depth contours along the coast, unless there is a disturbance in the form of coastal structures or dips in the bathymetry leading to rip currents.

Quasi-3D flow models, in this context, usually mean that the undertow profile is somehow accounted for. This can be just the depth-averaged undertow or a 1DV, analytical or numerical, vertical profile description similar to the ones we discussed in Chapter 3. Such a Q3D adaptation should in principle feed back into the bed shear stress, but this is not always done, as often the Q3D addition is carried out as a correction after the flow model.

Full 3D flow models in this kind of coastal area models, where horizontal scales are much larger than vertical scales, are normally based on the 3D shallow water equations, They can rather well represent the effects of breaking waves on the current profiles and at the same time deal with density-driven flows and deviations of the vertical due to strong curvature. Most 3D models apply $k-\varepsilon$ turbulence models or similar, 2-equation turbulence models, which can deal with simultaneous sources of turbulence due to e.g. wave breaking, bed shear, horizontal shear, buoyancy effects etc. and are therefore ideal 'integrators' of processes in complex situations. A nice example of a model/field data comparison where all such processes play a role is the Columbia River mouth, Wa, [Elias *et al.* (2011)].

9.3.2 *Sediment transport*

In most models, sediment transport is subdivided into bed load transport and suspended load transport. Bed load transport is always treated as a direct function of the near-bed velocity or bed shear stress; in 2DH the bed shear stress follows the depth-averaged flow, whereas in 3D it follows the near-bed flow.

The suspended transport in some models is still only a function of the local flow and wave conditions, but in most it is solved using the advection-diffusion equation in 2DH or 3D. In the case of 2DH, the source term is based on the difference between the equilibrium concentration and the actual (depth-averaged) concentration, as we discussed in Chapter 4.

The equilibrium concentration can be computed using one of many sediment transport formulations. In 3D, the sediment enters the model through the bed boundary condition and is further transported by the advection-diffusion equation, using the turbulence structure from the flow model or from some empirical formulation for the eddy diffusivity distribution.

The largest differences between 2DH and 3D sediment transports are in the surf zone and in strongly curved areas, as well as in areas with strong density gradients in horizontal and/or vertical.

Focusing on the surf zone, it is important to realize that if we decide to take into account the undertow, either by a Q3D addition or by running in 3D, we create a strong offshore directed transport component that will immediately start eating away at our beach. This has to be compensated by other cross-shore effects with a net onshore trend, such as wave skewness and asymmetry, as we discussed in Chapter 7. Conversely, if we are running in purely 2DH mode, we should turn such processes off.

In general, we may expect a more 'boring' coastal behaviour when running in 2DH mode: most erosion and sedimentation follows from longshore transport deviations, although interesting rhythmic patterns with rip circulation cells may develop. In 3D and Q3D, we may expect more profile variation, where during storms the upper profile is often eroded and during other conditions the beach may accrete. Finding the right long-term balance remains a major challenge; it certainly helps if the 3D model can also be run in 2DV profile mode, to enable calibration of just the profile behaviour.

9.3.3 *Bottom*

In a number of modern codes, the vertical profile of the bed composition can be tracked; the sediment is divided into a number of fractions, and the bed is divided into a number of layers, each of which has a different composition of these fractions. The top layer is often treated as a mixing layer, and different sediment fractions from this layer are picked up by the flow in proportion to their relative concentration and depending on their grain size and/or other characteristics. Some models include effects of hiding and exposure or mutual effects of sand and mud fractions.

The behaviour of models that take into account the changing bed composition can be quite different from single-fraction models. A typical effect is that channels and scour holes tend to get less deep because of a horizontal sorting that develops, since finer fractions can be transported more easily than coarser fractions. As was found in a recent study by [Dastgheib *et al.* (2009)], the surface sediment grain size in tidal inlets after some time tends to be correlated with the mean shear stress. Specifying the initial condition for the bed composition can be problematic, especially for hindcast studies, but several methods can be applied to get a fair estimate; using a reasonable initial condition considerably improves the model behaviour.

9.4 Grids and numerical aspects

The models discussed here use either *structured* grids, which can be rectilinear or curvilinear, or *unstructured* grids, which are typically built up from triangles, but can also contain combinations of triangles and quadrilaterals. The structured grids are used in *finite difference* methods, the unstructured grids in *finite element* or *finite volume* methods.

Traditionally, finite difference methods have been popular because they are relatively easy to understand and relatively fast per grid cell. Finite element methods are much more complicated mathematically and used to be much slower per grid cell. However, the finite element meshes are much more flexible and can easily cover very large differences in scale.

In our discussions here we will focus on the finite difference methods, since their use is most widespread in coastal morphology and our own experience is based on them. We will not go into numerical aspects in any detail, since these are very different in all models. It important, however, to realize that some models use *implicit* schemes, such as ADI (Delft3D, Mike21) while other models use *explicit* schemes (e.g. XBeach, ROMS). In implicit methods, systems of equations are set up that relate state variables (e.g. water level, velocities) *at the next time step* to each other. Since these variables are unknown except at the boundaries, the whole set of equations must be solved at each time step, by matrix inversion or iterative methods. On the other hand, in explicit methods, state variables at the next time step are directly solved as a function of the values at the previous time step, which is relatively simple. The implicit methods take more time per timestep, but can use much bigger timesteps because there is no hard stability limit; explicit schemes take little effort per timestep but have to keep to a strict time step criterion.

In implicit models, the user usually gets some guidance on how to set the time step, based on the grid cell sizes and the depth; this criterion is limited more by accuracy than by stability. The guidance is still given in terms of the Courant or CFL criterion, which is given by:

$$CFL = \frac{C \Delta t}{\Delta x} \ , \ C \approx \sqrt{gh} + u \qquad (9.1)$$

where in implicit methods, CFL values in the order of 10 are usual. In explicit methods, the user gets to specify a CFL criterion (less than 1) and the model computes an automatic time step each time, in order to guarantee stability.

For morphodynamic models, which typically cover relatively small domains, the wave model domain usually has to be much larger than the flow domain. This is due to the fact that most wave models have lateral boundaries that are not well specified and introduce a wedge of unreliable results at each lateral boundary. This is true for frequently used spectral wave models such as HISWA and SWAN. In Figure 9.1 an example is shown of a wave grid (blue) and a flow grid (red) for a Delft3D model of the beach at Egmond, NL. The flow grid is well clear of any boundary problems near the lateral boundaries, so wave information can be passed to the flow grid without problems. However, for data flowing back from flow model to wave model (water level, current field) is missing information outside the flow grid. Therefore it is necessary to extrapolate flow information to the wave grid for these regions.

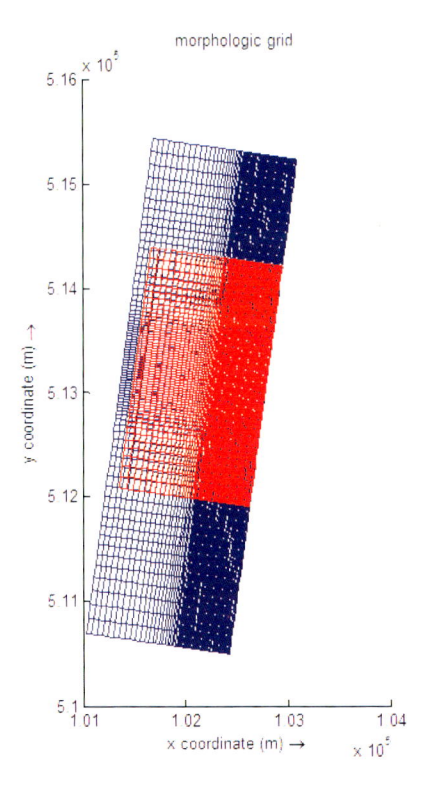

Fig. 9.1   Example of smaller flow grid embedded in large wave grid to avoid shadow zones.

9.4.1   *Overview of model components in some morphodynamic model systems*

9.5   Boundary conditions for coastal area models

9.5.1   *Flow model*

General

In general, boundary conditions are required at each boundary for:
(1) The water level, the velocity normal to the boundary *or* a combination of these such as a discharge or Riemann invariant, *and*
(2) The velocity along the boundary.

Model	Wave driver	Flow model	Sediment transport	Morphology updating method	Bed comp.	Grid system	Availability
Delft3D	Spectral Wave-averaged/ Short-wave averaged	2DH/3D	2DH/3D Sand and mud	Online with morfac, parallel online;	3D	Curvil. FD	Open Source
Mike21	Spectral/ parabolic/ mild slope eq.	2DH/Q3D	2DH/Q3D Sand or mud	Offline, online with morfac	3D	Rectil., Curvil. FD, Unstr.bFV	Licensed
Telemac	Spectral	2DH/3D	2DH/Q3D	Offline, no acceleration	3D	Unstr. FE	Licensed
ADCIRC	Spectral	2DH/3D	2DH/3D	Online	2DH	Unstr. FE/FV	Licensed
ROMS-SED	Spectral	3D	3D	Online with morfac	3D	Curvil. FD	Open Source
FINEL	Spectral	2DH	2DH	Online with morfac	2D	Unstr. FE	Not
XBeach	Spectral Wave-averaged/ Short-wave averaged	2DH	Q3D	Online with morfac	3D	Rectil. FD	Open Source

The boundary conditions for water level or velocity can be obtained by *nesting* in larger-scale models or by spatial interpolation between observation points. In the nesting process, time series of water level or current velocity at boundary support points are interpolated from surrounding points of a larger-scale model. Often a combination of water level and velocity boundaries is used; water level boundaries are necessary to fix the water level, but model behaviour is sensitive to small errors in the specified water level; velocity boundaries produce smoother velocity fields but leave the water level undetermined. The sensitivity of the model for small errors in the water level decreases for larger-scale models, which are often driven by water levels alone.

When nesting it is important to make sure that the forcing and bathymetry in overall and nested model match well. Matching the bathymetry is especially important in shallow water. The effect of a mismatch in wind forcing near a water level boundary will be that the cross-shore slopes in the overall and nested model are different, which will lead to spurious circulations. Near velocity boundaries, the velocity imposed based on the overall model will not match the wind-driven current according to the forcing of the nested model, which will again lead to deviations and spurious circulations.

Wave-driven currents are usually not accounted for at all in large-scale models, as their grid sizes are typically in the order of kilometres or more. This leads to strong boundary effects at the lateral (cross-shore) boundaries of small-sale coastal models. The next section will give a possible solution for this problem.

For the velocity along the boundary, approximations are usually made although it is possible in principle to prescribe them by nesting also. The approximations used are:
(1) On boundary sections across a river or tidal channel, the along-boundary velocity is set to zero on inflow; this prevents cross-channel oscillations.
(2) On boundary sections that are aligned with the flow or in open coast models, the *gradient* of the along-boundary velocity is set to zero; this prevents a horizontal boundary layer from forming due to advection of zero alongshore momentum into the model.

Cross-shore boundaries The problem of specifying boundary conditions at lateral boundaries is illustrated by a simple example. Let us consider the case of section 2.4 with 20 m/s wind blowing on a model with a plane sloping profile at 1:100 slope starting from 20 m depth. If we specify uniform water level boundaries around the model we get the result as in the right panel of Figure 9.2. Clearly the picture is very far from the correct stationary solution, which is a uniform velocity field over the whole model and a longshore uniform water level setup. The setup gradients due to the specified water level boundaries lead to flow towards the boundaries, which is compensated by an inward flow at the seaward boundary.

There are two ways to overcome this problem. The first is to try and predict the water level setup or the current velocity along the lateral boundary by solving a 1DH or 2DV problem along the boundary and to impose this. For simple cases this is possible but for more complex combinations of forcing conditions it is cumbersome. A better solution is to let the model determine the correct solution at the boundary by imposing the alongshore water level gradient (a so-called Neumann boundary condition) instead of a fixed water level or velocity. In many cases this can be assumed to be zero; only in tidal cases and in cases where a storm surge travels along a coast the alongshore gradient varies in time, but as we have seen in section 2.1, the alongshore gradient of the water level does not vary much in cross-shore direction.

The equations solved at the lateral boundaries now read, if s is the direction along the boundary and n the direction normal to the boundary:

$$\frac{\partial \eta}{\partial n} = f(t) \qquad (9.2)$$

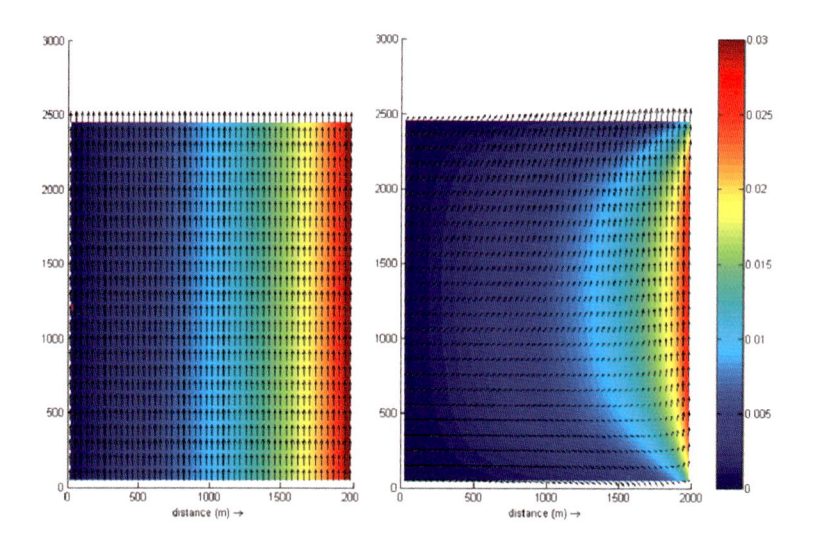

Fig. 9.2 Wind-driven current and water level set-up on a small-scale model. Left panel: uniform water level at seaward side, zero water level gradient boundary top and bottom. Right panel: uniform water level boundaries; Stationary solution, wind velocity 20 m/s, wind direction 225 deg. N.

$$\frac{\partial u_s}{\partial t} + u_s \frac{\partial u_s}{\partial s} + u_n \frac{\partial u_s}{\partial n} = -g\frac{\partial \eta}{\partial s} + f_{cor} u_n + \frac{\tau_{ws}}{\rho h} + \frac{F_s}{\rho h} + \frac{R_s}{\rho h} - \frac{\tau_{bs}}{\rho h} \quad (9.3)$$

$$\frac{\partial u_n}{\partial t} + u_s \frac{\partial u_n}{\partial s} + u_n \frac{\partial u_n}{\partial n} = -g\frac{\partial \eta}{\partial n} - f_{cor} u_s + \frac{\tau_{wn}}{\rho h} + \frac{F_n}{\rho h} + \frac{R_n}{\rho h} - \frac{\tau_{bn}}{\rho h} \quad (9.4)$$

Here, the greyed-out terms are being neglected. Note that the advection terms containing cross-shore gradients of the velocity are not neglected, as they are important during the spin-up of the model or in instationary conditions, and in the case of 3D flow.

These lateral boundary conditions can only be applied in combination with a water level boundary at the seaward boundary, which is needed to make the solution well-posed. At the seaward boundary, the water level is prescribed as a function of time and advection terms containing normal gradients of the velocity are set to zero.

In our example case, if we specify not the water level but the water level *gradient* normal to the boundary (equal to zero) then the wind-induced setup can freely develop at the lateral boundaries and the correct solution is found. We can see the uniform stationary solution on the left panel in Figure 9.2; in Figure 9.3 we see that the time-evolution in this case is also modelled correctly.

For wave-driven currents the situation is very similar in principle. However, as the wave-driven current only appears in a narrow strip along the coast, the effect of

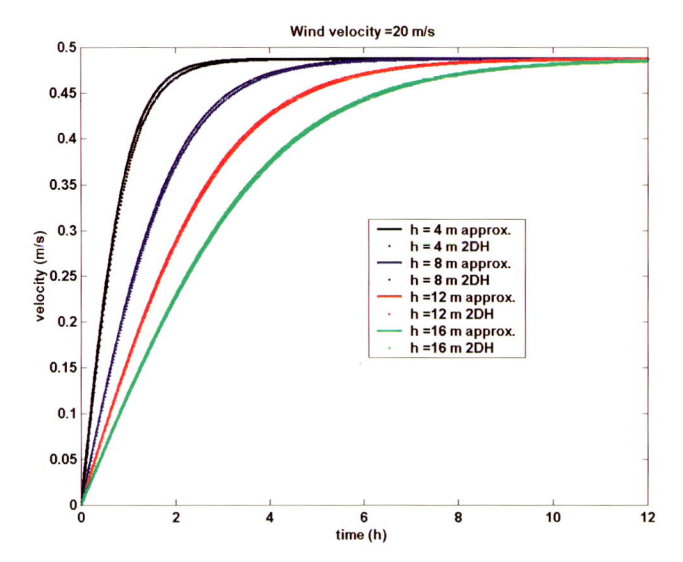

Fig. 9.3 Wind-driven velocity vs. time for different water depths; comparison of 2DH solution with gradient-type lateral boundary conditions and theoretical solution.

the boundary disturbances reaches less far into the model area. A typical example, for the same model now driven by obliquely incident waves, is shown in Figure 9.4.

For situations where a tidal wave propagates along the coast we may derive harmonic boundaries as follows. The propagation of the tidal water level along the coast can be described by:

$$\eta(s,t) = \sum_{j=1}^{N} \widehat{\eta}_j \cos\left(\omega_j t - k_j s - \varphi_j\right) \tag{9.5}$$

Here η is the water level, $\widehat{\eta}_j$ the amplitude of the j-th component, ω_j the angular frequency, k_j the alongshore wavenumber of the tidal component, s the alongshore distance and φ_j its phase relative to a fixed point in time and space. To obtain the alongshore gradient of the water level we can now simply differentiate with respect to s, to get:

$$\frac{\partial \eta}{\partial s}(s,t) = \sum_{j=1}^{N} k_j \widehat{\eta}_j \sin\left(\omega_j t - k_j s - \varphi_j\right) = \\ \sum_{j=1}^{N} k_j \widehat{\eta}_j \cos\left(\omega_j t - k_j s - \varphi_j - \pi/2\right) \tag{9.6}$$

If our model area is in between two water level stations where tidal amplitudes and phases are known, we can simply determine the local water level amplitudes, and phases by spatial interpolation; the alongshore wavenumber can be derived

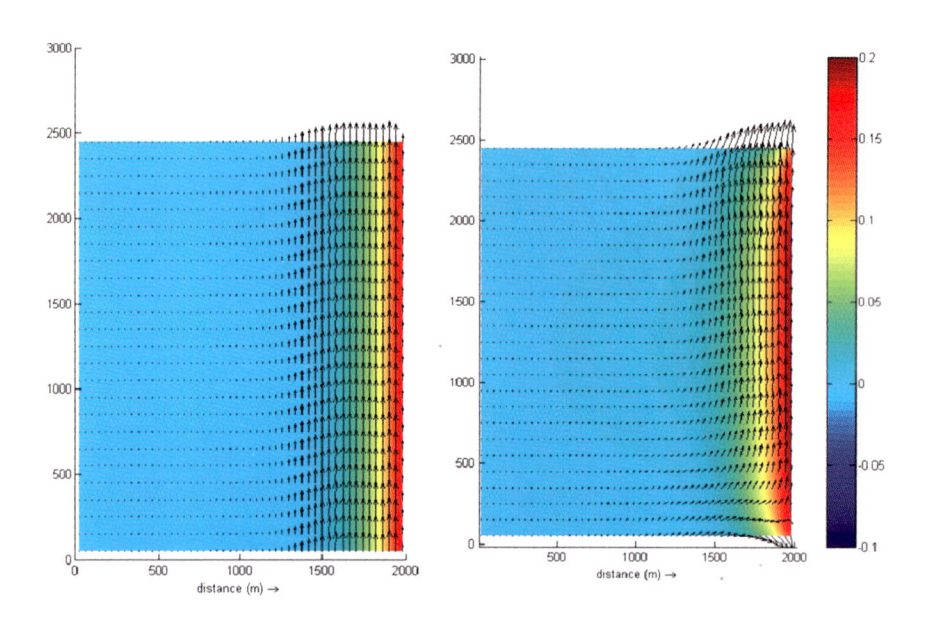

Fig. 9.4 Current velocity and water level pattern for waves H_s=2m, T_p=7s, direction 240 deg. N, with gradient-type lateral boundaries (left panel) and uniform water level boundaries (right).

for each component by analysing the phase difference between the two water level stations.

Similar procedures can be followed for output from larger-scale models. At the seaward boundary we now prescribe the water level and at the lateral boundaries we prescribe a uniform longshore pressure gradient as a function of time or as a combination of harmonic components with the right phases.

With tidal boundary conditions of this kind it is possible to add arbitrary forcing due to wind or waves, in 2DH or 3D without generating spurious circulations.

As an example, we show results for a 2D grid with dimensions of 2000 m in cross-shore direction and 2500 m in alongshore direction. The depth is 20 m at the seaward boundary and decreases linearly at a slope of 0.01. The tidal amplitude is 1 m and the tidal alongshore wave length is 400 km. The tidal alongshore wave number is then $1.57.10^{-5}$ rad/m and the phase difference between the southern and northern boundary is 2.25 degrees. The boundary conditions applied are then:

(1) Southern boundary: water level gradient amplitude $1.57.10^{-5}$ and phase 90 deg.;
(2) Northern boundary: water level gradient amplitude $1.57.10^{-5}$ and phase 92.25 deg.;
(3) Western boundary: water level amplitude 1 m, phase linearly varying from 0 deg at the southern end to 2.25 deg. at the northern end.

In Figure 9.5 the cross-shore distribution of the longshore velocity is shown at 6 time points in the tidal period. The distribution in longshore direction is quite

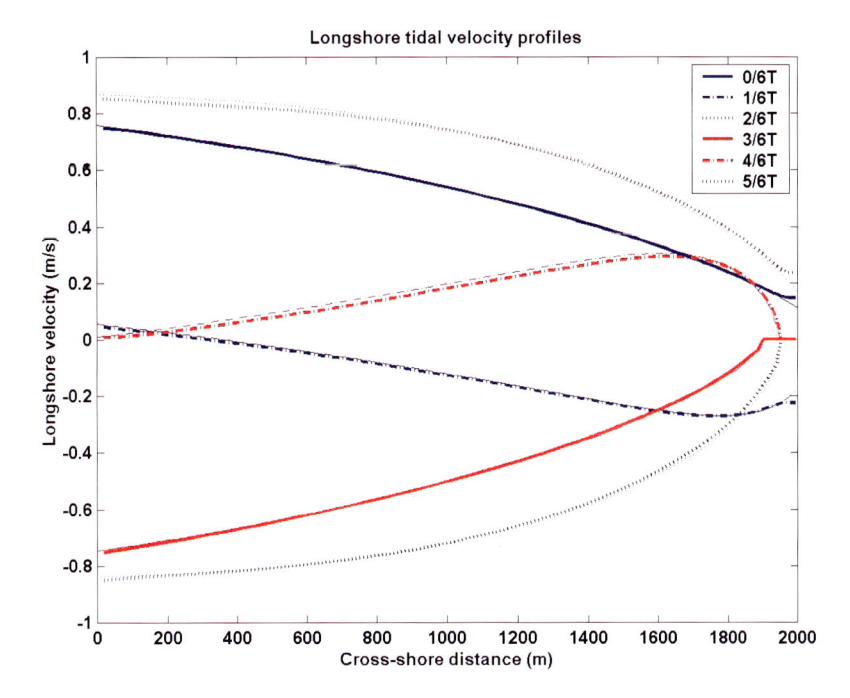

Fig. 9.5 Example tidal velocity profiles at 6 times during a tidal period. Thick lines: Delft3D using Neumann boundary conditions on a grid of 200 m cross-shore by 2500 m alongshore. Thin lines: offline numerical solution

uniform. In the figure the numerical solution using Delft3D is compared with a 1D solution of eqs. (9.3) and (9.4). The agreement is very good, indicating that in this case advection effects are apparently negligble as they are not taken into account in the 1D case.

9.5.2 *Waves*

Seaward boundary For spectral wave models the incoming spectral energy (divided over direction and frequency bins) is specified at the seaward boundary. In case on wave-group resolving models, these are time-varying as outlined in section 2.7.

Lateral boundaries Most spectral models, including HISWA and SWAN, have poor lateral boundaries, which exhibit a triangular wedge of unreliable results, increasing in width towards the shore. As is shown in XBeach, this is not necessary; simply setting the gradients in y-direction to zero at the lateral boundaries (so-called Neumann boundaries) solves this problem.

9.5.3 *Sediment transport*

In coastal models the boundary conditions for suspended sand transport are usually set to zero gradient across the boundary. This allows the concentration to vary in time without creating artificial gradients. It can be applied for sand since the adaptation length scales are usually very small compared to the model domain and there usually is an abundance of sand. For mud transport this is not generally the case since the mud fraction in the bed may be small and adaptation scales are very large. In this case concentrations have to be prescribed at the inflow boundaries. In cases of alternating flow so-called Thatcher-Harleman boundary conditions can be applied, which account for the fact that when the flow reverses the concentration at the inflow boundary for some time is still governed by what happened inside the model before.

9.5.4 *Bed level*

The bed level at the inflow boundaries must be prescribed. This can be done either by specifying no change, by specifying a time-series of bed levels or by specifying that the bed level change at the boundary is equal to the bed level change one row or column inside the model. The latter condition is convenient when modeling an area with relatively uniform conditions in longshore direction but with significant variations in cross-shore direction, e.g. bar migration.

9.6 Modeling strategies for wave-current interaction

In most morphodynamic models the wave-current motion is represented by a combination of an instationary flow model and a (quasi-)stationary wave model. The rationale behind this is that the wave models usually are wave-averaged and thus represent mean characteristics over a period in the order of half an hour; this is longer than the time required for waves to travel through a typical morphological model area that may cover some kilometres cross-shore.

The waves are modified by the flow through the water level and the current field. Changes in water level are reflected in changes in water depth, which lead to changes in wave propagation, shoaling, refraction and wave breaking. Changes in the current field modify the wave pattern due to current refraction and, in cases of strong opposing currents, wave blocking.

The flow in turn is modified by the waves: they lead to enhanced bottom friction, forcing of longshore currents, wave-induced set-up and horizontal and vertical circulations.

In the following sections we will discuss practical aspects and strategies of coupling waves and currents.

Data exchange Trivial as it may seem, a first requirement of the data exchange is to agree on the units of the parameters to exchange (e.g. forcing in m/s^2 or N/m^2) and on their exact definition (e.g. not just 'wave height' but H_{m0} or H_{rms}). Next, the wave and flow models typically operate on different grids, so we need to interpolate between grids. Especially when more modules are involved than just waves and flow, it is convenient to have a central grid that all modules interpolate to and from. In the case of Delft3D we have chosen the flow grid to be this central grid as it is the same as the grids for transport and morphological change, and we have put the responsibility of interpolating to and from this grid with the wave module. Within this wave module, multiple nested runs may have to be carried out at each time point, in order to get enough resolution in the areas of interest. This is especially the case for rectangular grids. With curvilinear grids as can be used in the SWAN model, this can usually be avoided. If we define the task of the wave module to 'get information from the flow grid, run the wave model(s) and send back wave information to the flow grid' this means that we will have to put the following functionality into the wave model, for a single time point:

(1) For all nested runs
   (a) Read water depth and velocity on wave input grids
   (b) Interpolate depth from flow model to wave input grid;
   (c) Interpolate water level from flow model to wave input grid;
   (d) Add interpolated depth and water level;
   (e) Transform velocity from flow grid direction to Cartesian direction;
   (f) Interpolate u and v components to wave input grid;
   (g) Update wave input grids for area within flow model;
   (h) Run wave computation module using nest information from previous runs (if any); output to nested runs;
   (i) Interpolate wave information from wave computational grid to flow grid for area within this wave computational grid;

The advantages of putting this complexity into the wave module are that the amount of data exchange is very limited, that all communication parameters are defined on the same grid and that the overall system is not burdened with these complexities. The wave module can be seen as a shell around the actual wave model (such as HISWA or SWAN); these models are called as they are from within the wave shell, which prepares their specific input and reads their output. This makes it relatively easy to add new wave models to the system.

Stationary conditions For stationary flow and wave conditions, a procedure that usually converges quickly is as shown in Figure 9.6. After allowing the flow to spin up without wave forcing, the flow model is suspended now and then and the wave model is called; the frequency of updating is either fixed or based on some criterion for the flow to become stationary after updating the waves. After a number of such cycles, the wave forcing does not change anymore and the whole system has converged to a stable solution.

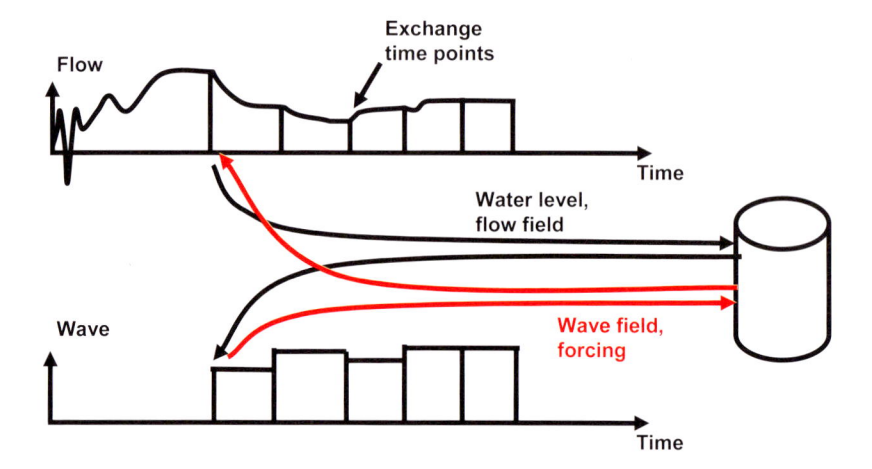

Fig. 9.6 Model set-up for wave-current interaction; stationary case

This method generally converges quickly if the effect of the current on the wave field is not too strong. An example of this is shown in Figure 9.7, which shows the computed current fields, with and without wave-current interaction, for the 'Hiswa-basin' test [Dingemans (1987)], where waves travelling in positive x-direction break over a submerged bar that covers part of the width of the basin. The concentrated wave breaking over the bar generates a strong circulation, which in turn has an effect on the wave propagation and hence on the forcing. The effect in this case is a moderate shift of the velocity pattern in positive y-direction. The measured current pattern is of a similar strength and extent, though the centre of the gyre is shifted a couple of metres towards the tip of the bar.

In cases where this effect is very strong, such as near concentrated rip currents, the system may 'flip-flop' between different states if the flow is allowed to go to a stationary state between wave updates. In such cases a solution may be reached by updating the waves very frequently, thus allowing flow and waves to reach their combined equilibrium together.

Iterative solution over a tidal cycle In many morphological scenarios waves, flow and sediment transport are evaluated over a representative, periodic tidal cycle, while the offshore wave conditions are kept constant. In order to limit the number of wave computations per tidal cycle, the following scheme is applied, as depicted in Figure 9.8.

(1) A flow computation is carried out over a number of tidal cycles, whereby output is generated on the communication-file, for a number of time points. At a

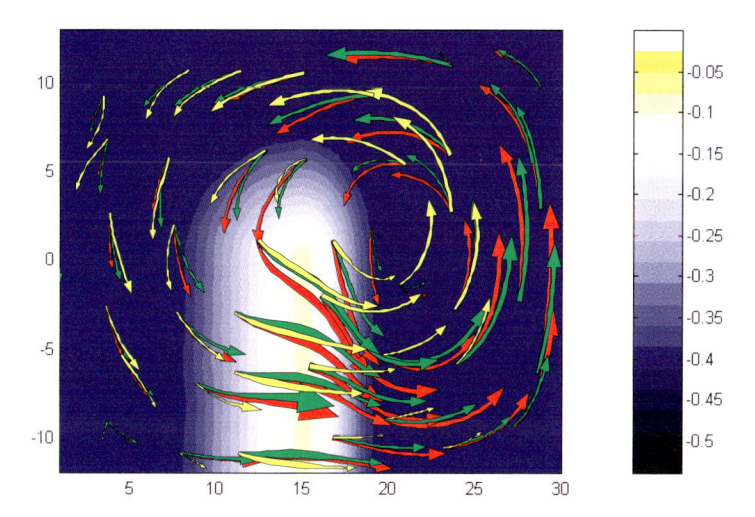

Fig. 9.7 Velocity field in 'Hiswa basin' test. Red arrows: no wave-current interaction; green arrows: converged wave-current interaction; yellow arrows: measurements. Arrows denote tracks following current for 50 seconds starting from randomly chosen points.

selection of these time points, wave computations are carried out and those results are stored on the communication file. The flow computation is repeated, while now interpolating wave forces and parameters in time from the fields stored. The dotted lines indicate how the wave fields can be assumed periodic in time in order to save on the number of computations. Steps 2 and 3 can be repeated until convergence is reached. However, with a strong tidal component one iteration is often enough.

Simultaneous simulation of waves and currents The iterative method described above has the advantage that the wave forcing applied in the flow model is a continuous function in time, because of the interpolation between previously stored wave fields. However, the method becomes cumbersome if long periods have to be simulated or if the bottom changes during the simulation. In such cases it is better to simulate the flow and waves (almost) simultaneously and to exchange information each time the wave field is updated. In this case, the wave forcing applied in the flow model changes abruptly each time the wave field is updated, so that it is necessary that the updating happens very frequently. This does not have to be a problem if a wave model is used that uses an iterative technique, such as the spectral wave model SWAN. Given that the model continues on its last solution (using a restart or 'hot' file), taking 6 steps of two iterations each is equivalent to taking one step that requires 12 iterations; in the first case the boundary conditions and flow field may vary smoothly and the details of the intra-tidal variation of the wave field are resolved much more accurately.

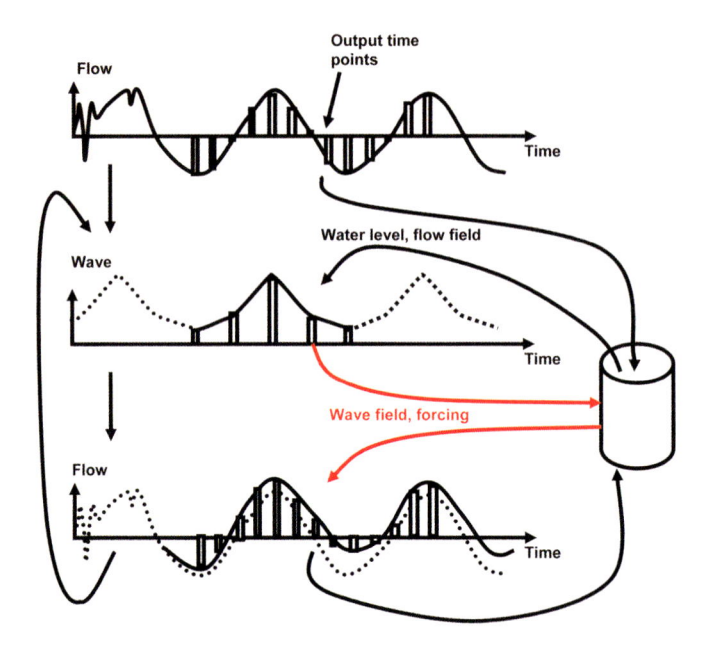

Fig. 9.8 Iterative solution of wave-current interaction over a (number of) tidal cycle(s)

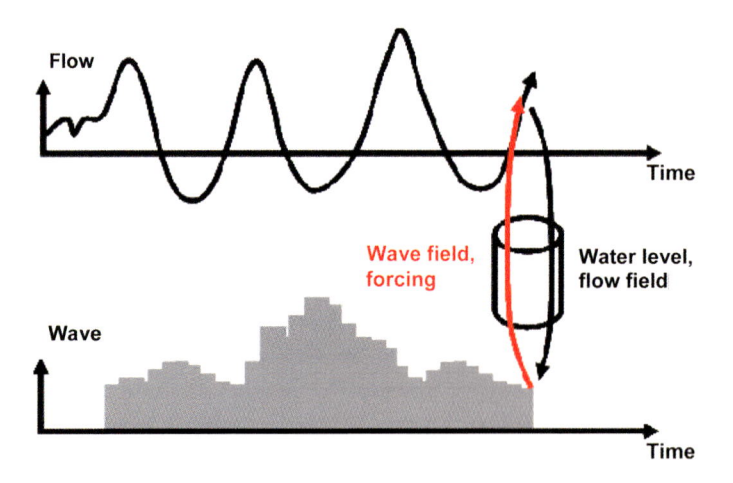

Fig. 9.9 Scheme for simultaneous simulation of waves and currents

9.7 Strategies for morphodynamic updating

In this section we will discuss various ways of accelerating the computation of morphological changes. Part of this discussion was published in [Roelvink (2006)].

9.7.1 Tide-averaging approach

This approach is based on the fact that morphological changes within a single tidal cycle are usually very small compared to the trends over a longer period, and that such small changes do not affect the hydrodynamics or sediment transport patterns. It is then acceptable to consider the bottom fixed during the computation of hydrodynamics and sediment transport over a tidal cycle. The rate of change of the bed level is computed from the gradients in the tidally averaged transport.

In the earlier morphological models this rate of change or Initial Sedimentation and Erosion (ISE) was an end product; it could be applied to assess sedimentation rates in a navigation channel or it could be used to assess large-scale changes to a sediment budget.

In morphodynamic models however, the bed level is updated using some (usually explicit) scheme and fed back into the hydrodynamic and transport models. As the transport pattern now continuously adapts to bed changes, this allows bed patterns to migrate along with the mean transport direction, as we discussed in Chapter 5.

In Figure 9.10 the flow scheme of the tide-averaging approach is shown. Starting from a given bathymetry, the wave-current interaction is solved over a tidal cycle, using the iterative approach as described in Figure 9.8. The resulting flow and wave fields are then fed into a transport model, which computes bed load and suspended load transports over the tidal cycle. The averaged result is applied to compute bed changes. The updated bathymetry is looped back to the transport model through 'continuity correction' (see next paragraph) or to the full hydrodynamics module. The same scheme is often applied to stationary situations, where the hydrodynamics and transport are made to converge at each full morphodynamic loop.

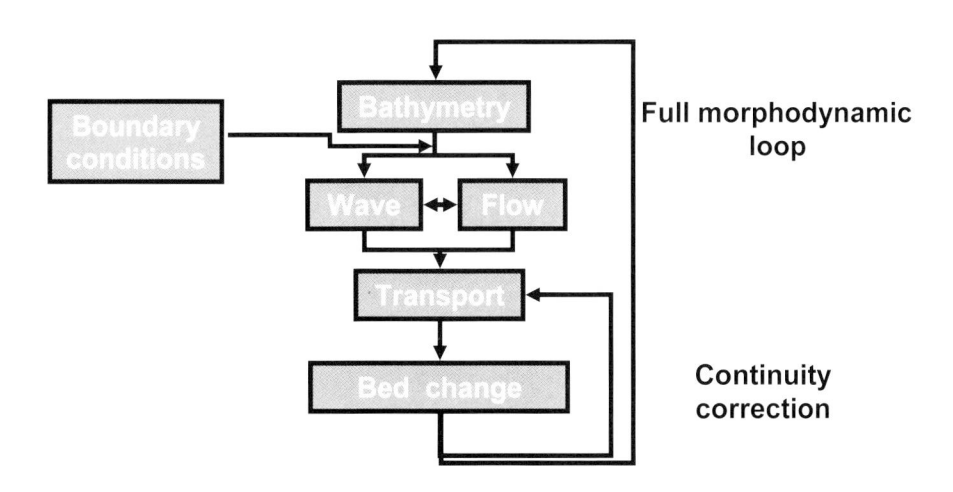

Fig. 9.10 Flow diagram of tide-averaging morphodynamic model setup

The morphological time step is numerically restricted by the bed courant number:

$$CFL = \frac{c\Delta t}{\Delta s} \qquad (9.7)$$

where the bed celerity c can be approximated by:

$$c = \frac{\partial S}{\partial z_b} \approx \frac{bS}{h} \qquad (9.8)$$

Here b is the power of the transport relation, S is the tide-averaged transport in s-direction and h the tide-averaged water depth.

Apart from this numerical restriction the morphological time step is limited by the accuracy of the time-integration method. This can be estimated as follows: the bed level change over a number n of tidal periods T is equal to:

$$\Delta z_b = \int_0^{nT} \frac{\partial z_b}{\partial t} dt = -\int_0^{nT} \left(\vec{\nabla}.\vec{S}\right) dt = -\vec{\nabla}.\int_0^{nT} \vec{S} dt \qquad (9.9)$$

Here the sediment transport vector changes both with time or phase of the tide and with the varying bed level. We may approximate the variation with bed level by a first-order Taylor expansion:

$$\vec{S}_{t,\Delta z_b} = \vec{S}_{t,\Delta z_b=0} + \frac{\partial \vec{S}_{t,\Delta z_b=0}}{\partial z_b} \Delta z_{b,t} + O(\Delta z_b^2) \qquad (9.10)$$

We can now time-integrate the transport vector, using the approximation in equation (9.8) and assuming an approximately linear increase or decrease of Δz_b per tidal cycle:

$$\Delta z_{b,t+nT} = -\vec{\nabla}.\int_t^{t+nT}\left(\vec{S}_{\tau,\Delta z_b=0} + \frac{\partial \vec{S}_{\tau,\Delta z_b=0}}{\partial z_b}\Delta z_{b,t} + O(\Delta z_b^2)\right)d\tau \approx$$

$$\approx -nT\vec{\nabla}.\frac{1}{T}\int_t^{t+T}\vec{S}_{\tau,\Delta z_b=0}d\tau - \vec{\nabla}.\int_t^{t+nT}\left(\frac{b\vec{S}_{\tau,\Delta z_b=0}}{h}z_{b,\tau}\right)d\tau + O(\Delta z_b^2) \approx$$

$$\approx -nT\left(1 + \tfrac{1}{2}\frac{b\Delta z_{b,t+nT}}{h}\right)\vec{\nabla}.<\vec{S}_{\Delta z_b=0}> + O(\Delta z_b^2)$$

$$(9.11)$$

Compared to a simple Euler scheme of updating, which will not reproduce the second term between the brackets, we find that the relative error is proportional to the ratio of bottom change over water depth and to the power b in the transport relation.

Because of these limitations on the morphological time step, it is necessary to update the transport regularly. In the next section we discuss the 'continuity correction', a cheap way of doing this in an approximate way.

9.7.2 Continuity correction

The sediment transport field is generally a function of the velocity field and the orbital velocity:

$$\vec{S} = f(\vec{u}, u_{orb}, ...) \tag{9.12}$$

When the bathymetry changes, the flow field and orbital velocity change, and have to be recomputed. The "continuity correction" is a frequently applied method to adjust the flow field after small changes in the bathymetry. The flow *pattern* is assumed not to vary for small bottom changes:

$$\vec{q} \neq f(t_{mor}) \tag{9.13}$$

where

$$\vec{q} = h\vec{u} \tag{9.14}$$

is the flow rate vector and h is the water depth. The same goes for wave pattern: wave height, period and direction are kept constant, and only the orbital velocity is adapted for the local water depth:

$$H_{rms}, T_p \neq f(t_{mor}) \tag{9.15}$$

Since:

$$\vec{u} = \frac{\vec{q}}{h} \tag{9.16}$$

and:

$$u_{orb} = f(H_{rms}, T_p, h) \tag{9.17}$$

adaptation of the sediment transport field is now simply a matter of adjusting the velocity and orbital velocity and recomputing the sediment transport using eq. (9.12).

In case of a tidal flow situation, a number of velocity and wave fields based on the original bathymetry are stored, and when the depth changes, the adapted transport field is computed for a number of time points in the tidal cycle and subsequently averaged. This averaged transport field is then used in the sediment balance.

The method still requires full transport computations through a tidal cycle, which can be time-consuming when suspended-load transport is to be accounted for. The morphological time step is often dominated by some shallow points, which are usually not of interest. This means that typically after some 5-20 continuity correction steps, the full hydrodynamic model has to be run on the updated bathymetry.

The main limitation to the continuity correction is the assumption that the flow pattern remains constant in time. In the case of a shallow area becoming shallower, the flow velocity will keep increasing under continuity correction, whereas in reality the flow will increasingly go around the shallow area.

9.7.3 *RAM approach*

In practical consultancy projects there is often a need to interpret the outcome of initial transport computations without having to resort to full morphodynamic simulations. One way of doing this is looking at initial sedimentation/erosion rates, but this method is flawed in many respects. Initial disturbances of the bathymetry lead to a very scattered pattern, and, as [De Vriend *et al.* (1993)] point out, sedimentation/erosion patterns tend to migrate in the direction of transport, a behaviour which is not represented in the initial sedimentation/erosion patterns.

The Delft3D-RAM module (Rapid Assessment of Morphology) is a simple method that overcomes these disadvantages. If we assume that for small bed level changes the overall flow and wave patterns do not change (an assumption also used in the "continuity correction" of many morphological models), the tide-averaged transport rates are a function of flow and wave patterns which do not vary on the morphological time-scale, and the local depth, which does vary on this time-scale. In other words: given a certain set of currents and waves, the transport at a given location is only a function of the water depth.

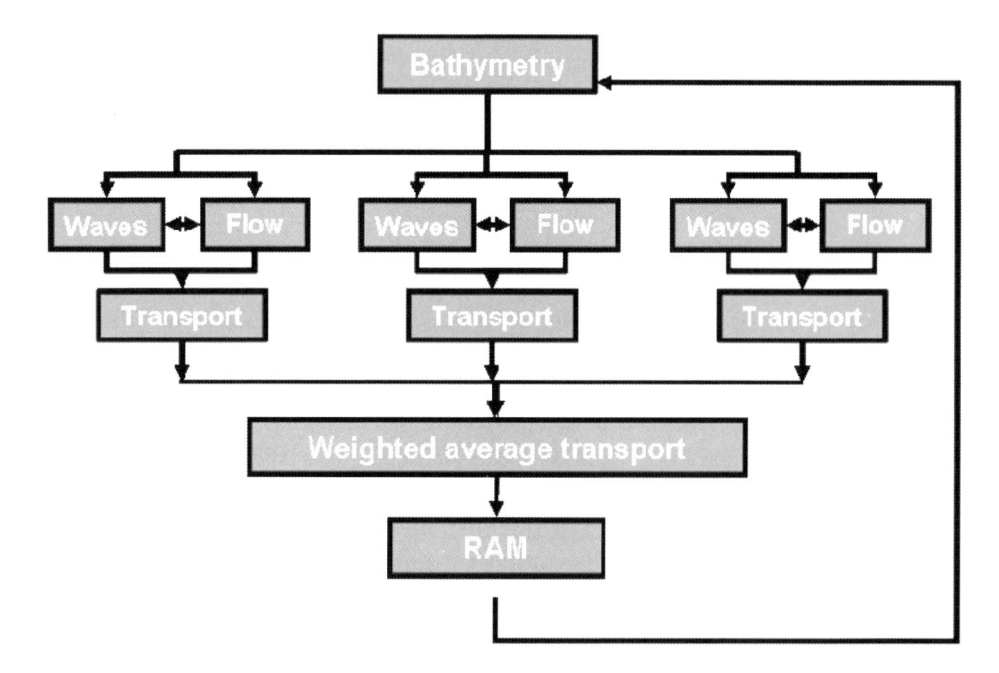

Fig. 9.11 Flow diagram of RAM simulations

If we can now approximate this function by some simple expression with coefficients which vary from place to place, we end up with a very simple set of two equations: the sediment balance which expresses bottom change in terms of sediment transport gradients:

$$\frac{\partial z_b}{\partial t} + \frac{\partial S_x}{\partial x} + \frac{\partial S_y}{\partial y} = 0 \qquad (9.18)$$

where z_b is the bed level and S_x, S_y are the sediment transport components, and:

$$\vec{S} = \frac{\vec{S}_{t=0}}{\left|\vec{S}_{t=0}\right|} f(z_b) \qquad (9.19)$$

This equation describes the reaction of sediment transport to bottom changes. The form of the function $f(z)$ can be estimated by considering that transport usually is proportional to the velocity to some power b:

$$\left|\vec{S}\right| \propto |u|^b \propto \left(\frac{|\vec{q}|}{h}\right)^b \propto |\vec{q}|^b h^{-b} \qquad (9.20)$$

where $\vec{q}$ is the discharge per unit width. Since a similar relationship with the orbital velocity can be assumed, a suitable function is:

$$\left|\vec{S}\right| = A(x,y) h^{-b(x,y)} \qquad (9.21)$$

where the water depth h is taken as $h = HW - z_b$ and HW is the high water level, which ensures that water depth is always positive.

As a further simplification b can be assumed constant throughout the field. In this case, the value of A in each point can be derived directly from the local water depth and the initial transport rate, which may be computed using a sophisticated transport model. The combination of equations (9.18) and (9.21) can be solved using the same bottom update scheme as in the full morphodynamic model and requires very little computational effort (in the order of minutes on a PC). The flow diagram is depicted in Figure 9.11.

In dynamic areas, such as estuaries and outer deltas, the RAM method may still work well enough to be applied as a quick updating scheme. As soon as bottom changes become too large, a full simulation of the hydrodynamics and sediment transport is carried out for a number of input conditions. A weighted average sediment transport field is then determined, which is the basis for the next RAM computation over, say, a year. The updated bathymetry is then fed back into the detailed hydrodynamic and transport model. An important point is, that the (costly) computations to update wave, flow and transport fields can be carried out in parallel, using different processors. The simplified updating scheme and the parallel computation for various input conditions together lead to a reduction in simulation time in the order of a factor of 20.

9.7.4 *Online approach with morphological factor*

The methods above have in common that the morphology is updated relatively infrequently compared to the number of flow time steps per tidal cycle (typically with less than one minute time steps) and the number of transport time steps per cycle (typically more than 20).

A completely different approach is to run flow, sediment transport and bottom updating all at the same small time steps. In case an advection-diffusion scheme is solved for the sediment transport this has to be done with comparable time steps as the flow solver anyway. The updating of the bottom only takes very little computation time. However, this 'brute force' method would not take into consideration the difference in time scales between the flow and morphology. Therefore a simple device is used, called the 'morphological factor'. This factor n simply increases the depth change rates by a constant factor, so that after a simulation over one tidal cycle we have in fact modelled the morphological changes over n cycles. This is similar to the concept of the 'elongated tide' proposed by [Latteux (1995)], except that there it was only used in combination with continuity correction. The idea is that nothing irreversible happens within an ebb or flood phase, even when all changes are multiplied by the factor n. The results obviously have to be evaluated after a whole number of tidal cycles.

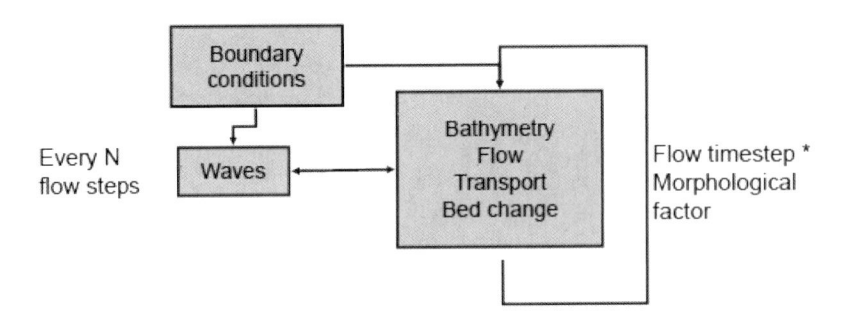

Fig. 9.12 Flow diagram of Online approach with morphological factor

An important difference with the previous methods is that the bottom evolution is computed in much smaller time steps, even when relatively large values of n are used. If we use an nof 60, this means that after completing 12 tidal cycles we have covered approximately one year of morphological change. In a typical flow model we would apply a time step of 1 minute; even with this high morphological factor this means we still update the bathymetry every hour. In comparison, using a tide-averaging approach we would have to take steps of a month in order to cover the same time period while calculating through the same number of tidal cycles. Even when applying 10 continuity steps in between each full morphological step this still means a 3-day time step; which then takes 10 times as many transport computations.

We can analyse the error made using this approach by time-integrating the transport vector multiplied by n as follows:

$$z_{b,t+nT} = -\vec{\nabla} \cdot \int_t^{t+nT} \left(\vec{S}_{\tau, \Delta z_b = 0} + \frac{\partial \vec{S}_{\tau, \Delta z_b = 0}}{\partial z_b} \Delta z_{b,t} + O(\Delta z_b^2) \right) d\tau =$$

$$\approx -\vec{\nabla} \cdot \int_t^{t+T} n\vec{S}_{\tau, \Delta z_b = 0} d\tau - \vec{\nabla} \cdot \int_t^{t+T} \left(\frac{nb \vec{S}_{\tau, \Delta z_b = 0}}{h} z_{b,\tau} \right) d\tau + O(\Delta z_b^2) \approx \quad (9.22)$$

$$\approx -nT \left(1 + \frac{1}{2} \frac{bnz_{b,t+T}}{h} \right) \vec{\nabla} \cdot <\vec{S}_{\Delta z_b = 0}> + O(\Delta z_b^2)$$

Compared to the expression in equation (9.11) we see that the second term between the brackets is not neglected as in the tide-averaging approach, but approximated using:

$$n \Delta z_{b,t+T} \approx \Delta z_{b,t+nT} \quad (9.23)$$

9.7.5 Tide-averaged approach vs. morphological factor

In order to make an honest comparison between the various methods we have devised a simple test case. Let us consider a tidal channel that is gradually widening over a typical length scale L. The flow diverges during flood (positive direction) and converges during ebb tide. A mean discharge is added. The transport gradient can then be approximated by the transport itself divided by a length scale L, so:

$$\frac{\partial z_b}{\partial t} = -\frac{\partial S_x}{\partial x} \approx -\frac{S_x}{L} \quad (9.24)$$

Let us now assume that the discharge per unit width through the channel has a mean component and an oscillatory component at the M2 frequency. This discharge is not sensitive to the depth, until the depth becomes so shallow that the flow chooses another channel and the discharge goes to zero. This effect is added by means of a smooth tapering function; the discharge including all effects is now described by:

$$q = \left(\bar{q} + \hat{q} \cos(\omega t) \right) \left(1 - \exp\left(-\left(\frac{h}{h_{sh}} \right)^2 \right) \right) \quad (9.25)$$

where h_{sh} is a depth scale that governs the tapering of the discharge to zero. We assume that the transport rate S_x is a simple function of the velocity and thus:

$$S_x = a \left(|u| \right)^{b-1} u = a \left(\left| \frac{q}{h} \right| \right)^{b-1} \frac{q}{h} \quad (9.26)$$

With this simple set of formulas we can now test the various time integration schemes. We have done this by integrating equation numerically for each of the schemes. We have used the same factor n for the time step in the averaging approach (multiplied by the the tidal period) and for the morphological factor; the value chosen in the example is 70. The intra-tide time step was equal to $T/50$, which is approx. 15 minutes.

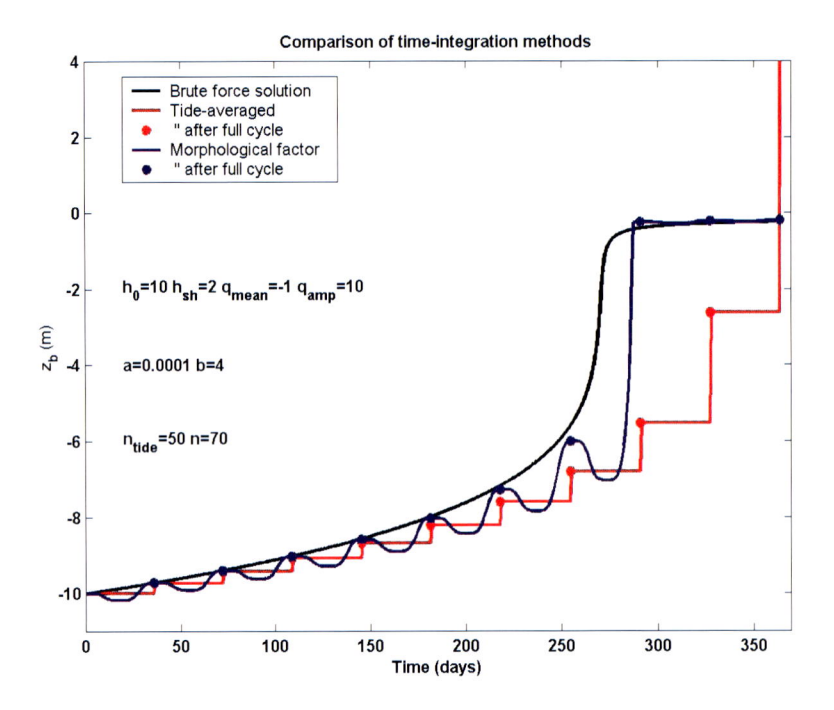

Fig. 9.13 Comparison of time-integration methods: morphological factor vs. tide-averaged, equal number of tidal cycles.

Figure 9.13 shows the evolution of the bed level in time. It is clear that with the 'brute force' integration, the intra-tidal bed changes are very small. The resulting bed change curve appears to be smooth. As the water depth decreases in the example, the transports increase and the rate of bed level change increases rapidly; as the depth becomes very shallow the tapering function starts to work and gradually nudges the water depth towards zero.

As long as the bed level change is approximately linear, both approximate methods do a reasonable job of following the 'brute force' line. The 'morphological factor' method wanders off within each (elongated) cycle, but returns very close to the right value after each full tidal cycle. As the transport is updated much more frequently than in the case of the 'tide-averaging' approach, this method is capable of following the curve even when the water depth becomes very shallow; it is also sensitive to the tapering of the transport for shallow depths. On the other hand, the 'tide-averaging' approach in this case misses the shallow part completely and shoots through the surface.

We can improve the 'tide-averaging' results without (in a realistic case) adding too much computational effort by using intermediate 'continuity correction' steps. If we keep the number of tidal cycles that we compute in a year constant (equal to

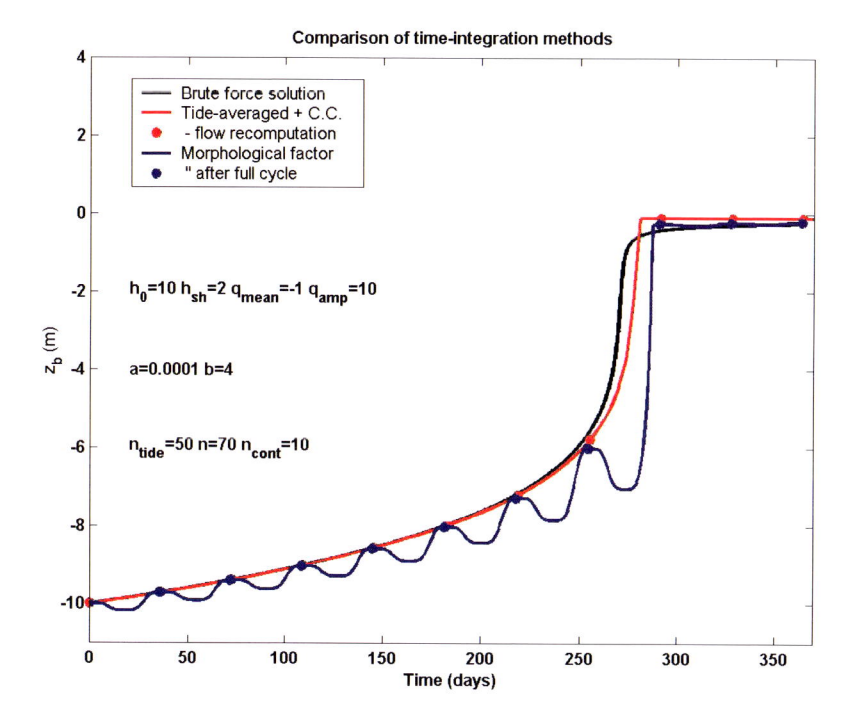

Fig. 9.14 Comparison of time-integration methods, morphological factor vs. tide-averaging + continuity correction

$700/70 = 10$) and use 10 continuity steps in between, we get the result as in Figure 9.14.

9.7.6 *Parallel online approach*

In this new approach we assume that hydrodynamic conditions vary much more rapidly than the morphology can follow. If the time interval within which all different conditions (ebb, flood, slack, spring tide, neap tide, NW storm, SW wind, etc) may occur is small relative to the morphological time-scale, these conditions may as well occur simultaneously. This leads to the idea that we may as well let simulations for different conditions run in parallel, as long as they share the same bathymetry that is updated according to the weighted average of the bottom changes due to each condition. The flow scheme of this approach is given in Figure 9.15

In this scheme the simulation is split into a number of parallel processes, which all represent different conditions; at a given frequency all processes provide bottom changes to the merging process, which returns a weighted average bottom change to all processes, which then continue the simulation. The parallel execution of the different processes lends itself to an efficient implementation on a series of PC's or Linux cluster.

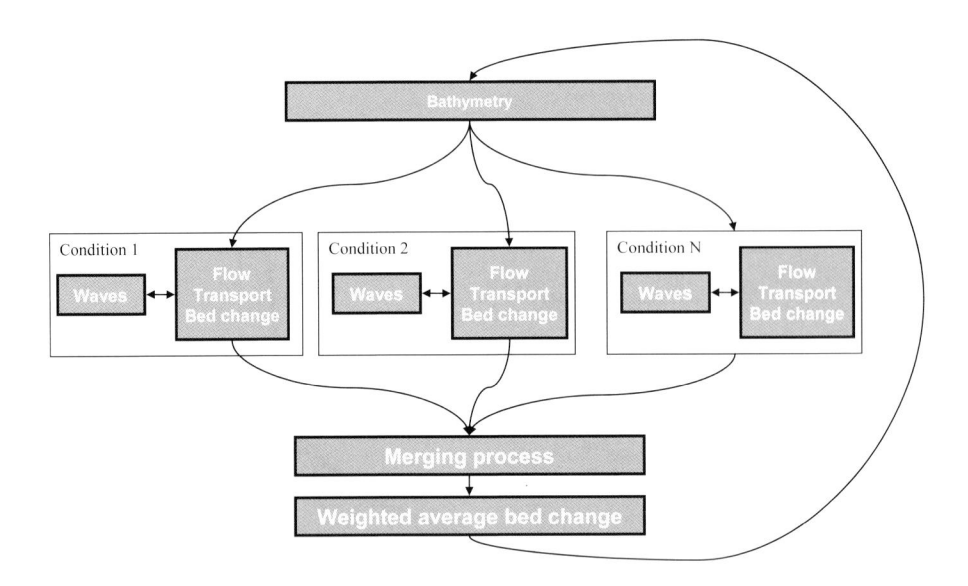

Fig. 9.15　Flow scheme of 'parallel online' approach

One can now design the different processes to keep each other in check, for instance by assigning a different tidal phase to different conditions, so that ebb and flood transports counteract each other at all times. This reduces the amplitude of short-term changes and thus allows the use of much higher morphological factors.

9.7.7　*Efficiency of the methods*

The relative efficiency of the methods is determined by three factors, described above: the numerical stability, the accuracy and the ability to cope with variable input conditions (wind, waves, discharges, etc.). The criteria for the different methods are summarised in Table 9.1. The parameter n is a good indicator of the run time in all cases, as one whole tidal cycle has to be run through to simulate n tidal cycles of morphological change. The parameter n_{RAM} denotes the number of RAM steps between full hydrodynamic updates. Under 'coping with variable input' we have assumed a number of conditions n_{cond} with equal probability of occurrence that have to be represented in a year (700 tides) of morphological change.

Stability criterion　The criterion of numerical stability is much more restrictive for the tide-averaging and RAM methods than for the online methods, as the flow time step is typically in the order of one minute, compared with the tidal period of approx. 745 minutes. For large-scale simulations with large grid sizes, deep water and/or low transport rates this need not be a problem, but in practice attention is shifted more and more towards shallow areas and detailed grid resolution. This can lead to a time-step restriction for the tide-averaging method of less than the tidal period. In such cases, the tide-averaging method has no advantages.

Table 9.1 Overview of criteria for different methods

Method	Stability	Accuracy	Coping with variable input
Tide-averaging	$\frac{h\,\Delta x}{b<S>T}$	$\frac{1}{2}\frac{b\Delta z_b}{h}$	$n < 700/n_{cond}$
RAM	$\frac{h\,\Delta x}{b<S>T}\,n_{RAM}$	$\frac{1}{2}\frac{b\Delta z_b}{h}$	No restriction
Online	$\frac{h\,\Delta x}{b\,S_{\max}\,dt_{flow}}$	$\frac{1}{2}\frac{b}{h}\left(\Delta z_{b,t+nT} - n\Delta z_{b,t+T}\right)$	$n < 700/n_{cond}$
Parallel Online	$\frac{h\,\Delta x}{b<S>_{\max}\,dt_{flow}}$	$\frac{1}{2}\frac{b}{h}\left(\Delta z_{b,t+nT} - n\Delta z_{b,t+T}\right)$	No restriction

Accuracy criterion The criterion of accuracy is easier to evaluate and more restrictive for the tide-averaging case than for the online methods. It is not grid-dependent but points out that relative bed level changes should be small, otherwise the assumption that the bathymetry is constant during the evaluation of the transport is violated.

Coping with variable input For the tide-averaging and standard online methods, which generally use a sequential scenario of varying input conditions, this restricts the value of n when a detailed input schematisation is chosen. This restriction does not hold for the RAM and parallel online methods, which both can use parallel processing to cope with a detailed input climate.

9.8 Strategies for longer-term simulations

9.8.1 *Beach profile extension*

In shallow coastal areas complex small-scale processes continually rework the beach profiles. Resolving such processes requires very advanced process modeling and very fine grids. This is not feasible in large-scale models, which may have in the order of five to ten grid cells across the surf zone, just enough to represent the wave-driven longshore current and transport. Another problem, even with the most advanced models, is the fact that the dry beach is generally not included and therefore dune erosion and accretion due to wind are not included.

On the other hand, a common observation both from field data and profile models is, that the upper part of the profile adapts very quickly to changes in the sediment balance and stays close to a dynamic equilibrium. This observation is widely used in coastline models and simple cross-shore models such as the Bruun rule: the profile is kept constant in shape but is allowed to move onshore and offshore in order to maintain the sand balance.

This concept has been implemented in RAM in the following way. First, the rate of bed level change is determined in all points using the usual scheme. For points that lie between a specified depth contour and the dune top, a special treatment is used, where we compute the bottom changes *per gridline* rather than per grid cell. The total volume change over the shallow grid cells in the grid line is determined by:

$$S_{in} = \frac{\partial V_{tot}}{\partial t} = \sum A_i \frac{\partial z_{b,i}}{\partial t} \qquad (9.27)$$

Here S_{in} is the net transport into the row of shallow grid cells, $\partial V_{tot}/\partial t$ the total volume change rate, A_i the individual grid cell areas and $\partial z_{b,i}/\partial t$ the bed level change rate of individual cells. We then treat the row of cells as if the net transport into the row of cells enters through the lowest cell and decreases towards the top of the dune according to:

$$S_i = S_{in} \frac{z_{top} - z_{b,i}}{z_{top} - z_{bottom}} \qquad (9.28)$$

Note that the transport rate now depends linearly on the bed level. The consequence of this is, that:

$$\frac{\partial S}{\partial z} = -\frac{S_{in}}{z_{top} - z_{bottom}} \qquad (9.29)$$

For uniform grid cell width $\Delta \eta$ along the gridline the transport rate per unit width is $s_\xi = S/\Delta \eta$; we can then write the one-dimensional sediment balance along the gridline direction ξ as:

$$\frac{\partial z_b}{\partial t} + \frac{\partial s_\xi}{d\xi} = \frac{\partial z_b}{\partial t} + \frac{\partial s_\xi}{dz_b}\frac{\partial z_b}{d\xi} = 0 \qquad (9.30)$$

Here we recognise a simple wave equation with a constant celerity; in other words, the profile propagates undistorted in cross-shore direction, which is exactly the desired behaviour. The volume balance per grid cell follows simply from eq. (9.28):

$$\frac{\partial V_i}{\partial t} = -\Delta S_i = +S_{in} \frac{\Delta z_{b,i}}{z_{top} - z_{bottom}} = \frac{\partial V_{tot}}{\partial t} \frac{\Delta z_{b,i}}{z_{top} - z_{bottom}} \qquad (9.31)$$

where the height difference over a grid cell $\Delta z_{b,i}$ is computed in an upwind manner in order to obtain a robust numerical behaviour. As we can see, the actual implementation boils down to a simple redistribution of the total volume change per profile, which makes it easy to ensure mass conservation.

In the present, first implementation of this procedure is carried out in the grid direction that gives most intersections with the waterline. The user can specify a polygon that confines the procedure to areas within the polygon.

9.8.2 *Representation of subgrid features*

In large-scale models, some features such as small groynes cannot be represented explicitly. Their effect can however be included in a schematic way by specifying polygons within which the transport rate is reduced by a specified factor.

As an example, along the Holland coast, groynes are found between km 2 - 31 (North-Holland groynes) and between km 98 - 118 (Delflandse Hoofden). Attempts to assess the effects of these groynes by looking at historical evolution of the beaches before and after construction have not been conclusive due to the variability of the wind and wave climate and the occurrence of sand waves along the coast. However, the fact that the construction of groynes always leads to leeside erosion is a strong indicator that they have some effect on longshore transport, even though there will still be a considerable rate of bypassing.

In a recent study on the effects of an airport island on the coast, reduction factors of 1 (no effect) and 0.5 have been applied to the validation case 1972-1980; based on the resulting pattern of shoreline accretion and erosion a value of 0.5 was chosen for further simulations

9.8.3 *Representation of dredging*

Dredging is an important aspect in many simulations of morphological change. The amount of dredging required to keep a navigation channel at its prescribed depth may be an important output of a study. Also, in longer-term simulations dredging is essential to keep the simulation realistic; without it entrances may silt up and bypassing rates may become too large, affecting the large-scale coastal behaviour.

In Delft3D-RAM the user can specify a number of dredging areas where the bottom is kept at a specified minimum depth. If the depth becomes less than this minimum depth, the excess sand is taken out of the cell and dumped in a dumping area specified together with the dredging area. The model then provides output of the amounts of dredged and dumped material per area as a function of time.

This is just an example of a dredging strategy implemented in a morphological model. It is relatively simple to implement other strategies of capital or maintenance dredging or of sand mining.

9.8.4 *Beach nourishments*

Long-term simulations of coastal morphological development would not be realistic if the coast were allowed to erode freely. The policy of "Dynamic Preservation" in the Netherlands for instance would prevent this from happening. In the model an approximation of this policy has been implemented, where eroding profiles are nourished immediately by just the amount of erosion per time step. The profiles are actually kept in place in such situations and the program keeps track of the cumulative amount of nourished sand per profile.

Chapter 10

Case Studies

10.1 Toy models of small coastal problems

10.1.1 *Introduction*

In this section we will play around with a relatively simple model set-up that we can use to discuss some processes taking place around hard and soft coastal structures. Such simple models have often been used to intercompare models, to create a basic understanding of processes around hard and soft engineering structures or to validate our models and understanding of coastal processes against lab or field data. [De Vriend et al. (1993)] analyzed processes and compared models for a semi-circular embayment and a straight coast with a river outflow. [Nicholson et al. (1997)] did a similar exercise for an offshore breakwater. [Zyserman and Johnson (2002)] summarized a number of model studies on the effects of offshore breakwaters. [Ranasinghe et al. (2006)] looked at the circulation patterns and shoreline response due to artificial surfing reefs, a special form of submerged breakwaters. [Roelvink and Walstra (2004)] studied the effect of a groyne on a straight coast and compared it with coastline modeling. [Grunnet et al. (2005)] studied the evolution of a large nearshore nourishment and its effect on the nearshore sediment balance in a relatively small model that included effects of tides, wind and waves.

To illustrate the kind of processes these models can capture we compare four runs with some engineering feature with a base case without any structure.

10.1.2 *Model setup*

The model grid and bathymetry was inherited from a study of coastal hydrodynamics as part of the Coast3D programme and was developed by [Elias et al. (2000)]. It represents a typical Dutch double-barred profile, which is rather uniform in longshore direction but does exhibit some nearshore variability giving rise to rip currents; see Figure 10.1. The grid resolution in the center area is in the order of 15 m by 20 m and coarsens towards the lateral and seaward boundaries.

At the time, the model was run using nested information from two larger regional models; nowadays we use a procedure as in [Roelvink and Walstra (2004)]

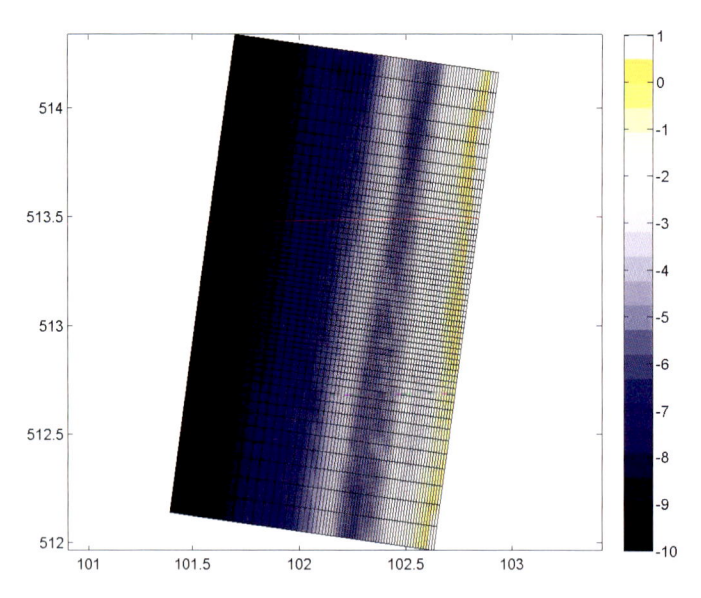

Fig. 10.1 Grid and bathymetry (depth i m indicated by the color bar) base case Egmond model (distance in km)

or [Grunnet et al. (2005)], where the boundary conditions are given as waterlevel boundary at the seaside and Neumann boundaries at the lateral boundaries. To further simplify the case we ignore tide and wind and use a single wave condition, with an H_{rms} wave height of 1 m, T_p peak period of 7 s and a wave direction of 240 deg. N. We use the Delft3D model in 2DH mode, as we will focus on the horizontal circulations and their effects on the coastal evolution due to the engineering measures. The wave propagation and dissipation is solved by the XBeach model in stationary mode, which is called by the Delft3D flow and morphology solver at regular intervals. The advantage of this setup is that the model grid for wave and flow solver can be identical and no disturbances are created at lateral boundaries.

We run the model for 30 days morphological time, with a morphological factor of 100. This means that in terms of hydrodynamic time we only run the model for 432 minutes, a little more than 7 hours. We allow the flow to spin up before we start the morphological updating; in this model this only takes about 10 minutes. Horizontal viscosity and diffusivity are set at a constant 0.5 m2/s and bed roughness is specified according to a Chezy value of 65 m1/2/s.

For sediment transport we use the standard [Van Rijn (1993)] settings in Delft3D, with a sediment diameter of 0.200 mm. Dry cell erosion effect is turned on fully, allowing the dry beach to erode when there is erosion just underwater.

The hard structures are resolved in the grid and are represented by areas of no available sediment for erosion. Elsewhere we have assumed a layer of 10 m thickness of erodible sand.

Five runs were carried out:

- A Base Case run without any engineering measure;
- A groyne with a length of approx. 200 m from the shoreline.
- An emerged offshore breakwater, with a length of approx. 200 m, at a distance of approx. 500 m from the shoreline;
- A submerged offshore breakwater, with the same length and location but a crest height 1 m below the water surface;
- A nourishment with the same dimensions of the submerged offshore breakwater;

10.1.3 Wave height patterns

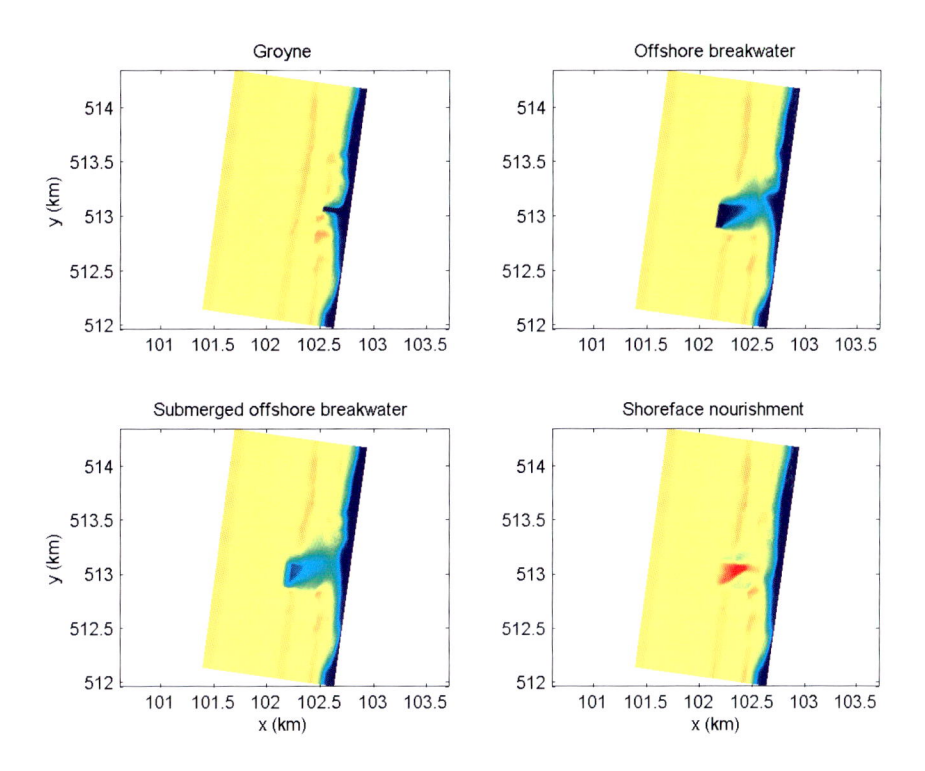

Fig. 10.2 Wave height patterns for different coastal engineering measures at the Egmond site.

In Figure 10.2 we see the wave height patterns due to the different schemes, at the end of the 30 days of morphological evolution. The effect of the groyne on the wave height pattern is mainly indirect, through the morphological evolution, since at the groyne location the waves are almost perpendicular to the shore. The shielding of the wave heights due to the offshore breakwater is very obvious; also for the submerged breakwater this effect is substantial. For the nourishment the effect is initially the same as for the submerged breakwater but what we see here is the effect after the nourishment has been topped off and has migrated towards the shore. Interestingly, the waves are increased in height because of shoaling and wave focusing due to refraction.

10.1.4 Current patterns

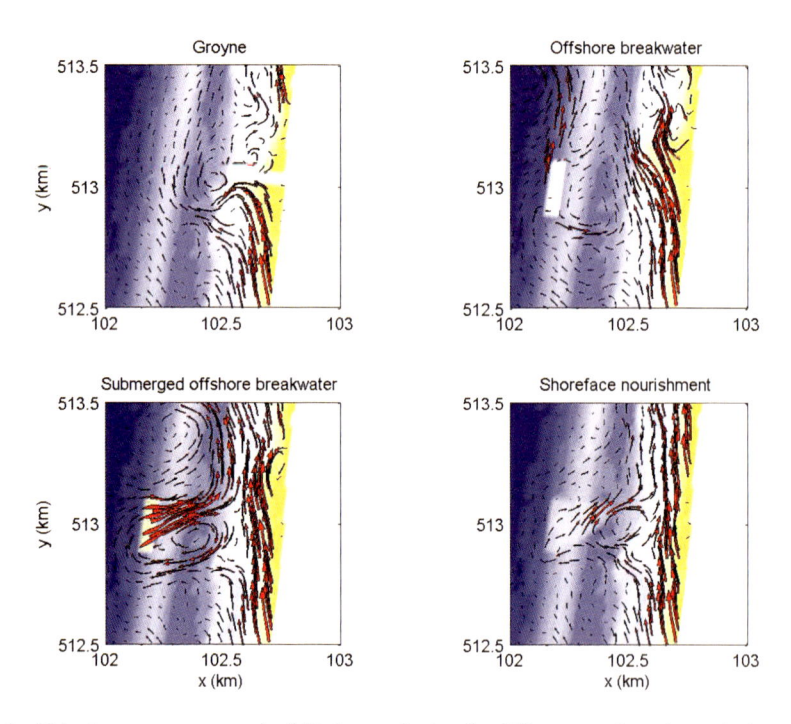

Fig. 10.3 Velocity patterns at end of 30-day evolution for different engineering solutions (detail)

As we can see in Figure 10.3, the different coastal structures have a strong effect on current patterns and thereby on the local morphology. In the groyne simulation we see that there is a strong longshore current that is diverted away from the coast and suddenly decreases strongly in magnitude, with complex little patterns around the tip; at the downstream end, it takes quite a distance for the current to pick up again. In the offshore breakwater case a rather similar pattern is seen, which also leads to a rather similar shoreline behavior, as we'll see in the next paragraph. The submerged breakwater leads to a strong onshore current induced by wave breaking over the structure. As [Ranasinghe et al. (2006)] have shown, this current pattern can cause a problem when the structure is too close to shore, in which case it can even lead to erosion behind the structure; for this case the structure seems to be far enough from the shore to avoid this. However, there is a rather strong rip current south of the structure that could cause safety problems for swimmers. In the nourishment simulation a somewhat similar circulation pattern is visible, but it is much less pronounced because in the 30 days of simulation the nourishment has reduced in height and has spread out over a larger area.

10.1.5 Effect on bathymetry

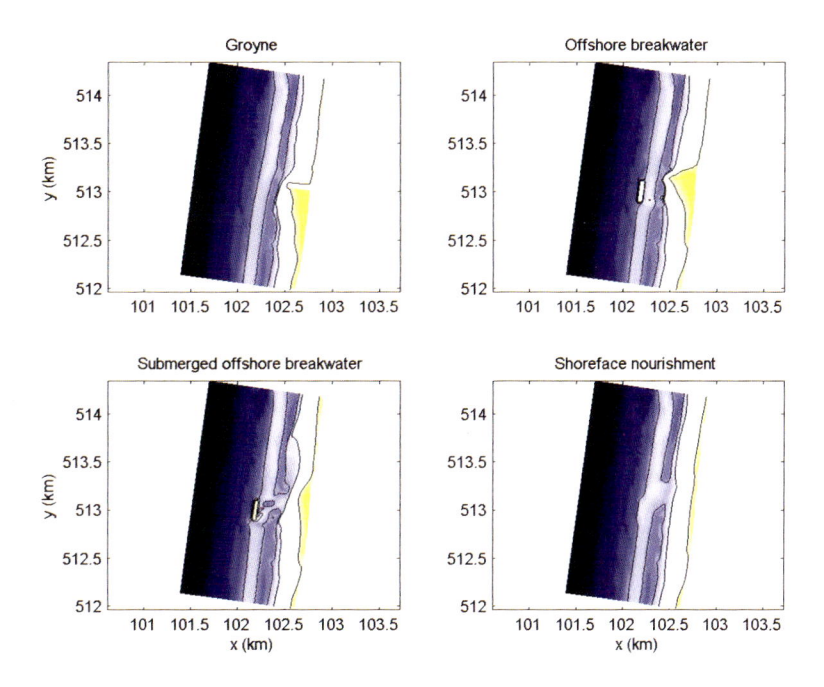

Fig. 10.4 Final bathymetries for different engineering options after 30 days of simulation

The effect of the groin and the emerged offshore breakwater is very similar: the shoreline upstream of the structure accretes rapidly and the coast downstream experiences rapid erosion (see Figure 10.4). On the updrift side the behavior is rather similar to that according to coastline theory, but on the downdrift side the situation is different; here aspects like wave setup differences and advection terms lead to a slow buildup of the longshore current, which leads to a rather more extended erosion area than the classical solutions suggest, where the downdrift area is a mirror image of the updrift area. Part of this may be due to the profile response in the downdrift area. If the eroded profiles react by shifting horizontally, the longshore transport can be established more quickly than when they scour down to below depth of breaking. This depends both on the treatment of the 'dry cell erosion' and on the cross-shore transport processes in this area. A minimum requirement is that there is enough beach and dune area within the model domain to allow for erosion; if there is not, the erosion shadow will propagate along the coast too fast.

For the case of the submerged breakwater, the effect appears like a weakened version of the effect of the emerged breakwater, but some additional complexity is evident, which is not surprising given the strong circulation patterns. For the shoreface nourishment the effect on the coastline is limited on the timescale considered; however, there is a clear accretion in the trough area, which is likely to have a beneficial effect in the longer run.

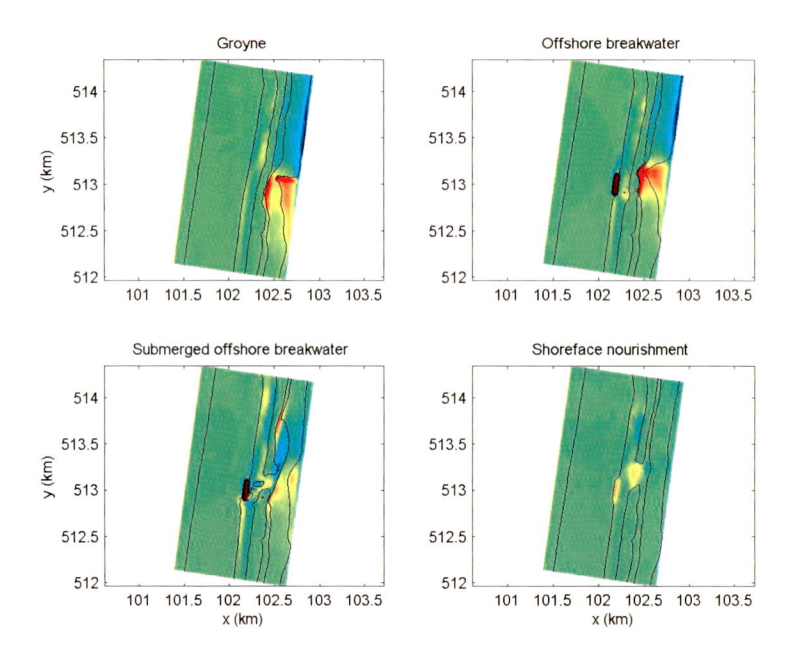

Fig. 10.5 relative erosion/sedimentation patterns for different coastal engineering options after 30 days of simulation.

10.1.6 *Relative erosion/sedimentation patterns*

Finally, in Figure 10.5 we show the differences in bed level between the situation with engineering measures and the base case without any measures. Clearly, the groyne and the emerged offshore breakwater have the strongest effects, both negative and positive. The submerged breakwater in this case has a similar but watered-down effect. Only in the case of the nearshore nourishment the effect is positive overall, which is logical since this is the only option where sediment is actively brought into the system.

10.1.7 *Discussion*

Though there may be a number of interesting and realistic-looking features that come out of these model runs, any aspect of these runs could do with vast improvements. The resolution is only just sufficient to represent some rip current patterns (but not any on a scale of, say, some meters in cross-section); the domain is on the small side given how far the effects may extend; the beach and dune profiles do not extend far enough to allow for much erosion; the flow is only 2DH, so misses much of the return flow profiles; there are hardly any cross-shore effects included; there is only a single sediment size; we only took a constant wave condition and forgot about tides and wind, and we could go on for much longer. We leave it to you, the reader, to do better on all these points and a few more before you present it to an innocent client.

10.2 Long-term modeling of tidal inlets, estuaries and deltas

10.2.1 *Introduction*

Tidal inlets in many parts of the world are valuable natural resources, which are threatened by human interference, bed level subsidence and sea level rise. Based on a wealth of empirical data, we know that intra-tidal flats and supra-tidal marshes tend to follow the relative mean sea level or mean high water level, by trapping part of the large amounts of sediment that enter and leave the inlets every tidal cycle. This means that inlets facing relative sea level rise act as sinks of sediment, often at the cost of adjacent coasts. Over long time-scales this may lead to a shoreward transgression of barrier islands or, in case of insufficient sediment supply, to loss of valuable intertidal areas and marshes. Both types of behaviour pose serious coastal management problems: houses and villages are threatened and governments are often committed to preserving valuable habitat areas.

We've seen that the overall behaviour of tidal inlets facing sea level rise can be modelled using semi-empirical models, but detailed measures and effects are difficult to represent in that approach. A process-based approach could be an alternative, but until recently this was not believed to be feasible.

10.2.2 *How far can upscaling lead us?*

In the early '90s, the general belief was that upscaling of small-scale processes to large space- and time-scales could only be done over a short period, after which everything would diverge and collapse; indeed, we have seen this behaviour in many cases, when we simply took a model that was calibrated over a period of some years and continued it for a much longer period.

On the other hand, many natural systems tend towards some kind of equilibrium, and if we have some idea of the counteracting forces leading to this equilibrium, why should we not be able to make process-based models with similar behaviour? Some groundbreaking work was carried out by [Wang et al. (1995)], [Hibma et al. (2003)] and [Marciano et al. (2005)], who produced promising and realistic-looking patterns based on a Delft3D models using classical tide-averaged approaches, for

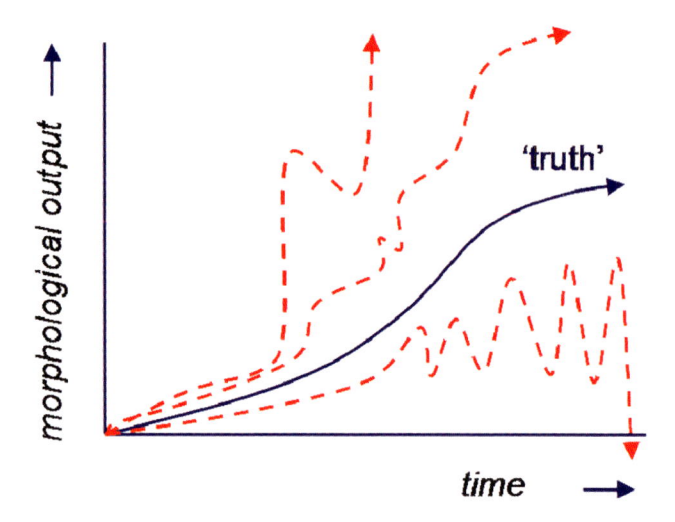

Fig. 10.6 Morphosceptics' view of bottom-up models

schematic cases resembling the Western Scheldt and the Ameland Inlet. While these simulations covered long time periods, they could not start from just any initial bathymetry, they tended to fail when shoals emerged and generally led to channel patterns that were too sharp.

10.2.3 Necessary model improvements for long-term modeling

Various model improvements and schematization 'tricks', sometimes very simple, have been developed over the last couple of years, which have led to much more stable and realistic solutions:

- The boundary conditions on the open sea used to pose serious problems, especially in the case of a propagating tide. Even small disturbances at the boundaries travel through the model domain in morphological runs, and eventually ruin our simulations. With the use of the Neumann boundary conditions on lateral boundaries, as described in section 9.5, we can have simulations over thousands of years without these problems, see e.g. [Dastgheib et al. (2008)] and [Dissanayake et al. (2009)].
- The "morphological factor" (MORFAC) approach generally leads to numerically much smoother and accurate results compared to the tide-averaging approach
- In simulations like this, shoals develop that are relatively flat and have sharp fronts. Dominantly central schemes such as Lax-Wendroff may develop oscillations on such fronts, which very often lead to instability problems and/or very much reduced time steps. Schemes with a more upwind character (e.g.) [Lesser et al. (2004)] are much better able to handle such shocks. Though in hypothetical tests of hump migration upwind schemes look terribly diffusive, in realistic

situations with flat shoals they correctly describe purely horizontal motion of the shoals.

- When in a tidal situation shoals develop above low tide level, tide-averaged simulations run into problems; the morphological factor approach deals with these shoals in a smooth fashion. In [Hibma et al. (2003)], who used the classical tide-averaging approach, this limited the simulation of developing shoal patterns to where they would actually emerge;
- In most models, after a shoal has developed above high tide level, it gets out of reach of the active model domain and becomes fixed, which in turn fixes the adjacent parts of the model. In the end, the model becomes less and less dynamic. This can be avoided by introducing some form of bank erosion, either through 'dry cell erosion', e.g. [Lesser et al. (2004)], or through an avalanching mechanism. In these cases the developments underwater are leading, and dry shoals and banks can follow; erosion next to dry areas leads to sideways migration rather than vertical scouring.
- Even with these improvements, the morphology can still tend to develop into too narrow and deep channel patterns. Again such deeply incised channels become too fixed and the simulations lose their dynamic behaviour. Even if the exact causes of these developments are not known, a pragmatic approach of increasing the coefficients regulating bed slope effects can ensure that a model exhibits much more natural dynamic behaviour, as was shown by [Lesser (2009)] for a model of Willapa Bay on the US West coast. [Van der Wegen and Roelvink (2008)], [Van der Wegen et al. (2008b)] using the morphological factor approach, 'dry cell erosion' and increased bed slope factors find their model to be able to run almost indefinitely while retaining realistic patterns. Figure 10.7 shows an impression of such results for the case of a long estuary with freely erodible banks.

Two well-known empirical relation are chosen to compare the results of this simulation with the empirical relations: cross-sectional area (A) – tidal prism (P) [Jarrett (1976)] – and channel volume – tidal prism [Eysink (1990)] (Figure 10.8). As it is shown in the figures the trend of the P/A relationship along the basins (i.e. not only at the mouth) for different points in time shows a similar trend as the empirical P/A relationship. Also the result of the model for the channel volume – tidal prism is in good agreement with [Eysink (1990)] suggestions. Detailed discussion about the pattern formation and results can be found in [Van der Wegen et al. (2008b)].

10.2.4 How much of the morphology of an estuary is forced by its boundaries?

For very schematic cases we have seen patterns emerge that look similar to patterns found in nature, and we have seen that these patterns are different when we have a short and wide basin than when it is long and narrow; also, if the boundaries are

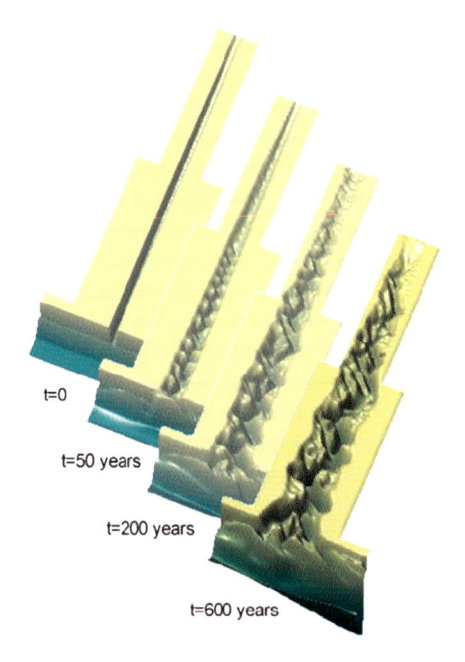

Fig. 10.7 Impression of bathymetry of 80 km long basin after 0, 50, 200 and 600 years. After [Van der Wegen *et al.* (2008b)]

fixed we get deeper and narrower channels than when the banks can be eroded. A next question then arises: how much of the morphology of an estuary is actually forced by its boundaries (dikes, headlands, unerodible layers and other geological or man-made constraints)? If we're given the outline of an estuary as seen from Google Earth, and given the tidal amplitudes outside it, can we guess the channel pattern and characteristic cross-sections? This would be very handy in areas where data is scarce (as in many parts of the world). Besides, it provides an excellent test of the skill of our models in predicting the *morphology* rather than *morphology change*. In practice so far, we have always tried to reproduce observed erosion and sedimentation patterns, but predicting the morphology itself is often more useful, for instance when you need to predict the depth and extent of a scour hole or when asked to predict what happens when you reopen a closed estuary mouth or wetland.

[Van der Wegen *et al.* (2008a)] and [Van der Wegen (2010)] carried out long-term simulations for the Western Scheldt, starting from a flat bathymetry with a depth representing the average depth of the whole basin, specifying the present-day dikes as hard boundaries and using schematized tidal boundary conditions at the seaside. Even with just a simple transport formula and depth-averaged flow equations, remarkably realistic patterns are computed; using 3D hydrodynamics and including known unerodible layers, the simulated channel pattern is remarkably similar to the present observed channel geometry, and hypsometry (wetted area as function of wa-

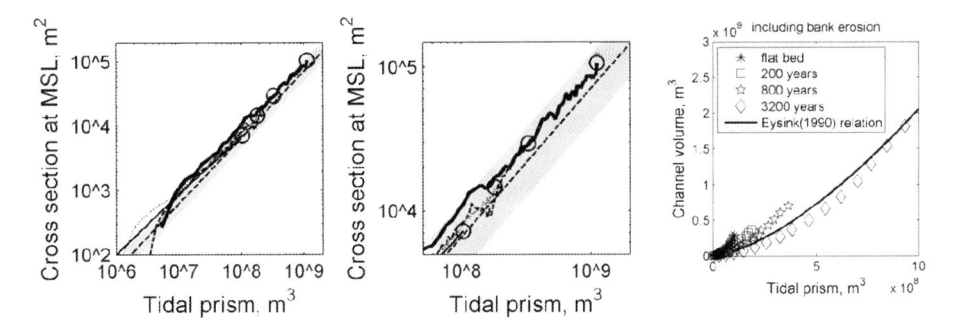

Fig. 10.8 Relation between Tidal Prism and Cross-sectional area. (left) whole 80 km long domain; (middle) In detail; unjettied (straight dashed line); gray area represents 95% confidence intervals [Jarrett (1976)]. Model results after 1 year (thin solid line); after 200 years (dotted line); after 800 years (thick dashed line); after 3200 years (thick solid line); open circles indicate values at the mouth; after [Van der Wegen et al. (2010b)]. (right) Relation between Tidal Prism and Cross-sectional area and channel volume; Model results are for different locations along the estuary at different times.

ter level) and P-A relations at different cross-sections are reproduced very well. The simulations show a quick generation of channel patterns in the first decades, after which a much slower redistribution of sediment along the estuary takes place. The best fit with the actual geometry was found after some 50-100 years. Figure 10.9 shows initial and final simulated channel patterns in comparison with the present-day geometry, for some different model settings: 2DH vs. 3D hydrodynamics and with or without unerodible layers. Because of the tide propagating from the south we see the development of a dominant channel on the southern side of the mouth; the 3D simulation show sharper features in the channel pattern due to the spiral flow effect in the bends.

[Dissanayake et al. (2009)] investigated channel patterns in a schematized version of the Ameland inlet and found patterns that are rather similar to the existing patterns. Especially the main channels are also similar to the real ones in location and size; he shows that again the tide characteristics are very important in steering the orientation of the main channel. Further work on the effects of sealevel rise and on comparing process-based modeling with semi-empirical modeling using ASMITA is currently underway.

[Dastgheib et al. (2008)] carried out very long-term simulations of the first three (interconnected) tidal inlets in the Western Wadden Sea using different initial conditions. He found that the three would develop channel patterns of similar strength and size when given the same initial flat bed but that once the initial depth of the inlet basins was different, very different evolution would take place. Deeper basins would become dominant with respect to the more shallow basin, which is exactly what has happened in this area, where the Marsdiep and Vlie basins used to be connected to the large Zuyderzee and therefore have much larger average depth than the smaller Eijerlandse Gat. Figure 10.10 shows the modelled patterns with

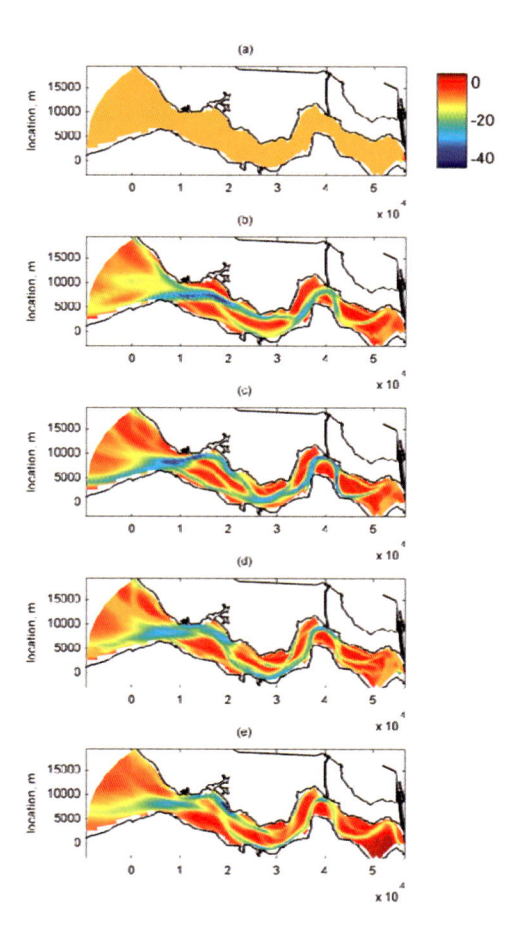

Fig. 10.9 Simulated development of Western Scheldt bathymetry after 200 years starting from flat bed. (a) initial bathymetry; (b) 2DH hydrodynamics;(c) 3D hydrodynamics; (d) 3D hydrodynamics and unerodible layers; (e) observed 1998 bathymetry. After [Van der Wegen (2010)]

different initial conditions in comparison with the present-day morphology. Again, the main channel patterns and even the ebb delta that is generated in the model are surprisingly similar to the present-day pattern and some equilibrium relations are reproduced beautifully, while others are followed in a qualitative sense. However, the simulated patterns are generally too sharp, even though exaggerated bed slope transport coefficients are used. A likely culprit is the single grain diameter used, as we will see in the next section.

10.2.5 *Effect of sediment sorting*

In rivers, where the bed material often varies from mud through sand to gravel over short stretches, taking into account horizontal grain size variations has long

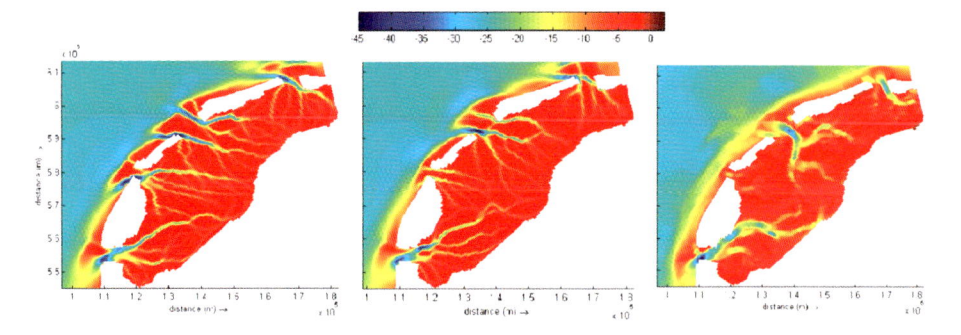

Fig. 10.10 The resulting bathymetry of Wadden Sea after 2100 Morphological years simulation from different initial bathymetries : Flat bathymetry, $d=$ 4.54m (left), sloping bathymetry (middle) , compared to bathymetry of 1998 (right panel) After [Dastgheib et al. (2008)].

been common practice, even though it generates a wealth of complexity. In the simulations so far this effect was ignored, because the range of sediment diameters is generally less than in rivers. However, in systems like the Marsdiep, even if it is mostly sandy, there is still a large range of grain sizes, with typically the largest grain sizes (D50 in the order of 0.6 mm or more) in the channels and the smallest (less than 0.1 mm to mud) on the tidal flats. Recently Dastgheib et al. (in prep.) and [Van der Wegen et al. (2010b)] have been studying the effects of taking these grain size variations into account in different ways:

- to use a spatially varying D_{50} grain size;
- to separately model the transport of different size fractions and to have a book-keeping of the changes in bed composition, using a number of sediment under-layers and a mixing layer at the sediment-water interface.

The first option is the easiest, but loses its validity when channels move too much. One may overcome this by adapting the spatial grain size distribution every now and then. The second option is physically much more correct but costs more computation time and is more complex to program.

An important difficulty is how to prescribe the initial condition for the bed composition, as usually there is insufficient data available. Again there are different options. One is to use a multiple-fraction model to update the bed composition without updating the bathymetry. Such a method was applied by [Van der Wegen et al. (2010a)] for San Pablo Bay north of San Francisco Bay. Another option is to use observed data of the spatial sediment distribution and to link that to hydrodynamic parameters. It turns out that there is a clear relation between the sediment diameter and the tide-averaged bed shear stress or velocity variance, as fine sediment tends to be winnowed out from areas with high shear stresses. Dastgheib et al. (in prep.) applied such a relationship to create initial conditions for the bed composition.

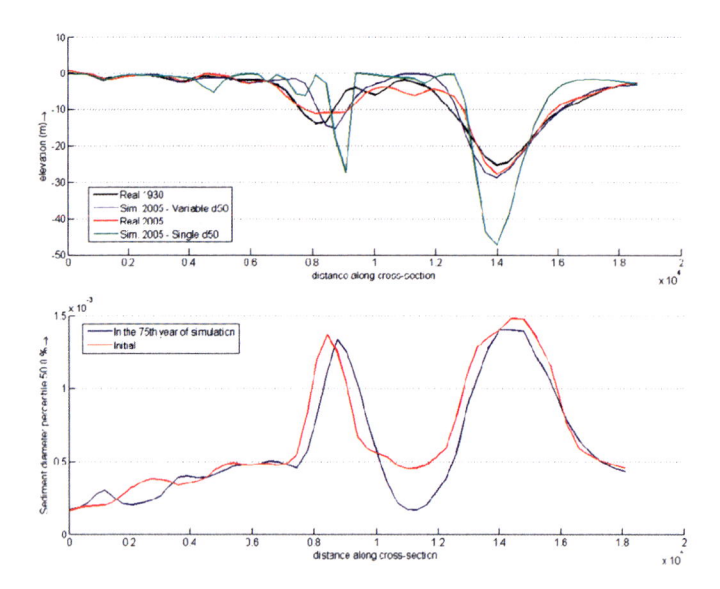

Fig. 10.11 Upper Panel : Resulting cross section after 75 years of morphological simulation for the model with spatial d50 distribution (Blue) and model with single D_{50} (Green) compared with the measured cross-section in 1930 (Black) and 2005 (Red). Lower Panel : Distribution of D_{50} in a lateral cross-section of Marsdiep basin, Initial (Red) and after 75 years of morphological simulation (Blue).

Regardless of the details of the method chosen, it turns out that taking into account both the horizontal distribution of sediment sizes and using a reasonable initial estimate for the sediment distribution leads to much less initial disturbance or 'morphological spinup' and much more realistic channel configurations. Dastgheib et al. (2010) found typical channel depths in the Marsdiep almost halved when using realistic spatial distributions of sediment size (see Figure 10.11)

10.2.6 *Some sample simulations*

The results we discussed just now were based on simulations that were relatively simple compared to how complex they might be, but still too complex to describe in a few words. Therefore we thought it useful to discuss some cases that are easy to set up and describe, and which can be run in a computation time of hours on a PC. We start with the tidal basin simulation described in [Roelvink (2006)] and vary the following parameters:

- Tidal propagation: perpendicular or alongshore, with a phase speed of 1°/km.
- Coriolis forcing: off (equator) or at latitude 51° (Holland)
- Waves: off or on (H_{rms}=1m, T_p=7s, direction 330 oN, directional distribution according to cos10 function)
- Coastal profile: relatively flat (1:500 down to -10 m) or relatively steep (1:100 down to -20 m)

The model grid is rectangular with (recklessly coarse) grid cells of 100 m by 100 m and 150 by 150 cells. The gorge width is 2 km. For this model we use Delft3D in 2DH mode with default sediment settings ([Van Rijn (1993)] sediment transport, sediment D50 of 0.200 mm). For the wave forcing we use a simple stationary refraction model with directional spreading, based on the XBeach wave solver [Roelvink et al. (2009)] used in stationary mode. This solver can be called from Delft3D and has the advantage that it is fast, can operate on the same grid and domain as Delft3D and does not have the lateral boundary problems common in models such as SWAN.

The tidal boundary conditions are given as a single harmonic component with an amplitude of 1 m (tidal range of 2 m) and a phase difference of 15 deg. (0 deg.) over the seaward boundary for propagating (perpendicular) tide. The lateral sea boundaries are Neumann boundaries with amplitude of the water level gradient of 1.75*10-5 for propagating tide and 0 for perpendicular tide.

Because the model is extremely dynamic in the beginning, with its shallow depth and high velocities, we only use a morphological factor of 10. We run it for 20 days (hydrodynamic time), which amounts to 200 days in morphological time.

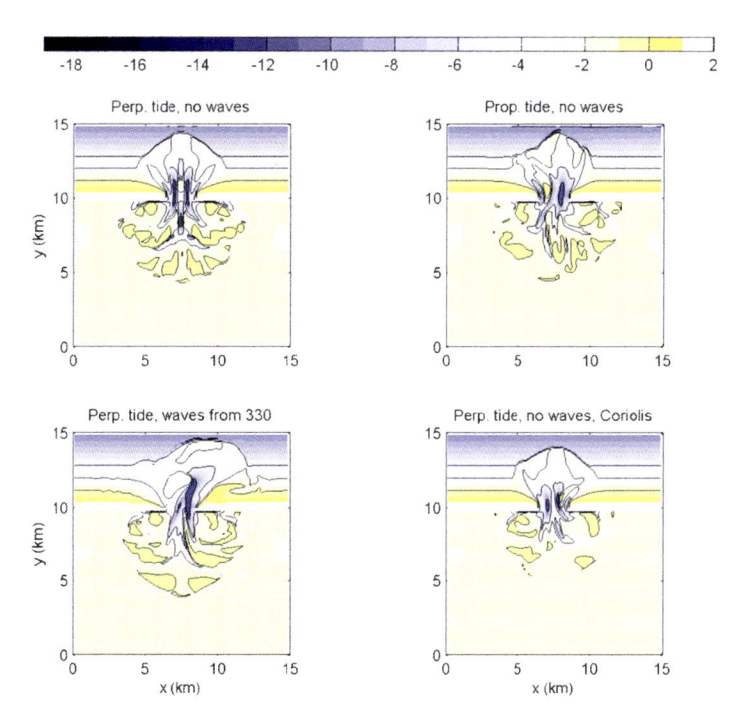

Fig. 10.12 Bathymetry (m) after 200 days of simulation, relatively mild seaward slope; effect of tide propagation, waves and Coriolis forcing.

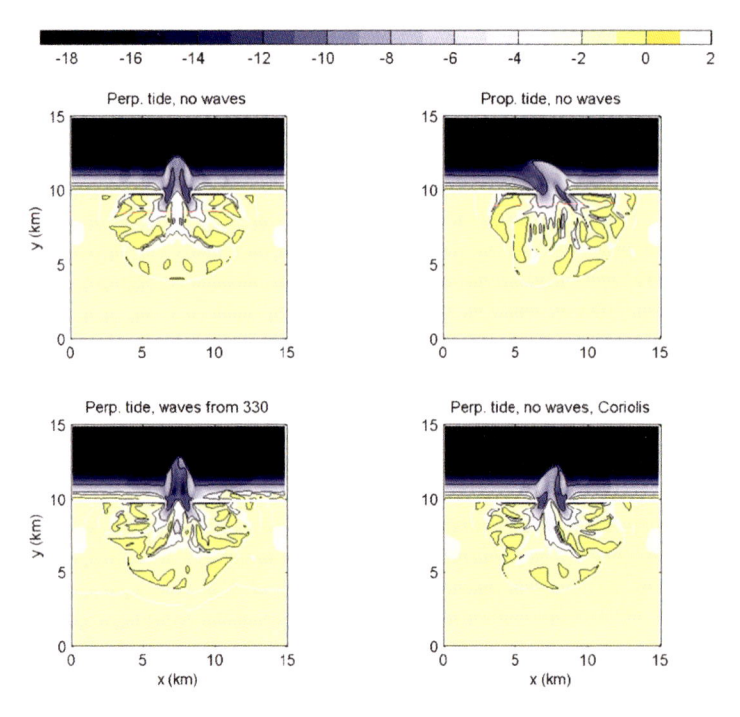

Fig. 10.13 Bathymetry (m) after 200 days of simulation, relatively steep seaward slope; effect of tide propagation, waves and Coriolis forcing.

In Figure 10.12 we show the results for a relatively mild coastal slope and in Figure 10.13 for a relatively steep slope. We see a big horizontal development of the ebb tidal delta for the mild slope and much less so for the steep slope; in the case without Coriolis and with a perpendicularly incident tide, we get a perfectly symmetrical solution in both cases. Interestingly, the response to a propagating tide is rather different for different profiles; for the steep slope we get a clear orientation of the ebb channel to the Northwest, so against the tide propagation direction, as expected. For the milder profile this is less clear; because there is so much sand around the shallow channels are easily choked and shift around more.

The local Coriolis effect (with perpendicular tide) is clearly another mechanism creating an asymmetry, this time in the other direction. Especially the ebb jet is pushed to the right by the Coriolis force (at this Northern latitude) and this leads to the ebb channel orientation towards the Northeast. Again, the picture is easier to interpret with the steeper coastal profile.

The wave effects on the ebb delta formation are spectacularly different for the mild and steep profiles. On the mild profile, we see the classical response of the obliquely incident waves pushing the ebb delta and ebb channel to the downdrift direction, because of the longshore transport entering the area from the West and because of the radiation stress gradients over the ebb delta acting in the same direction. A striking feature often observed in reality, for example at the western part of Ameland in the Netherlands, is the small spit growing on the downdrift coast. Though the tidal currents in this area are very strong, this feature only occurs in combination with (obliquely incident) waves. In the simulation with the steeper profile, the ebb delta is below the depth where waves are breaking so this effect is much less present. What is evident now is that the waves mainly act to stir up more sediment in the ebb flow, which leads to a more extended ebb delta compared to the situation without waves.

Although these simulations are very coarse and simplified, they do illustrate the potential of modeling morphological processes in complex environments.

10.3 Dune erosion

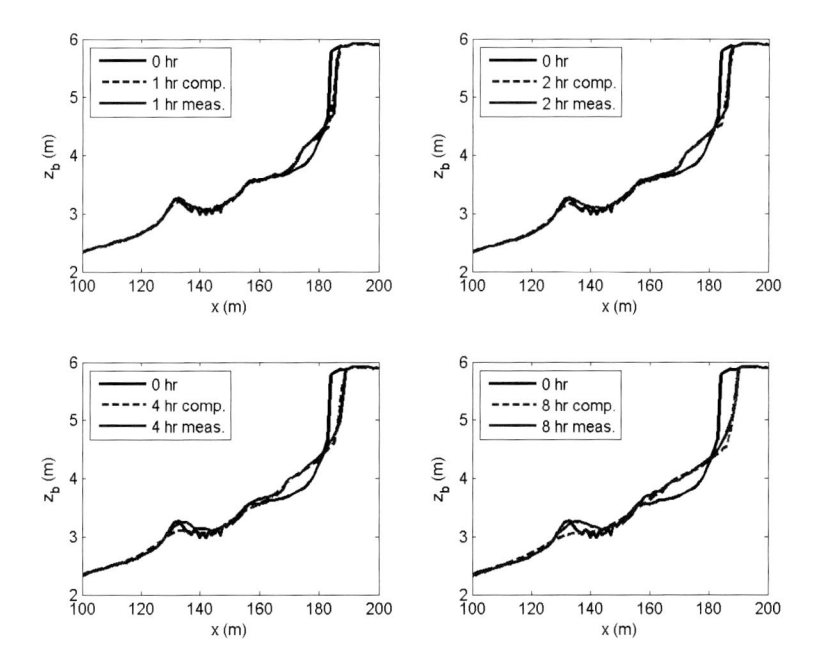

Fig. 10.14 Measured and modeled profile evolution after 1, 2, 4 and 8 hours for the LIP-11D dune erosion test.

Here we discuss an application of Xbeach to dune erosion ([Roelvink et al. (2009)]). To demonstrate the effects of both the infragravity swash and the avalanching a comparison with the LIP11D Delta Flume 1993 - test 2E is made ([Arcilla et al. (1994)]). This model test concerned extreme conditions with a raised water level at 4.58 m above the flume bottom, a significant wave height, H_{m0}, of 1.4 m and peak period, T_p, of 5 s. Bed material consisted of sand with a D_{50} of approximately 0.2 mm. During the test substantial dune erosion took place.

Based on the integral wave parameters H_{m0} and T_p and a standard Jonswap spectral shape, time series of wave energy ware generated and imposed as boundary condition (see Section 2.7). Since the flume tests were carried out with first-order wave generation (not accounting for bound long waves (see Section 3.6.2) or bound superharmonics (see Section 4.4.1), the hindcast runs were carried out with the incoming bound long waves set to zero ('first order wave generation'). Active wave reflection compensation was applied in the physical model, which has a result similar to the weakly reflective boundary condition used in the model [Verboom and Slob (1984)], namely to prevent re-reflecting of outgoing waves at the wave paddle (offshore boundary). A grid resolution of 1 m was applied and the sediment transport settings were set at default values. For the morphodynamic testing the model was run for 0.8 hours of hydrodynamic time with a morphological factor of 10, effectively representing a morphological simulation time of 8 hours.

Comparison of predicted and observed bed profiles (see Figure 10.14) shows a very good match in the swash and dune area. The mismatch mainly occurs within the surfzone, where intra-wave sediment transports become increasingly important, resulting in the initial formation of a bar. The calibrated model is used next to eximane the sensitivity to the avalanching and the wave group forcing.

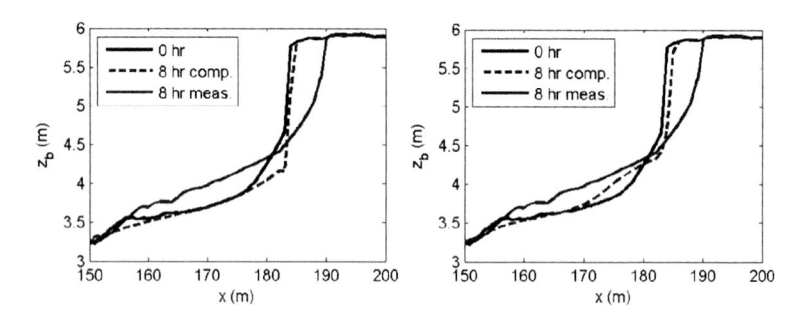

Fig. 10.15 Left panel: Predicted bottom profile after 8 hours without avalanching. Right panel: Predicted bottom profile after 8 hours without wave group forcing.

Without avalanching the dune erosion is severely underpredicted (see left panel of Figure 10.15). Hence, even though surfbeat waves running up and down the upper beach are fully resolved by the model, without a mechanism to transport sand from the dry dune face to the beach the dune face erosion rate is substantially underestimated. Similarly, if the avalanching is included but the wave group forcing is excluded a similar underprediction of the erosion is obtained (see right panel of Figure 10.15). This is explained by the fact that the avalanching mechanisms only kicks in when the dune face gets inundated by the incident infragravity swash. This demonstrates that both mechanisms should be present to correctly simulate dune erosion.

10.4 Overwash

Fig. 10.16 Part of the Florida Panhandle showing the locations of Santa Rosa Isand and Beasly Park. Picure courtesy of Google.

A rapid assessment of overwash is discussed next using Xbeach to predict the bed elevation changes observed after the impact of Hurricane Ivan at Beasly Park, Florida, USA (see Figure 10.16). The hurricane impact is simulated with a constant surge level of 2.0 m present for 15 hours at which time the offshore incident significant wave height is kept at 10 m with a mean wave period of 12.5 s. This is an over-simplified representation of the actual conditions (see [Wang and Horwitz (2007)] for a description of the conditions and morphological response during hurricane Ivan).

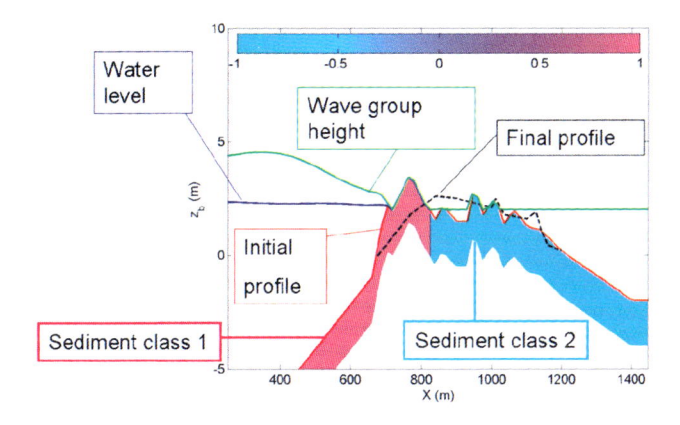

Fig. 10.17 Intial model configuration for Beasly Park with two sediment classes separating the dune and beach area from the barrier island and back-barrier bay. Observed profile after hurricane Ivan indicated by the dashed black line.

We have defined two types of sediment corresponding to dune and beach sand and a class representing that sand on the island and the back barrier bay (see Figure 10.17). Grain sizes for both sand composites are the same with a D_{50} of 0.003 m and a D_{90} of 0.006 mm. The initial sediment class distribution is presented in the top panel of Figure 10.17, where an intensity of 1 corresponds to sediment class one only and -1 to the presence of sediment class 2 only. The sand on the barrier island is mostly vegetated by grass which mitigates the erosion. This effect has been simulated with an increased friction coefficient $C = 40$ m$^{\frac{1}{2}}$/s.

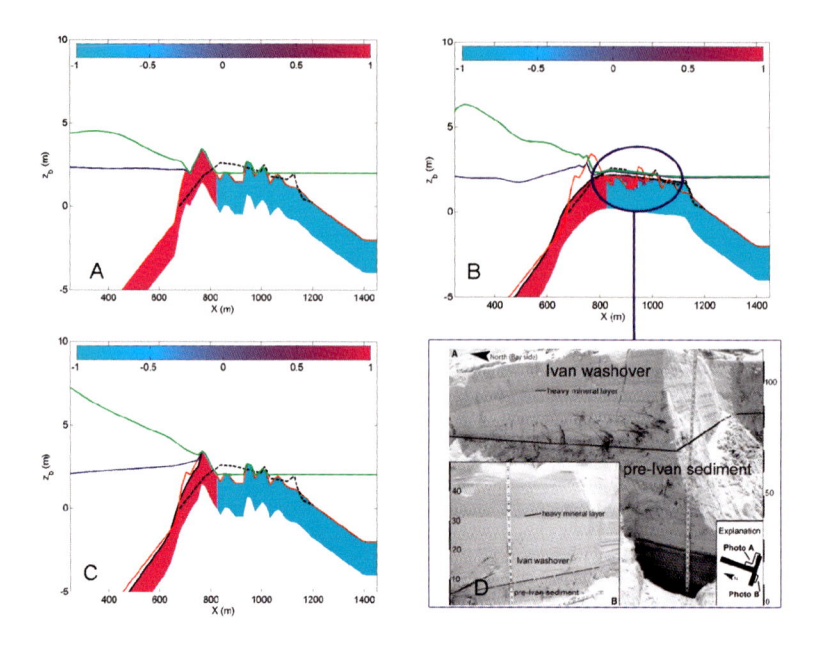

Fig. 10.18 Panel A: initial model configuration. Panel B: Model output after 15 hours simulating the impact of hurricane Ivan. Panel C: Similar but without wave group forcing. Panel D: Observed sand distribution after hurricane Ivan courtesy of [Wang and Horwitz (2007)].

The bed-elevation and sediment class distribution after 15 hours are shown in panel B of Figure 10.18. The calculated bed-level is similar to the observations although differences are apparent. These differences can be related to the fact that the hurricane impact is simply modeled (i.e. constant conditions) and the fact that the post-survey was performed approximately 10 months after the hurricane had past. Still the overall evolution is consistent with the observations. The calculated changes in the sediment classes, compare panels B and D in Figure 10.18, are also consistent with the observations of [Wang and Horwitz (2007)] based on a number of cores showing that the intersection of the new washover with the pre-hurricane sediment occurs approximately at the original bed level. Finally, calculating the hurricane impact without wave group forcing leads in this case to dune erosion only (see panel C in Figure 10.18).

As mentioned earlier, the two dimensional nature of the barrier islands can be important in the morphological response. This was examined recently by [McCall et al. (2010)] for a stretch of Santa Rosa Island (see Figure 10.16). The initial

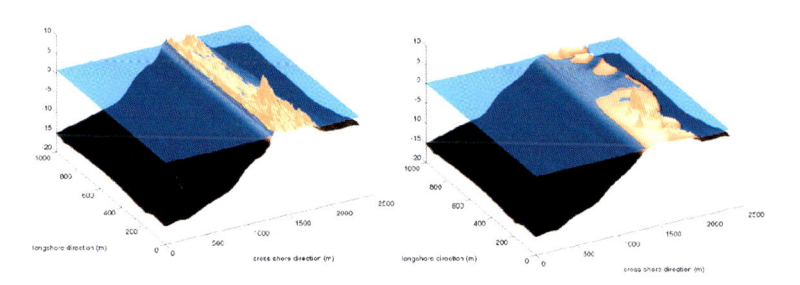

Fig. 10.19 Left panel: Pre-hurricane topography of part of Santa Rosa Island obtained with LIDAR (courtesy of the USGS). Right panel: Predicted post-hurricane topography after severe overwashing. Figures couretsy of [McCall (2008)].

topography shows significant variation in dune heights along the barrier island (see left panel of Figure 10.19 which clearly results in a spatially varying morphological response to the hurricane impact (compare panels in Figure 10.19). Variability of morphological impact to forcing conditions, initial conditions and model configurations are discussed extensively by [McCall et al. (2010)], where the largest uncertainty is related to the (limitation of) sediment transport during overwash conditions.

10.5 Sand bars and rip channels

Beach states have been observed to display a wide variety as reported by [Wright and Short (1984)] and [Lippmann and Holman (1989)], ranging from alongshore uniform dissipative beaches to intermediate beaches with a longshore bar-trough, quasi-rithmic bar and beach, shore connected transverse bars intersected by (obliquey oriented) rips, a low tide terrace and a reflective beach. The transition from one beach state to the next is defined as up-state or down-state. A large number of modeling efforts have been made to explain these transitions of which the down-state transitions have been reasonably successful. These include analyses based on the linear (in-)stability of an initially alongshore uniform barred or planar profile after the pioneering work of [Hino (1974)] as described by [Dodd et al. (2003)] and references therein. These methods yield information on the emergence of alongshore periodic patterns that are most likely to occur, corresponding to the fastest growing mode (see Section 3.5.4 on longshore current instabilities), subject to the (perturbed) sediment transport patterns over the initial bathymetry. As a result the outcome is sensitive to initial bottom profile, wave height, period and direction, tidal elevation, sediment grain size (distribution), etc, limiting the predictive capability of these models in a deterministic sense (e.g. [Calvete et al. (2007)]). The model results are valid for the initial evolution of rhythmic patterns only, i.e. corresponding to infinitesimal changes in the bed elevation. As the bed-perturbations grow a non-linear analysis taking into account the finite amplitude effects on the wave transformation and flow circulation has to be performed, which can be achieved with the morphodynamic process models presented in the following.

The transition process occurs within a relatively short time span, typically in the order of days to weeks, and the corresponding length scales of the evolving bathymetric features are typically a hundred metres or more in the alongshore and of order surfzone width in the cross-shore (see Figure 5.8). These observed spatial scales suggest that hydrodynamic processes at similar scales may be important agents in determining the morphodynamic response. This notion lead [Holman and Bowen (1982)] to propose that the streaming within the wave boundary layer of phase-locked edge waves could lead to alongshore rhithmic features. Other hydrodynamic processes that have similar length scales are associated with the shear instability of the longshore current (Section 3.5.4) as well as wave-group generated surfzone circulations or VLFs (Section 3.6.5).

To adequately represent the bathymetric features an alongshore spacing of O(10) m is required, i.e. 10 grid points to represent a single rip channel shoal combination as well as the corresponding wave group variation at those alongshore scales. The cross-shore grid can be varying, with larger grid spacing offshore, and smaller grid spacing within the surfzone. The offshore spacing is dictated by the need to adequately represent the cross-shore structure of the wave groups:

$$\Delta x \leq 0.1 c_g T_{group} \qquad (10.1)$$

where T_{group} corresponds to a typical wave-group/long-wave period of $O(25)$ s in prototype conditions. The surfzone grid spacing on the other hand is governed by a correct representation of the rapid transitions at wave breaking over the shallow shoals, which typically occurs at shorter than wave group spatial scales:

$$\Delta x \leq 0.1 \frac{H_b}{\tan \beta} \qquad (10.2)$$

where H_b is the wave height at breaking and $\tan \beta$ is the bed slope in the surfzone. The time step is determined by a correct representation of the temporal evolution of the wave groups:

$$\Delta t \leq 0.1 T_{group} \qquad (10.3)$$

In the following we will consider a case where an initially alongshore uniform bar, representative of a reset event, is subsequently subject to normally incident directionally spread waves, taking into account the variation of wave energy at the wave-group time scale (see sections 2.7 and 3.6). The bathymetry is a loose fit to the observed bathymetry at Palm Beach, Australia [Reniers et al. (2004a)] , with a single barred profile located approximately 75 m from the shore line and an average bed slope of 1 in 50 (upper left panel of Figure 10.20). Wave conditions at the site are characterized by the arrival of both sea and swell waves [Holman et al. (2006)], with at time significant sheltering of the beach by the protruding headlands [Smit (2010)]. Here we consider normally incident waves with a constant root mean square wave height of 1 m with a peak period of 10 s (upper left panel of Figure 10.20). The corresponding cross-shore grid spacing ranges from O(20) m offshore to O(5) m within the surfzone based on eq.s 10.1 and 10.2, a fixed 10 m spacing in the alongshore direction and a time step of 2.4 seconds.

The alongshore model domain is 1500 m (see lower left panel of Figure 10.20) which corresponds to the length of the beach within view of the Palm Beach Argus camera (see Figure 5.8) and allows for the potential generation of a range of rip-channels and shoals at all observed length scales. The cross-shore domain is 1200 m with a corresponding depth of 8 m at the offshore boundary. The reasoning behind this is that the depth is large enough that the bound long waves at the offshore boundary, (Section 3.6.2), are small and can be neglected in the formulation of the flow boundary condition. At times when the incident wave height over depth is large, the incoming bound long wave has to be specified to avoid (large) spurious long waves [van Dongeren and Svendsen (1997)] or the cross-shore domain has to be extended to larger depths thus increasing the computational time.

The presence of the headlands results in the blocking of the nearshore flows and reflection of long waves which is simulated with no-flux boundary conditions at the lateral sides of the model domain. Reflection of long waves at the shore line is also accounted for with a zero flux boundary condition. The waterline is defined by the interface between cells that are dry and those that have a minimum water depth of 0.1 m calculated with a drying and flooding procedure [Stelling (1984)]. The offshore boundary is given by a weakly reflective Riemann boundary which allows offshore propagating long waves to pass undisturbed with only weak reflections for obliquely exiting long waves [Verboom and Slob (1984)].

The frequency distribution of the corresponding energy density is given by a Jonswap spectrum with a $\cos^m$-directional distribution with $m = 20$. Using a single summation random phase method to generate the time series of the surface elevation at the offshore boundary and applying a low-pass filtered Hilbert transform (see Section 2.7) the wave-group varying energy time series at the offshore boundary are obtained. These are propagated into the model domain with the directionally and frequency averaged wave energy equation accounting for refraction, shoaling and wave breaking at the wave group time scale (see Section 2.7).

The bed shear stress takes into account both the waves and currents using the parameterization of [Soulsby et al. (1993)] for the wave-current boundary layer model of [Fredsoe (1984)] (see section 3.2.5). The current-only friction is calculated with the expression by Manning with $n = 0.02$ and the wave-only friction parameter is obtained from [Swart (1974)]. Turbulent mixing is modelled with eq. 3.49 using the dissipation of roller energy to calculate the turbulent kinetic energy with α equal to 1.

The sediment response is calculated with the 2DH-advection diffusion equation (Section 4.2.2)., with the source term given by the Soulsby-van Rijn sediment transport equation (see Section 4.3.2). We have used a D_{50} of 200 μ and a z_0 of 0.006 m. Intra-wave sediment transport is incorporated using the method outlined by [Reniers et al. (2004a)], with an efficiency factor α_w of 0.75.

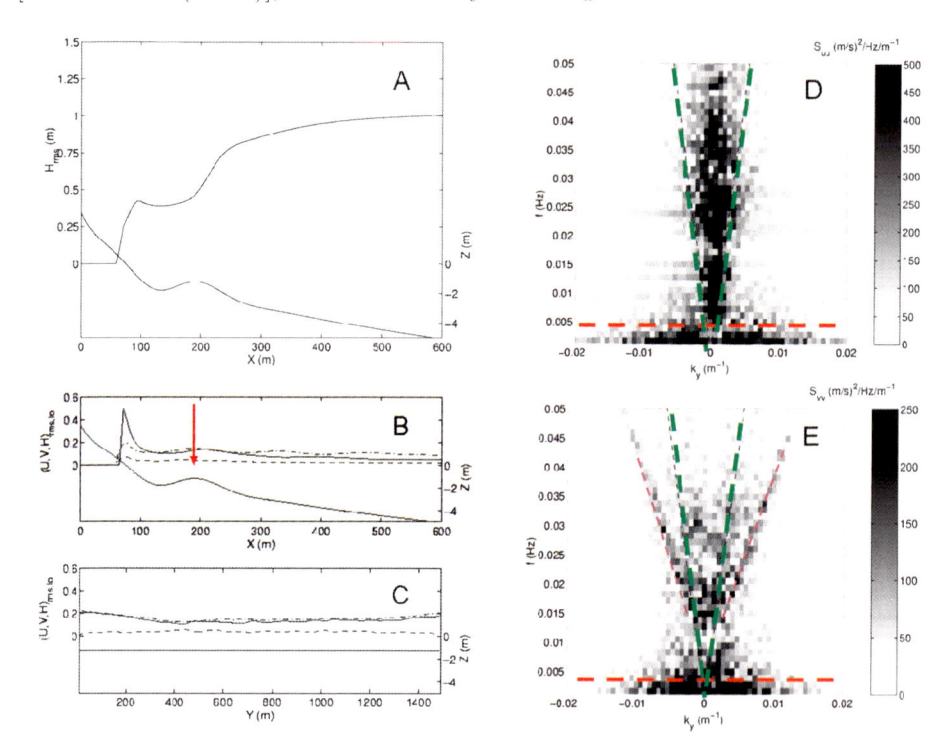

Fig. 10.20 Panel A: Initial wave-group-averaged short-wave transformation over alongshore uniform barred bathymetry. Panel B: Corresponding infragravity wave height $H_{rms,lo}$, cross-shore velocity, $U_{rms,lo}$, and alongshore velocity, $V_{rms,lo}$. Panel C: Alongshore infragravity response at the bar-crest (location indicated by the red arrow). Panel D: Calculated f-k_y-energy density of cross-shore velocity along the bar crest. Leaky wave regime enclosed by the green dashed lines. Vlf regime below the dashed magenta line (see Section 3.6.5 for a description of the different f-k_y regimes). Panel E: Similar for the alongshore velocity with edge wave ridges indicated by the dashed magenta lines.

Initial model calculations over the fixed alongshore uniform beach show negligible alongshore variation in the infragravity response (see panel C of Figure 10.20), a prerequisite for the edge wave template model proposed by [Holman and Bowen (1982)]. A 2D-FFT analysis of the model calculated velocities along the ar crest show that, although the leaky waves are dominant (panel D in Figure 10.20, there is significant edge-wave energy present (see panel E of figure 10.20), but this is smoothly distributed over a large range of alongshore wave numbers, i.e. no preferential length scale. In contrast, there is significant variation in the very low frequency (VLF) velocity field with VLF energy density concentrated around $k_y = 0.005$ m^{-1} corresponding to a length scale in the order of 200 m (see panels D and E of Figure 10.20). As descibed in Section 3.6.5, this alongshore variation in the VLF response is associated with the alongshore variability of the incident wave groups.

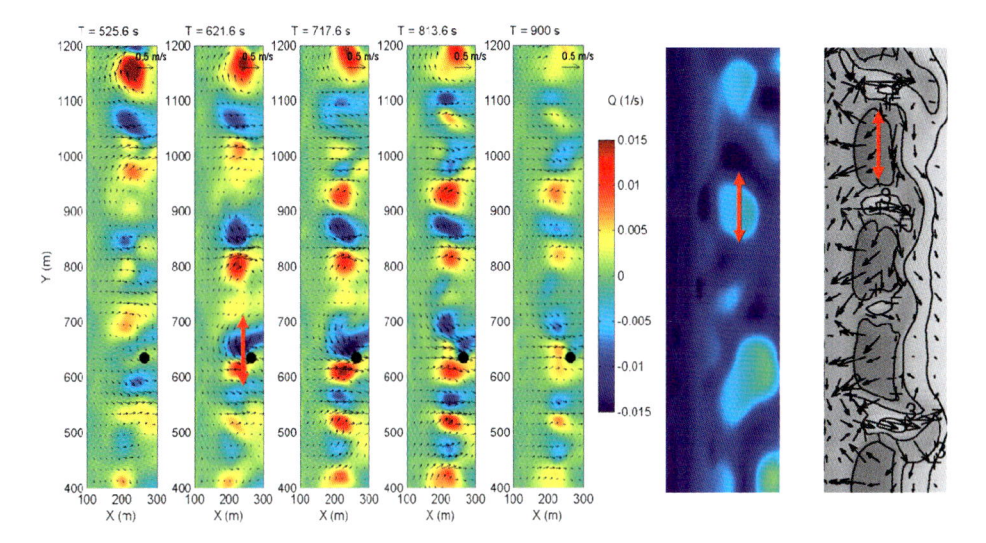

Fig. 10.21 Left 5 panels show the low-pass filtered, $f < 0.004$ Hz, VLF velocity field with wave-group breaking induced vortical circulations (vorticity indicated by the color scale). The sixth panel shows a snapshot of the wave group energy with the alongshore spacing corresponding to a cos-directional spreading factor of 20. The right panel shows the final predicted rip-channel shoal bathymetry (depth contour in meters), lighter (darker) areas are deeper (shallower). Typical lengthscales for the eddies, wave groups and resulting rip channel spacing are similar as indicated by the red arrow.

An illustration of the low-pass filtered, $f < 0.004$ Hz, velocity time series in the vicinity of the bar crest shows the persistence of the VLF-surfzone eddies with narrow offshore directed transient rip-flows distributed approximately 200 m apart in the alongshore direction. The fact that these VLF have relatively long hydrodynamic time scales and attain significant velocities makes them important agents in creating alongshore variability within the initially alongshore uniform bathymetry. Once this variability is created positive feedback (see Section 5.3.2) can lead to further erosion (deposition) leading to the generation of rip channels and shoals (see right panel of Figure 10.21). The corresponding length scales are a function of the alongshore wave group length, which in turn is a function of the directional spreading of the incident waves, and can deviate significantly from the FGM obtained with linear stability analysis. The alongshore rip-spacing is irregular consistent with the stochastic wave group forcing (see Figure 10.21).

This also suggests that in nature, where both the wave (group) forcing and the pre-existing variability in the nearshore bathymetry along an open coast are stochastic, it is unlikely to be able to predict the rip channel positions in a deterministic sense ([Stive and Reniers (2003)]). This is supported by the observations ([Holman et al. (2006)], [Turner et al. (2007)]). Instead, the expected variability in bed elevation and rip channel spacing can be used to verify model predictions of down state transitions initialized after a reset (e.g. [Smit (2010)]). In contrast, if the initial model bathymetry has significant variation such as rip channels and shoals the prediction of the down state beach evolution can succesfully be executed in a deterministic sense ([Ranasinghe et al. (2004)], [Smit (2010)]).

At present the modeling of up-state transitions, most notably the transition from a bathymetry with significant alongshore variation to a reset bathymetry with a uniform longshore bar-trough system still poses a great challenge. In this case the stirring and transport related to wave-induced turbulence inserted by the vigourishly breaking waves may play an important role (similar to the case of dune erosion discussed earlier) as suggested by [Smit (2010)]. Also the presence of a strong alongshore current rectifying and elongating the shallow shoals thereby filling in the rip channels is likely to be important and deserves further examination.

Chapter 11

Modeling Procedure

11.1 Introduction

In this chapter we attempt to define a sound morphological modeling procedure, based on the experience in the model studies described in the previous chapters. Obviously, following the full procedure is not always possible, given constraints of time and budget. In such cases explicit choices must be made to leave out certain elements. Also, our desciption may be biased by our own experience and environments we have worked in. Still, we think it is useful as a list of things to think about.

11.2 Data collection and analysis

11.2.1 *Bathymetry data*

The analysis of bathymetric data can comprise the following steps:

(1) Conduct a literature review of the problem area.
(2) Find historical maps and digitise key features and contours so they can be easily combined and displayed.
(3) Interpolate digital depth data to one common grid. Make difference maps.
(4) Determine relevant areas for which volume changes must be determined.
(5) Compute volume changes for these areas.
(6) Draw a number of cross-sections and analyse behaviour of cross-sections.
(7) Select morphological units for an analysis of the growth or decay and migration rate of these units.
(8) Make animations of evolution of bathymetry.

11.2.2 *Wave and wind data*

Analyse wave and wind climate; divide into sectors of, for instance, 30 degrees, 0.5 s T_p, 0.5 m H_s. Plot wave roses. Determine dominant wave directions.

11.2.3 *Tidal data*

Check availability of regional tidal models. Collect water level data from neighbouring stations. Analyse current and discharge measurements. Select time periods for calibration of a tidal model.

11.2.4 *Longshore current data*

These are usually not available, since it takes major field campaigns to collect useful data. The only option is to test your model against these datasets and hope that it will be applicable in the specific situation too.

11.2.5 *Sediment transport data*

Collect data on sediment properties. Analyse (if available) sediment concentration data and select data for model calibration/validation. Estimate longshore transport rates from local accretion/erosion near structures.

11.3 Conceptual model

Analyse current patterns from regional model, current atlas or previous model studies. Analyse grain size distribution over area. Estimate dominant wave-driven current patterns. Estimate transport paths and set up hypotheses about causes of bottom changes. Draw a picture of this estimate, to be updated in the real study.

11.4 Setting up modeling strategy

Define the morphological elements that need to be resolved by the model. Estimate the grid sizes required to represent them. Define model boundaries. Inside tidal inlets, choose boundaries over the tidal shoals as natural boundaries. On the seaward side, choose boundaries perpendicular to or parallel with flow direction. Put them as far away from the problem area as possible. Coarsen grid towards the boundaries.

Determine which wave directions will be included in the simulations and sketch wave grids. Estimate runtime for wave run.

Estimate flow and morphological time steps. Estimate probable run times for a single tide and a single morphological step. Estimate how long it will take to simulate the desired number of years and lower your standards and expectations if necessary. Take into account that you'll need about five to ten runs in the calibration and validation phase to arrive at a single run you can live with.

Define necessary sensitivity runs.

11.5 Setting up model grid and bathymetry

11.5.1 *Flow and morphology grid*

Boundaries and grid resolutions have already been determined. Draw spline grid as a first sketch. In a number of steps, refine grid, globally or locally, orthogonalise, repair local glitches, refine further. Check orthogonality (< 0.05-0.10), smoothness (< 1.3), resolution (conform requirements).

11.5.2 *Wave grids*

Define one overall rectangular bottom grid for all coarse-grid wave computations. Try to ensure that the nested wave grids are within the flow model domain, so they can use the bathymetry of the flow model, which is updated in a morphological run. Select the nested grid in such a way that:

(1) resolution is better than flow model in most areas
(2) grid direction is close to local wave direction
(3) the order of nested computations is such that disturbances at the boundaries are avoided

For each grid set-up, plot wave heights, dissipation rates and wave forces at least once for a relevant condition.

11.5.3 *Bathymetry*

Interpolate digital data to grid. Use triangulation if data points are few, grid cell averaging if you have many points per grid cell. Check important cross-sections. Produce clear and detailed figures of the interpolated bathymetry.

11.6 Boundary conditions

An extensive discussion of schematization approaches for the boundary conditions already given in Section 6.3. Here we just list some activities for the simplest of approaches.

11.6.1 *Wave schematisation*

Select number of wave conditions by which the full climate is represented. Select criteria (longshore transport rate at some coast sections, stirring of sediment in deeper water at some locations)

Group wave conditions. Determine average of the criteria per group. For each group, select the condition for which most of the criteria match the average of the group. Try to avoid using a weight factor for a condition that differs from the probability of occurrence of the group.

Make sure that you use the right parameters for input. Don't confuse H_s with H_{rms}, T_p with T_z, T_{m01} or T_{m02}. Use appropriate relations between these parameters.

11.6.2 *Representative tide*

Select a month with an average spring/neap amplitude ratio. Run tidal model with time history output in a number of representative locations. At each point, estimate the transport averaged over 59 tidal cycles. Starting at each flow reversal, determine average transport over two consecutive tides, at each point. Select the period for which the transport matches the monthly averaged transport most closely. Generate boundary conditions for this period. Carry out a Fourier analysis over the two selected tidal periods for each of the boundary support points. Take out the diurnal and odd components, in order to get a single representative tide. *(Note: the reason for first selecting two consecutive tides and then removing the diurnal and odd components is, that in this way we avoid the situation that the mean component is polluted by the diurnal component.)*

11.6.3 *Sediment transport*

Usually, equilibrium transport is prescribed at the open boundaries, which is computed based on the local flow and wave conditions.

11.6.4 *Bottom change*

For models of tidal inlets, where the boundaries should be chosen reasonably far from the area of interest, the appropriate type of boundary condition is a fixed bed level.

11.7 Calibration

Calibration is the process of tuning all parts of the model system based on local data and common sense. Typically, the following checks are carried out:

11.8 Validation

In the validation phase, no further adjustments are made to the model. First we look at the results at the end of the calibration phase and assess how well the model is capable of simulating a number of morphological processes *simultaneously*. Even in a model that has been calibrated to its limit, it is still very well possible that some features are not represented well. The number of degrees of freedom in the morphological model is limited, and the calibration parameters and coefficients are global parameters, which are not allowed to vary over the domain.

Calibration of flow model		
Criterion	Check	Adjustments to make
Smooth flow fields without boundary disturbances	Vector plots of velocity at some points in time, vector plot of tide-averaged velocity	Type, values of boundary conditions
Smooth time series, small enough time step	Comparison of time series plots for different time steps.	Time step
Periodic solution	Time series plot	Longer initialisation, use of restart file
Matching with overall model in case of nesting	Comparison of time series of water level and velocity in collocated points.	Type of boundary conditions, location of support points
Water levels, flow velocity, total flow rates match measurements	Comparison of time series, determining rms error and phase errors	Roughness, viscosity
Wave-driven currents represented well	Vector plots showing at least five grid rows in surf zone	Grid resolution

When such data is available, a more strict validation may consist of simulating a different time period in the same area, preferably one in which different things happen. The criteria against which the model performance is judged are the same as in the calibration phase of the morphological model.

11.9 Preparing scenarios

Once the model has been calibrated and validated to satisfaction, it can be applied to evaluate relative effects of various scenarios, such as different port layouts, dredging scenarios, nourishment schemes or structures aimed at mitigating coastal erosion or improving navigability. Since these schemes are usually yet to be built, the starting bathymetry is usually the most recent one. In order to enable a clean evaluation of the effect of the scheme, a so-called T0 or reference simulation must be carried out first: a simulation starting from the latest bathymetry and covering the desired number of years.

Calibration of wave model		
Criterion	Check	Adjustments to make
No disturbances at grid boundaries	Contour plot of wave heights, plot of dissipation and wave forces	Changes nested grids and order of nested computations
Water level and velocity inside flow grid match overall values outside flow grid.	Check disturbances at boundaries flow grid in contour plot of wave heights on overall wave grid	Change overall values of water level and velocity per time point
Wave heights match measurements	Compare wave heights as function of time for measurement locations	Wave breaking parameters, bottom friction.

Calibration of transport model		
Criterion	Check	Adjustments to make
Smooth time series, small enough time step suspended transport	Time series of concentration and transport at some locations; comparison of different time steps	Time step, time interval, number of initial steps
Smooth, consistent transport fields	Vector plots at some points in the tidal cycle and residual transport	Remove errors in bathymetry or wave grids
Overall transports through some cross-sections in accordance with observations or conceptual model	Integral of residual transport through cross-sections	Change coefficients in transport formulation

Next, for each of the schemes the bathymetry must be adapted and if necessary structures have to be added to the schematisation. The simulations are then carried out sing exactly the same settings as in the reference computation.

Calibration of morphological model		
Criterion	Check	Adjustments to make
Sedimentation-erosion pattern in agreement with measurements	Contour plots of measured and computed sedimentation and erosion	Transport coefficients or formulation, wave climate, sediment parameters
Volume changes over control areas in accordance with soundings or dredging figures	Area integrals of bed level changes	As above
Cross-section shape	Compare shape, migration and area change of measured and computed cross-sections	Side slope effect, (spiral flow), transport formulation
Plan view of morphology after some years	Contour and isoline plots of measured and computed bathymetry	All of the above

11.10 Defining output

By this time, the amount of output can be reduced considerably and can be limited to outputs directly relevant to the engineering questions, and to a minimum set of standard plots which allow the modeller to ascertain that the process is running correctly.

It is important to realise at this point all the steps necessary to translate the model results to the design criteria that are important to the client. He or she may not be interested at all in nice colour pictures, but mainly in the effect of various layouts on the annual dredging volume. Make sure that you give that type of information in a clear way, besides the necessary information to substantiate it.

11.11 Running and postprocessing

Since the running of various scenarios is vary similar to the reference simulation, it pays off to automate this process by using shell scripts or batch files. Preferably, in these scripts everything is arranged from running the scenarios to making graphs and doing volume computations and integration of transports along sections. This is even useful if not too many scenarios are run, because very often errors are found at a late stage, and sometimes all runs must be redone. When this has been well organised, it can be done quickly. Additionally, these scripts offer a clear insight to the experienced user into how results were obtained. This is vital for quality checking.

It is advised to use one machine and disk section for a morphological project, and to use separate directories for separate activities, and for each run. The directory for each run can be further subdivided into input and output directories per simulated condition.

11.12 Interpretation

The end results must be interpreted carefully, with the deficiencies encountered in the validation phase in mind. Part of the interpretation consists of combining and reducing data, for instance from erosion and sedimentation patterns to volume changes over control areas, and from transport vector fields to integrated transports over a selection of cross-sections. A useful way of showing the effects of a certain scheme is by plotting on one page the bathymetry, the bottom changes in the reference run, the bottom changes for the scheme concerned, and the difference in bottom change between the reference run and the scheme concerned.

At the end, the main findings related to the various schemes to be compared must be summarised in a few tables and / or figures showing the effect of the schemes in terms of parameters relevant to the client, such as overall dredging volumes, nourishment needs depth of scour.

11.13 Reporting

Typically, the kind of reporting carried out in these studies falls into the following categories:

(1) A background report or study report in which the full model set-up is given and an extensive description is given of the whole modeling process, the choices made in schematisations and detailed explanations of the results. Such a report should explain in enough detail *how* a model was built, so that the results may be reproduced by someone else using the same model. It must explain *why* the inevitable choices were made.
(2) An executive summary (sometimes written by the client) in which the main findings are given in layman's terms and only those figures are given that are necessary to illustrate the conclusions.
(3) Sometimes a CD-ROM is provided with a large number of graphs and animations, which can be browsed using an Internet browser.

11.14 Archiving

The contents of the project directory and all subdirectories must be stored on tape and on CD-ROM's and DVD's. The latter is preferable since CD's and DVD's follow a very clear world standard, contrary to tape devices.

Chapter 12

Modeling Philosophy

12.1 Virtual reality or realistic analogue?

As we have seen in the previous chapters, process-based morphodynamic modeling of coastal areas has taken off in recent years, due to robust numerical schemes, increased computer power and clever tricks to speed up the simulations. It is allowing engineers and scientists to use models as numerical laboratories of which the conceptual handling is similar to that in physical scale models: create bathymetries, fill with sediment ('define initial sediment thickness'), turn on pumps ('discharge boundaries'), create weirs ('clamped water level boundaries') and place wave paddles ('incident spectral wave conditions') to create boundary conditions and analyze what happens in terms of evolving bed patterns and sedimentation/erosion rates. Now that we have these numerical labs, an important question is how best to use them to solve real-life problems. In this section we will discuss pros and cons of two fundamentally different approaches.

Virtual reality... The first, most common method applied today is to try and create a 'virtual reality': we reproduce in the best possible detail the geometry, sedimentology, bathymetry, 3D flow, wave and sediment processes and simulate the evolution of the sea or river bed and associated processes under various future scenarios, in order to find out likely effects of something we intend to do (deepen a channel, build a new port, reclaim land...). The success of this approach depends critically on our ability a) to prove that our model accurately represents the morphological processes going on at present and b) to prove that it is still valid when the situation changes drastically in the future. The method to validate a) is usually to carry out a hindcast over a recent period where our model results are calibrated against the observed hydrodynamics, sediment concentration patterns, sedimentation/erosion patterns and dredging quantities. Here we encounter the problem that, no matter how good our model is, it is not reality and it may not accept the initial conditions we offer it as the near-equilibrium conditions that they are. This will lead our model to quickly try to adjust to a more fitting bathymetry. This will show up in the results as sedimentation/erosion patterns that are different from the observations, which can pose serious problems, for example, if one investigates ways to optimize dredging and the model predicts erosion where most of the dredging takes place. Unless one can identify and fix the root cause for such a discrepancy, 'tuning' the model may seemingly improve the hindcast but actually worsen its longer-term behaviour, as illustrated in Figure 12.1. Of course, this pessimistic scenario need not always be true, but there is a real danger here that we need to be aware of. Taking this route we can only slowly improve the confidence in our models by making sure and proving that they contain all the relevant physics and by assessing them against case after case with as little adjustments as possible.

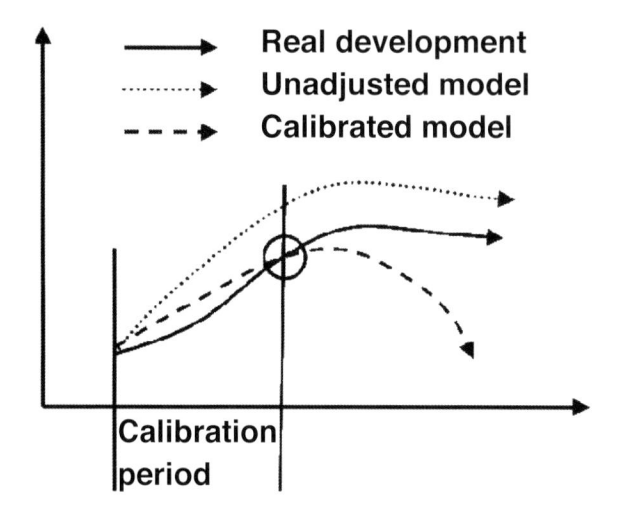

Fig. 12.1 Pessimistic scenario for effect of calibration

...or realistic analogue? In the second approach we use our models differently: we do not try to replicate the situation exactly but we use the model as an analogue to reality, with which we can investigate processes and effects in relative isolation. Instead of asking the question: 'where will this particular channel be located next year?' we may ask: 'why is there a channel here, with these particular dimensions?' We may investigate which hydrodynamic forcing creates and maintains it, in how far it is self-organized or forced by geometric constraints, how important is the geological setting, by carrying out numerical experiments where we vary such parameters. Given some necessary simplifications such experiments can be run over very long time periods, allowing us to study the equilibrium behaviour of the model for specific situations (delta formation, channel structure in tidal inlets and estuaries). Once a near-equilibrium situation is established, various integral parameters of the simulated pattern can be compared with empirical relationships. While such studies offer valuable insight into what causes different bed patterns, there is a practical application, too. Once an 'establishment scenario' has been set up and a situation has been created that shows realistic behaviour and patterns, this provides a good starting point for sensitivity studies investigating in principle the effects of, say, sea level rise, land reclamations, dredging or nourishment strategies etc.

Our aim in discussing these two approaches is not to select one or the other; rather, we believe that they complement each other. The detailed 'virtual reality' models may gain much from insight gained in the 'realistic analogue' models and can gain public acceptance when it is shown that the physics contained in them leads to realistic long-term results, whereas the accuracy of the more schematic models obviously may profit from improved process descriptions validated at detailed scale.

12.2 Process-based or data-driven?

This book has been mostly about process-based modeling. We've seen that this kind of modeling can explain a lot of morphological processes that we see, but this does not mean that we can always predict what we want to. If we take the example of predicting the behavior of longshore bars as we discussed in Chapter 7, we've shown that we can hindcast the evolution of the bars, with both onshore and offshore motions, in a realistic way. However, predicting the bar behavior into the future is another thing; it suffers from the fact that the wave climate is not deterministic and from the build up of errors over the months and years. To represent the longer-term behavior we need to carefully select the most relevant processes, be careful to maintain only the necessary adjustable coefficients and to carry out an extensive calibration. By this time, you could argue that our model has become to a large extent data-driven, so why not directly use a data-driven approach such as a (linear or nonlinear) neural network? [Pape (2010)] has compared process-based (UNIBEST) model results of long-term evolution of longshore bars with neural networks and concludes that the latter outperform the process-based model in predicting parameters such as the bar location. His claim is that this is due to the accumulation of errors in process-based models, but it may well be that it is just easier to train a neural network model than a process-based model. On the other hand, once sufficiently trained for longer-term evolution, process-based modeling can be used to predict things that are not in the training dataset, such as nourishments and coastal structures, and can still be applied, although with more uncertainty, in data-scarce environments, especially when we use the model to predict the overall behavior and the relative effects of engineering measures, rather than absolute locations of the bars.

12.3 Top-down or bottom-up?

The distinction between top-down approaches (e.g. [Stive and Wang (2003)], [Plant et al. (1999)] and bottom-up approaches is much less sharp than it seems in the kind of applications like predicting bar behavior or the evolution of tidal systems. A real bottom-up approach would mean to calibrate a model against data over a short period and then extrapolating the resulting model over years or centuries. This obviously leads to a big accumulation of errors and poor results on the long term, not only quantitatively but also qualitatively. But we can do better than that: if we examine the model behavior on longer time scales by doing many runs, we slowly find out which processes and parameters govern the longer-term trends and how to set them. In other words, we then also work from the top (the desired end result) down (to the processes and parameters needed to get there). Important success factors in this is that the model must be robust enough to survive all these runs and that the processes it contains allow enough richness of behavior.

12.4 More physics, better model?

Though it sounds reasonable that adding more physics will improve the ability to represent certain processes, the overall model performance does not always have to improve. The problem is that our representations of physical processes, especially around breaking waves and sediment transport, are not exact and every process we add will come with at least one extra uncertain coefficient. We may then end up with lots of adjustable coefficients, for which there is no clear guidance and for which it is unclear how they influence the end result. A simpler model with fewer coefficients that have a predictable effect on the outcome will then be preferable. A nice example is given by [Ruessink *et al.* (2007)] who, based on a very extensive sensitivity analysis, threw out a lot of processes in a profile model and ended up with a much more manageable set of processes and coefficients, with which they could hindcast profile evolution for different sites with considerable skill. Of course we should keep in mind the Einstein Principle, which says that "a scientific theory should be as simple as possible, but no simpler". The same holds for models.

12.5 How to judge model skill?

It is perhaps a sign of the growing maturity of morphological models that quantifying the skill of morphological simulations is becoming a serious issue. In the early days of intercomparing models in EU projects this was carefully avoided, but especially during the Coast3D project important steps were taken to develop useful metrics for our models in [Sutherland *et al.* (2004a)], which were then tested for a complex field site in [Sutherland *et al.* (2004b)]. For hydrodynamic data, usually time series of values, it is not so hard to come up with meaningful relative measures of error. Evaluating the skill in predicting bottom changes is harder. Often we see encouraging visual agreement but if we try to quantify this it turns out to be not so good. A notorious skill measure is the Brier Skill Score (BSS), which indicates if your model does better than just stating that nothing will change. Negative scores indicate that you are worse than that, a perfect score is 1. In many practical cases it is very difficult to come up with skill socres above 0, let alone to get close to 1. This should not discourage us from using this skill score; first of all, it will still tell us if model improvements actually improve results and secondly, it is no use pretending the models are better than they are. In the meantime, a bad skill score for absolute bottom changes does not have to mean the model is all bad; a slight shift in channel location can kill a BSS score. Then again, this shift might be just what the client was interested in...

12.6 Absolute vs. relative skill

Finally we come back to the question: what do we use our model for? We spend a lot of time hindcasting observed morphology changes, with varying success, but is predicting the past what we are after? These simulations mainly serve to establish whether or not our models behave realistically, with a variability in the right order of magnitude and with natural features generated or preserved in roughly the right shape. Once this is the case, we can start feeling somewhat confident in using the model for evaluating scenarios, 'what if?' questions. If the model does not look like reality, it is unlikely that it can predict even relative effects correctly, although this is often suggested. If it does, it is still no guarantee that the effect of a particular scenario we're interested in will be modeled correctly; we will need a collection of case studies on real executed projects to be able to make that judgement. That way we can slowly make progress towards having useful morpological models with known skills.

Bibliography

Allen, J., Newberger, P. and Holman, R. (1996). Nonlinear shear instabilities of alongshore currents on plane beaches, *J. Fluid Mech.* **310**, pp. 181–213.
Andrews, D. and McIntyre, M. (1978). An exact theory of non-linear waves on a lagrangian-mean flow, *J. Fluid Mech.* **89**, pp. 609–646.
Apotsos, A., Raubenheimer, B., Elgar, S. and Guza, R. (2008). Testing and calibrating parametric wave transformation models on natural beaches, *J. Geophys. Res.* **90**, pp. 9159–9167.
Arcilla, A. S., Roelvink, J. A., O'Connor, B. A., Reniers, A. and A.Jimenez, J. (1994). The delta flume '93 experiment, in *Proc. Coastal Dynamics* (Barcelona, Spain), pp. 488–502.
Ardhuin, F., Herbers, T., Jessen, P. and O'Reilly, W. (2003). Swell transformation across the continental shelf, part 2: Validation of a spectral energy balance equation, *J. Phys. Oceanogr.* **33**, pp. 1940–1953.
Ardhuin, F., Rascle, N. and Belibassakis, K. (2008). Explicit wave-averaged primitive equations using a generalized lagrangian mean, *Ocean Modelling* **20**, pp. 35–60.
Ashton, A. and Murray, B. (2006). High-angle wave instability and emergent shoreline shapes: 1. Modeling of sand waves, flying spits, and capes, *J. Geophys. Res.* **111**, F04011.
Bailard, J. (1981). An energetics total load sediment transport model for a plane sloping beach, *J. Geophys. Res.* **86**, pp. 10938–10954.
Bakker, W. and De Vroeg, J. (1988). Is de kust veilig? (is the coast safe?, in dutch), Tech. Rep. GWAO 88.017, Rijkswaterstaat.
Baldock, T., Holmes, P., Bunker, S. and van Weert, P. (1998). Cross-shore hydrodynamics within an unsaturated surf zone, *Coastal Eng.* **34**, pp. 173–196.
Barber and Ursell (1948). Propagation of ocean waves and swell, 1. wave periods and velocities, *Phil. Trans. of the Royal Soc. of London* **240**, 824, pp. 527–560.
Battjes, J. (1975). Modeling of turbulence in the surf zone, in *Proc. Symp. Model Techniques* (San Francisco, USA), pp. 1050–1061.
Battjes, J., Bakkenes, H., Janssen, T. and van Dongeren, A. (2004). Shoaling of subharmonic gravity waves, *J. Geophys. Res.* **109**, pp. 1–15.
Battjes, J. and Janssen, P. (1978). Energy loss and set-up due to breaking of random waves, in *Proc. 16th Int. Conf. Coastal Eng.* (Hamburg, Germany), pp. 569–587.
Battjes, J. and Stive, M. (1985). Calibration and verification of a dissipation model for random breaking waves, *Coastal Eng.* **55**, pp. 224–235.
Beji, S. and Battjes, J. (1993). Experimental investigation of wave propagation, *Coastal Eng.* **19**, pp. 151–162.

Benoit, M., Frigaard, P. and Schaffer, H. (1997). Analysing multidirectional wave spectra: A tentative classification of available methods, in *IAHR Seminar Multidirectional waves and their interaction with structures* (San Francisco, USA), pp. 131–158.

Biesel, F. (1952). Equations generales au second ordre de la houle irreguliere, *Houille Blanche* **7**, pp. 372–376.

Bijker, E. (1988). Some considerations about scales for coastal models with movable bed, Tech. Rep. Publication No. 50, Delft Hydraulics Laboratory.

Bouws, E. and Battjes, J. (1982). A Monte-Carlo approach to the computation of refraction of water waves, *J. of Geophys. Res.* **87**, pp. 5718–5722.

Bowen, A. and Holman, R. (1989). Shear instabilities of the mean longshore current, 1. Theory, *J. of Geophys. Res.* **94**, pp. 18023–18030.

Brevik, I. and Aas, B. (1980). Fume experiments on waves and currents, 1. Rippled bed. *Coastal Eng.* **3**, pp. 149–177.

Broker-Hedegaard, I., Deigaard, R. and Fredsoe, J. (1991). Onshore/offshore sediment transport and morphological modeling of coastal profiles, in *Proc. ASCE Specialty Conf. Coastal Sediments* (Seattle, USA), pp. 643–657.

Broker-Hedegaard, I., Roelvink, J., Southgate, H., Pechon, P., Nicholson, J. and Hamm, L. (1992). Intercomparison of coastal profile models, in *Proc. 23rd Int. Conf. Coastal Eng.* (Venice, Italy), pp. 2108–2121.

Brown, J., MacMahan, J., Reniers, A. and Thornton, E. (2009). Surf zone diffusivity on a rip-channeled beach, *J. of Geophys. Res.* **114**, pp. 1–20.

Buijsman, M., Ruggiero, P. and Kaminsky, G. (2001). Sensitivity of shoreline change predictions to wave climate variability along the southwest Washington coast, usa, in *Proc. Coastal Dynamics* (Lund, Sweden), pp. 617–626.

Calvete, D., Cocco, G., Falques, A. and Dodd, N. (2007). (un)predictability in rip channel systems, *Geophys. Res. Letters* **34**, L05605.

Castelle, B., Ruessink, B. G., Bonneton, P., Marieu, V., Bruneau, N. and Price, T. D. (2010). Coupling mechanisms in double sandbar systems. Part 2: Impact on alongshore variability of inner-bar rip channels, *Earth Surface Processes and Landforms* **35**, 7, pp. 771–781.

Chawla, A. and Kirby, J. (2002). Monochromatic and random wave breaking at blocking points, *J. of Geophys. Res.* **107**, C7.

Cooley, J. W. and Tukey, J. W. (1965). An algorithm for the machine calculation of complex fourier series, *Math. Comput.* **19**, pp. 297–301.

Dally, W. and Dean, R. (1984). Suspended sediment transport and beach profile evolution. *J. Waterw. Port Coastal Eng.* **110**, 1, pp. 15–33.

Dalrymple, R., MacMahan, J., Reniers, A. and Nelko, V. (2011). Rip currents, *Ann. Review of Fluid Mech.* **43**, pp. 551–581.

Dalrymple, R. A. (1975). A mechanism for rip current generation on an open coast, *J. of Geophys. Res.* **80**, C24, pp. 3485–3487.

Damgaard, J., Dodd, N., Hall, L. and Chesher, T. (2002). Morphodynamic modelling of rip channel growth, *Coastal Eng.* **45**, 3-4, pp. 199–221.

Dastgheib, A., Roelvink, J. and Van der Wegen, M. (2009). Effect of different sediment mixtures on the long term morphological simulation of tidal basins, in *Proc. RCEM* (Santa Fe, Argentina), pp. 913–918.

Dastgheib, A., Roelvink, J. and Wang, Z. (2008). Long-term process-based morphological modeling of the Marsdiep tidal basin. *Marine Geology* **256**, pp. 90–100.

De Jong, H. and Gerritsen, F. (1984). Stability parameters of Western Scheldt estuary, in *Proc. 19nd Int. Conf. Coastal Eng.* (Houston, USA), pp. 3078–3093.

De Vriend, H. and Kitou, N. (1990). Incorporation of wave effects in a 3d hydrostatic mean current model, in *Proc. 22nd Int. Conf. Coastal Eng.* (Delft, The Netherlands), pp. 855–865.

De Vriend, H., Zyserman, J., Nicholson, J., Roelvink, J., Pechon, P. and Southgate, H. (1993). Medium-term 2dh coastal area modelling, *Coastal Eng.* **21**, pp. 193–224.

De Vriend, T., H. J.and Louters, Berben, F. M. L. and Steijn, R. C. (1989). Hybrid prediction of a sandy shoal evolution in an estuary, in *Proc. Int. Conf. Hydraulic and Environmental Modeling of Coastal, Estuarine and Rivers* (Bradford,UK), pp. 145–156.

Dean, R. (1973). Heuristic models of sand transport in the surf zone, in *Proc. of Conference on Engineering Dynamics in the Surf Zone* (Sydney, Australia Institution of Engineers), pp. 208–214.

Dean, R. (1992). Beach nourishment: Design principles, in *Short Course on Design and Reliability of Coastal structures attached to the 23th Int. Conf. Coastal Eng.*

Dean, R. and Dalrymple, R. (1991). *Water wave mechanics* (World Scientific, Singapore).

Deigaard, R. (1993). A note on the three dimensional shear stress distribution in the surf zone, *Coastal Eng.* **20**, pp. 157–171.

Deigaard, R., Christensen, E., Damgaard, J. and Fredsoe, J. (1994). Numerical simulation of finite amplitude shear waves and sediment transport, in *Proc. 24th Int. Conf. Coastal Eng.* (Kobe, Japan), pp. 1919–1933.

Deigaard, R. and Fredsoe, J. (1989). Shear stress distribution in dissipative water waves, *Coastal Eng.* **13**, pp. 357–387.

Dingemans, M. (1987). Verification of numerical wave propagation models with laboratory measurements : Hiswa verification in the directional wave basin, Tech. Rep. H228, Waterloopkundig Laboratorium.

Dingemans, M. (1997). *Water wave propagation over uneven bottoms* (World Scientific, Singapore).

Dingemans, M., Radder, A. and de Vriend, H. (1987). Computations of the driving forces of wave-induced currents, *Coast. Eng.* **11**, pp. 539–563.

Dissanayake, D., Roelvink, J. and van der Wegen, M. (2009). Modelled channel patterns in a schematized tidal inlet. *Coastal Eng.* **56**, pp. 1069–1083.

Dodd, N., Blondeaux, P., Calvete, D., de Swart, H., Falques, A., Hulscher, S., Rozynski, G. and Vittori, G. (2003). Understanding coastal morphodynamics using stability methods, *J. of Coast. Res.* **19**, 4, pp. 849–865.

Dodd, N., Iranzo, V. and Reniers, A. (2000). Alongshore-current shear waves, *Rev. of Geophys.* **38**, 4, pp. 16075–463.

Dodd, N. and Thornton, E. (1992). Longshore current instabilities: growth to finite amplitude, in *Proc. 23rd Int. Conf. Coastal Eng.* (Venice, Italy), pp. 2655–2668.

Dodd, N. and Thornton, E. (1993). Growth and energetics of shear waves in the nearshore, *J. Geophys. Res.* **95**, pp. 16075–16083.

Donelan, M., Haus, B., Reul, N., Plant, W., Stiassnie, M., Graber, H., Brown, O. and Saltzman, E. (2004). On the limiting aerodynamic roughness of the ocean in very strong winds, *Geophys. Res. Letters* **31**, L18306.

Drake, T. and Calantoni, J. (2001). Discrete particle mode for sheet flow sediment transport in the nearshore, *J. Geophys. Res.* **106**, C9, pp. 19859–19868.

Dronen, N. and Deigaard, R. (2007). Quasi-three-dimensional modelling of the morphology of longshore bars, *J. Geophys. Res.* **54**, 3, pp. 197–215.

Eckart, C. (1951). Surface waves on water of variable depth, Tech. Rep. Wave Report 100, S10 Ref 51-12, Scripps Institution of Oceanography.

Eldeberky, Y. and Battjes, J. (1995). Parameterization of triad interactions in wave energy models, in *Proc. Coastal Dynamics* (Gdansk, Poland), pp. 140–148.

Elgar, S. and Guza, R. (1985). Observation of bispectra of shoaling surface gravity waves, *J. of Fluid Mech.* **167**, pp. 425–448.

Elgar, S., Raubenheimer, R. G. B. and Gallagher, E. (1997). Spectral evolution of shoaling and breaking waves on a barred beach, *J. Geophys. Res.* **102**, pp. 15797–15805.

Elias, E., Gelfenbaum, G., van Ormondt, M. and Moritz, H. (2011). Predicting sediment transport patterns at the mouth of the Columbia River, in *Proc. Coastal Sediments* (Miami, USA).

Elias, E., Walstra, D., Roelvink, J., Stive, M. and Klein, M. (2000). Hydrodynamc validation of Delft3D with field measurements at Egmond, in *Proc. 27th Int. Conf. Coastal Eng.* (Sydney, Australia).

Escoffier, F. (1940). The stability of tidal inlets, *Shore and Beach* **8**, 4, pp. 114–115.

Eysink, W. (1990). Morphologic response of tidal basins to changes, in *Proc. 22nd Int. Conf. Coastal Eng.* (Delft, the Netherlands), pp. 1948–1961.

Falques, A. and Calvete, D. (2004). Large scale dynamics of sandy coastlines. diffusitivy and instability, *J. Geophys. Res.* **101**, C03007.

Falques, A., Coco, G. and Huntley, D. (2000). A mechanism for the generation of wave-driven rhythmic patterns in the surf zone, *J. Geophys. Res.* **105**, pp. 24071–24088.

Falques, A., Iranzo, V. and Caballeria, M. (1994). Shear instability of longshore currents: Effects of dissipation and non-linearity, in *Proc. 24th Int. Conf. Coastal Eng.* (Kobe, Japan), pp. 1983–1997.

Feddersen, F., Guza, R., Elgar, S. and Herbers, T. (2000). Velocity moments in alongshore bottom shear stress parameterizations, *J. Geophys. Res.* **105**, pp. 8673–8688.

Foster, D., Bowen, A., Holman, R. and Natoo, P. (2006). Field evidence of pressure gradient induced incipient motion, *J. Geophys. Res.* **111**, C05004.

Fowler, R. E. and Dalrymple, R. A. (1990). Wave group forced nearshore circulation, in *Proc. 22nd Int. Conf. Coastal Eng.* (Delft, the Netherlands), pp. 729–742.

Fredsoe, J. (1984). Turbulent boundary layer in wave-current motion, *J. Hydr. Engrg.* **110**, 1103, pp. 1103–1120.

Fredsoe, J. and Deigaard, R. (1992). *Mechanics of coastal sediment transport* (World Scientific, Singapore).

Friedrichs, C. (1995). Stability shear stress and equilibrium cross-sectional geometry of sheltered tidal channels, *J. Coastal Res* **11**, 4, pp. 1062–1074.

Friedrichs, C. and Aubrey, D. G. (1988). Non-linear tidal distortion in shallow well-mixed estuaries: a synthesis, *Estuarine, Coastal and Shelf Sciences* **27**, 2, pp. 521–545.

Galappatti, R. and Vreugdenhil, C. (1985). A depth integrated model for suspended transport, *J. Hydraul. Res.* **23**, 4, pp. 359–377.

Gallagher, E., Elgar, S. and Guza, R. (1996). Observations of sand bar evolution on a natural beach, *J. of Geophys. Res.* **103**, pp. 3203–3215.

Garcez Faria, A., Thornton, E., Lippmann, T. and Stanton, T. (2000). Undertow over a barred beach, *J. of Geophys. Res.* **105**, pp. 16999–17010.

Gelci, R., Cazale, H. and Vassal, J. (1956). Utilization des diagrammes de propagation'a la prevision energetique de la houle, *Bulletin dinformation du Comite Central doceanographie et detudes des cotes* **8**, 4, pp. 160–197.

Gonzalez-Rodriguez, D. and Madsen, O. (2007). Seabed shear stress and bedload transport due to asymmetric and skewed waves, *Coastal Eng.* **54**, 12, pp. 914–929.

Grant, W. and Madsen, O. (1978). Combined wave and current interaction with a rough bottom, *J. Geophys. Res.* **84**, C4, pp. 1979–1808.

Grunnet, N. M., Ruessink, B. and Walstra, D. (2005). The influence of tides, wind and waves on the redistribution of nourished sediment at Terschelling, The Netherlands. *Coastal Eng.* **52**, 7, pp. 617–631.

Guza, R. and Thornton, E. (1985). Velocity moments in nearshore, *J. Waterway, Port, Coastal Ocean Eng.* **111**, 2, pp. 235–256.

Haller, M. C., Putrevu, U., Oltman-Shay, J. and Dalrymple, R. A. (1999). Wave group forcing of low frequency surf zone motion, *Coastal Eng.* **41**, 2, pp. 121–136.

Hasselmann, K. (1962). On the non-linear transfer in a gravity wave spectrum, part 1. general theory, *J. Fluid Mech.* **12**, pp. 481–500.

Hasselmann, K. (1970). Wave-driven inertial oscillation, *Geophys. Fluid Dyn.* **1**, pp. 463–502.

Hasselmann, K., Barnett, T., E.Bouws, H.Carlson, D.E.Cartwright, Enke, K., Ewing, J., Gienapp, H., Hasselmann, D., Kruseman, P., Meerburg, A., Miller, P., Olbers, D., Richter, K., Sell, W. and Walden, H. (1973). Measurements of wind-wave growth and swell decay during the Joint North Sea Wave Project (JONSWAP)', *Ergnzungsheft zur Deutschen Hydrographischen Zeitschrift Reihe* **8**, 12, p. 95.

Henderson, S. M., Allen, J. and Newberger, P. (2004). Nearshore sandbar migration predicted by an eddy-diffusive boundary layer model, *J. Geophys. Res.* **109**, C06024.

Herbers, T., Elgar, S. and Guza, R. (1994). Infragravity-frequency (0.005-0.05 hz) motions on the shelf, part 1, Forced waves, *J. Phys. Oceanogr.* **24**, pp. 917–927.

Herbers, T., Elgar, S. and Guza, R. (1995). Generation and propagation of infragravity waves, *J. Geophys. Res.* **100**, C12, pp. 24863–24872.

Hibma, A., De Vriend, H. and Stive, M. (2003). Numerical modelling of shoal pattern formation in well-mixed elongated estuaries. *Estuarine, Coastal and Shelf Science* **57**, pp. 981–991.

Hino, M. (1974). Theory on the formation of rip-current and cuspidal coast, in *Proc. 14th Int. Conf. Coastal Eng.* (Brisbane, Australia), pp. 901–911.

Hjelmfelt, A. and Lenau, C. (1970). Nonequilibrium transport of suspended sediment, *J. Hydraul. Div.* **96**, (HY7), pp. 1567–1586.

Hoefel, F. and Elgar, S. (2003). Wave-induced sediment transport and sandbar migration, *Science* **299**, 1, pp. 1885–1887.

Hoitink, A., Hoekstra, P. and van Maren, D. (2003). Flow asymmetry associated with astronomical tides: Implications for the residual transport of sediment, *J. Geophys. Res.* **108**, C10.

Holman, R. and Bowen, A. (1982). Bars, bumps and holes: Models for the generation of complex beach topography. *J. of Geophys. Res.* **87**, pp. 457–468.

Holman, R. and Stanley, J. (2007). The history and technical capabilities of argus, *Coastal Eng.* **54**, pp. 477–491.

Holman, R. A., Symonds, G., Thornton, E. B. and Ranasinghe, R. (2006). Rip spacing and persistence on an embayed beach, *J. of Geophys. Res.* **111**, C01006.

Holthuijsen, L. (2007). *Waves in oceanic and coastal waters* (Cambridge University Press, Cambridge).

Holthuijsen, L., Booij, N. and Herbers, T. (1989). A prediction model for stationary, short-crested waves in shallow water with ambient currents, *Coastal Eng.* **13**, pp. 23–54.

Hume, T. and Herdendorf, C. (1993). On the use of empirical stability relationships for characterising estuaries, *Journal of Coastal Res.* **9**, 2, pp. 413–422.

Huntley, D., Guza, R. and Thornton, E. (1984). Field observations of surf beat. 1. Progressive edge waves, *J. Geophys. Res.* **86**, C7, pp. 6451–6466.

Ilic, S., van der Westhuysen, A., Roelvink, J. and Chadwick, A. (2007). Multidirectional wave transformation around detached breakwaters, *Coastal Eng.* **54**, 10, pp. 775–789.

Janssen, C. M., Hassan, W. N., v. d. Wal, R. and Ribberink, J. S. (1998). Grain-size influence on sand-transport mechanisms, in *Proc. Coastal Dynamics* (Plymouth, UK), pp. 58–67.

Janssen, T. and Battjes, J. (2007). A note on wave energy dissipation over steep beaches, *Coastal Eng.* **54**, 9, pp. 711–716.

Janssen, T., Battjes, J. and van Dongeren, A. (2003). Long waves induced by short-wave groups over a sloping bottom, *J. Geophys. Res.* **108**, C8.

Janssen, T., Herbers, T. and Battjes, J. (2006). Generalized evolution equations for non-linear surface gravity waves over two-dimensional topography, *J. Fluid Mech.* **552**, pp. 393–418.

Jarrett, J. (1976). Tidal prism-inlet relationships, Gen. Invest. tidal inlets rep. 3, Tech. rep., AUS Army Coastal Engineering and Research Centre, Fort Belvoir, Va.

Johnson, D. and Pattiaratchi, C. (2006). Boussinesq modelling of transient rip currents, *Coastal Eng.* **53**, 5-6, pp. 419–439.

Komar, P. (1976). *Beach Processes* (World Scientific, Singapore).

Koster, L. (2006). *Humplike nourishing of the shoreface; A study on more efficient nourishing of the shoreface*, Master's thesis, Delft University of Technology, Delft, The Netherlands.

Kraus, N. (1998). Inlet cross-section area calculated by process-based model, in *Proc. 26th Int. Conf. Coastal Eng.*, pp. 3265–3278.

Latteux, B. (1995). Techniques for longterm morphological simulation under tidal action, *Marine Geology* **126**, pp. 129–141.

Leatherman, S., Williams, A. and Fisher, J. (1977). Overwash sedimentation associated with a large-scale northeaster, *Marine Geology* **24**, pp. 109–121.

LeConte, L. (1905). Discussion on the paper, notes on the improvement of river and harbor outlets in the united states by D. A. watt, paper no. 1009, *Trans. ASCE* **55**, pp. 306–308.

Lentz, S., Fewings, M., Howd, P., Fredericks, J. and Hathaway, K. (2008). Observations and a model of undertow over the inner continental shelf, *J. Phys. Oceanogr.* **38**, pp. 2341–2357.

Lesser, G. (2009). *An approach to medium-term coastal morphological modelling*, Ph.D. thesis, Delft Univ. of Technology, Delft.

Lesser, G., Roelvink, J., van Kester, J. and Stelling, G. (2004). Development and validation of a three-dimensional morphological model, *Coastal Eng.* **51**, 8-9, pp. 883–915.

Lippmann, T. C. and Holman, R. A. (1989). Quantification of sand bar morphology: a video technique based on wave dispersion, *J. Geophys. Res.* **94**, pp. 995–1011.

Long, J. W. and Özkan-Haller, H. T. (2009). Low frequency characteristics of wave group-forced vortices, *Journal Geophysical Research* **114**, C08004, pp. 1–21.

Long, W., Kirby, J. T. and Hsu, T. (2006). Cross shore sandbar migration predicted by a time domain boussinesq model incorporating undertow, in *Proc. 30th Int. Conf. Coastal Eng.* (San Diego, USA), pp. 2655–2667.

Longuet-Higgins, M. (1953). Mass transport in water waves, *Trans. Royal Soc. London Ser.* **A.**, 245, pp. 535–581.

Longuet-Higgins, M. (1960). Mass transport in the boundary layer at a free oscillating surface, *J. Fluid Mech.* **8**, pp. 293–305.

Longuet-Higgins, M. (1970). Longshore currents generated by obliquely incident sea waves, *J. Geophys. Res.* **76**, pp. 6778–6801.

Longuet-Higgins, M. and Stewart, R. (1962). Radiation stress and mass transport in surface gravity waves with application to surf beats, *J. Fluid Mech.* **29**, pp. 481–504.

Longuet-Higgins, M. and Stewart, R. (1964). Radiation stresses in water waves: a physical discussion, with applications, *Deep Sea Res.* **11**, pp. 529–562.

Longuet-Higgins, M. and Turner, J. (1974). An entrainment plume model of a spilling breaker, *J. Fluid Mech.* **63**, pp. 1–20.

Lygre, A. and Krogstad, H. (1986). Maximum entropy estimation of the directional distribution in ocean wave spectra, *J. Phys. Oceanogr.* **16**, pp. 2052–2060.

MacMahan, J., Brown, J., Brown, J., Thornton, E., Reniers, A., Stanton, T., Henriquez, M., Gallagher, E., Morrison, J., Austin, M., Scott, T. and Senechal, N. (2010a). Mean lagrangian flow behavior on an open coast rip-channeled beach: A new perspective, *Marine Geology* **268**, 1-4, pp. 1–15.

MacMahan, J., Thornton, E., Stanton, T. and Reniers, A. (2005). Ripex: Observations of a rip current system, *Marine Geology* **218**, 1-4, pp. 113–134.

MacMahan, J. H., Reniers, A. and Thornton, E. (2010b). Vortical surf zone velocity fluctuations with O(10) min period, *J. Geophys. Res.* **115**, C06007.

Madsen, O. (1975). Stability of a sand bed under breaking waves, in *Proc. 14th Int. Conf. Coastal Eng.* (Copenhagen, Denmark), pp. 776–794.

Madsen, O. (1978). A note on mass transport in deep water waves, *J. Phys. Oceanogr.* **8**, 6, pp. 1009–1015.

Marciano, R., Wang, Z., Hibma, A., de Vriend, H. and Defina, A. (2005). Modelling of channel patterns in short tidal basins, *J. Geophys. Res.* **100**, F01001.

McCall, R. (2008). *The longshore dimension in dune overwash modelling. Development, verification and validation of XBeach*, Master's thesis, Delft University of Technology, Delft, The Netherlands.

McCall, R., van Thiel de Vries, J., Plant, N., van Dongeren, A., Roelvink, J., Thompson, D. and Reniers, A. (2010). Two-dimensional time dependent hurricane overwash and erosion modeling at Santa Rosa Island, *Coastal Eng.* **57**, 7, pp. 1–18.

McWilliams, J., Restrepo, J. and Lane, E. (2004). An asymptotic theory for the interaction of waves and currents in coastal waters, *J. Fluid Mech.* **511**, pp. 135–178.

Mei, C. (1989). *The applied dynamics of ocean surface waves* (World Scientific, Singapore).

Mei, C. and Benmoussa, C. (1984). Long waves induced by short wave groups, *J. Fluid Mech.* **139**, pp. 219–235.

Mellor, G. (2003). The three-dimensional current and surface wave equations, *J. Phys. Oceanography* **33**, pp. 1978–1989.

Melville, W. and Matusov, P. (2002). Distribution of breaking waves at the ocean surface, *Letters to Nature* **417**, pp. 58–63.

Meyer-Peter, E. and Muller, R. (1948). Formulas for bed-load transport, in *Proceedings of the 2nd Meeting of the International Association for Hydraulic Structures Research* (Delt, The Netherlands), pp. 39–64.

Miche, R. (1944). Mouvements ondulatoires des ners en profondeur constante ou decroissante, in *Annales des Ponts et chausses*, pp. 25–78, 131–164,270–292,369–406.

Miles, J. (1957). On the generation of surface waves by shear flows, *J. Fluid Mech.* **3**, pp. 185–204.

Miles, M. and Funke, E. (1989). A comparison of methods for synthesis of directional seas, *J. Offshore Mech. Polar Eng* **111**, pp. 43–48.

Monismith, S., Cowen, E., Nepf, H., Magnaudet, J. and Thais, L. (2007). Laboratory observations of mean flows under surface gravity waves, *J. Fluid Mech* **573**, 111, pp. 131–147.

Munk, W. H. (1949). Surf beats, *Trans. Am. Geophys. Union* **30**, pp. 849–854.

Nadaoka, K. and Yagi, H. (1993). A turbulent flow modelling to simulate horizontal large eddies in shallow water, *Adv. Hydrosci. Eng.* **1B**, pp. 356–365.

Nairn, R., Roelvink, J. and Southgate, H. (1990). Transition zone width and applications for modeling surfzone hydrodynamics, in *Proc. 22nd Int. Conf. Coastal Eng.* (New York, USA), pp. 68–81.

Newberger, P. and Allen, J. (2007). Forcing a three-dimensional, hydrostatic primitive-equation model for application in the surf zone, part 1: Formulation, *J. Geophys. Res.* **112**, C08018.

Nicholson, J., Broker, I., Roelvink, J., Price, D., Tanguy, J. and Moreno., L. (1997). Intercomparison of coastal area morphodynamic models, *Coast. Eng.* **31**, pp. 97–123.

Nielsen, P. (1992). *Coastal bottom boundary layers and sediment transport* (World Scientific, Advanced series on ocean engineering).

Nielsen, P. and Callaghan, D. (2003). Shear stress and sediment transport calculations for sheet flow under waves, *Coast. Eng.* **47**, pp. 347–354.

Nishi, R. and Kraus, N. (1996). Mechanism and calculation of sand dune erosion by storms, in *Proc. 25th Int. Conf. Coastal Eng.*, pp. 3034–3047.

O'Brien, M. (1931). Estuary and tidal prisms related to entrance areas, *Civil Eng.* **1**, 8, pp. 738–739.

O'Brien, M. (1969). Equilibrium flow areas of inlets on sandy coasts, *J. Waterway, Port, Coastal and Ocean Eng.* **95**, 1, pp. 43–52.

Oltman-Shay, J., Howd, P. and Birkemeier, W. (1989). Shear instabilities of the mean longshore current, 2. Field observations, *J. Geophys. Res.* **94**, C12, pp. 18031–18042.

Orzech, M. D., Thornton, E. B., MacMahan, J. H., O'Reilly, W. C. and Stanton, T. P. (2010). Alongshore rip channel migration and sediment transport, *Marine Geology* **271**, 3-4, pp. 278–291.

Overton, M. and Fisher, J. (1988). Laboratory investigation of dune erosion, *J. of Waterway, Port, Coastal and Ocean Eng.* **114**, 3, pp. 367–373.

Özkan-Haller, H. and Kirby, J. (1999). Non-linear evolution of shear instabilities of the longshore current: A comparison of observations and computations, *J. Geophys. Res.* **104**, pp. 25953–25984.

Pape, L. (2010). *Predictability of nearshore sandbar behavior*, Ph.D. thesis, University of Utrecht, Utrecht.

Pawka, S. (1983). Island shadows in wave directional spectra, *J. Geophys. Res.* **88**, C4, pp. 2579–2591.

Pelnard-Considere (1954). Essai de theory de l'evolution des formes de rivage en plages de sable et de galets, *Quatrieme Journees de l'Hydraulique, Les Energies de la Mer* **3**, pp. 289–298.

Peregrine, D. (1976). Interaction of water waves and currents, *Adv. Appl. Mech.* **16**, pp. 9–117.

Phillips, O. (1957). On the generation of waves by turbulent wind, *J. Fluid Mech.* **2**, pp. 417–445.

Phillips, O. (1977). *The dynamics of the upper ocean* (Cambridge University Press, Cambridge).

Plant, N., Edwards, K., Kaihatu, J., Veeramony, J., Hsu, L. and Holland, K. (2009). The effect of bathymetric filtering on nearshore process model results, *Coastal Eng.* **56**, 4, pp. 484–493.

Plant, N., Holman, R. and Freilich, M. (1999). A simple model for interannual sandbar behavior, *J. Geophys. Res.* **104**, C7, pp. 15,755–15,776.

Powell, M., Thieke, R. and Mehta, A. (2006). Morphodynamic relationships for ebb and flood delta volumes at floridas entrances, *Ocean Dynamics* **56**, pp. 295–307.

Powell, M., Vickery, P. and Reinhold, T. (2003). Reduced drag coefficients for high wind speeds in tropical cyclones, *Nature* **422**, pp. 279–283.

Putrevu, U. and Svendsen, I. (1992). Shear instability of longshore currents: A numerical study, *J. Geophys. Res.* **97**, pp. 7283–7303.

Rakha, K. A. (1998). A quasi-3d phase-resolving hydrodynamic and sediment transport model, *Coastal Eng.* **34**, 3-4, pp. 277–311.

Ranasinghe, R., Symonds, G., Black, K. and Holman, R. (2000). Processes governing rip spacing, persistence, and strength in a swell dominated, microtidal environment, in *Proc. 27th Int. Conf. Coastal Eng.* (Sydney, Australia), pp. 454–467.

Ranasinghe, R., Symonds, G., Black, K. and Holman, R. (2004). Morphodynamics of intermediate beaches: a video imaging and numerical modelling study, *Coastal Eng.* **51**, pp. 629–655.

Ranasinghe, R., Turner, I. and Symonds, G. (2006). Shoreline response to multi-functional artificial surfing reefs: A numerical and physical modelling study, *Coastal Eng.* **53**, 7, pp. 589–611.

Raubenheimer, B. and Guza, R. (1996). Observations and predictions of run-up, *J. of Geophys. Res.* **101**, C10, pp. 25575–25587.

Renger, E. and Partenscky, H. W. (1974). Stability criteria for tidal basins, in *Proc. 14th Int. Conf. Coastal Eng.*, pp. 1605–1618.

Reniers, A. and Battjes, J. (1997). A laboratory study of longshore currents over barred and non-barred beaches, *Coastal Eng.* **30**, pp. 1–22.

Reniers, A., Battjes, J., Falques, A. and Huntley, D. (1997). A laboratory study on the shear instability of longshore currents, *J. Geophys. Res.* **102**, C4, pp. 8597–8609.

Reniers, A., Groenewegen, M., Ewans, K., Masterton, S., Stelling, G. and Meek, J. (2010a). Estimation of infragravity waves at intermediate water depth, *Coastal Eng.* **57**, pp. 52–61.

Reniers, A., MacMahan, J., Beron-Vera, F. and Olascoaga, J. (2010b). Rip-current pulses tied to lagrangian coherent structures, *Geophys. Res. Letters* **37**, L05605.

Reniers, A., MacMahan, J., Thornton, E. and Stanton, T. (2007). Modeling of very low frequency motions during RIPEX, *J. Geophys. Res.* **112**, C07013.

Reniers, A., MacMahan, J., Thornton, E., Stanton, T., Henriquez, M., Brown, J., Brown, J. and Gallagher, E. (2009). Surf zone retention on a rip-channeled beach, *J. Geophys. Res.* **114**, C10010.

Reniers, A., Roelvink, J. and Thornton, E. (2004a). Morphodynamic modeling of an embayed beach under wave group forcing, *J. Geophys. Res.* **109**, C01030.

Reniers, A., Symonds, G. and Thornton, E. (2001). Modelling of rip-currents during RDEX, in *Proc. Coastal Dynamics* (Lund, Sweden), pp. 493–499.

Reniers, A., Thornton, E., Stanton, T. and Roelvink, J. (2004b). Vertical flow structure during sandy duck, *Coastal Eng.* **51**, pp. 237–260.

Reniers, A., van Dongeren, A., Battjes, J. and Thornton, E. (2002). Linear modeling of infragravity waves during delilah, *J. Geophys. Res.* **107**, C10.

Ribberink, J. (1998). Bed-load transport for steady flows and unsteady oscillatory flows, *Coastal Eng.* **34**, 1-2, pp. 59–82.

Ribberink, J. and Chen, Z. (1993). Sediment transport of fine sand in asymmetric flow, Tech. Rep. Delft Hydraulics Report, Publ. H840, Part VII, WL-Delft Hydraulics.

Rienecker, M. and Fenton, J. (1981). A fourier approximation method for steady water waves, *J. Fluid. Mech* **104**, pp. 119–137.

Rijkswaterstaat (1988). Handboek zandsuppleties (in dutch, beach nourishment manual), Tech. rep.

Rivero, F. and Arcilla, A. (1995). On the vertical distribution of $< \tilde{u}\tilde{v} >$, *Coastal Eng.* **25**, pp. 137–152.

Rodi, W. (1984). Turbulence models and their application in hydraulics, in *State-of-the-art paper article sur letat de connaissance. Paper presented by the IAHR-Section on Fundamentals of Division II: Experimental and Mathematical Fluid Dynamics* (The Netherlands).

Roelvink, J. (1993). Dissipation in random wave groups incident on a beach, *Coastal Eng.* **19**, pp. 127–150.

Roelvink, J. (2006). Coastal morphodynamic evolution techniques, *Coastal Eng.* **53**, 2-3, pp. 277–287.

Roelvink, J. and Broker, I. (1993). Coastal profile models, *Coastal Eng.* **21**, pp. 163–191.

Roelvink, J., Meijer, T., Houwman, K., Bakker, R. and Spanhoff, R. (1995). Field validation and application of a coastal profile model, in *Proc. Coastal Dynamics* (Gdansk, Poland), pp. 818–828.

Roelvink, J. and Reniers, A. (1995). LIP 11D delta flume experiments, Tech. Rep. H2130, WL-Delft Hydraulics, Delft, The Netherlands.

Roelvink, J., Reniers, A., van Dongeren, A., van Thiel de Vries, J., McCall, R. and Lescinski, J. (2009). Modeling storm impacts on beaches, dunes and barrier islands, *Coastal Eng.* **56**, pp. 1133–1152.

Roelvink, J. and Stive, M. (1989). Bar generating cross-shore flow mechanisms on a beach, *J. Geophys. Res.* **94**, C4, pp. 4485–4800.

Roelvink, J. and Walstra, D. (2004). Keeping it simple by using complex models, in *Advances in Hydro-Science and Engineering* (Brisbane, Australia), pp. 1–11.

Ruessink, B., Kuriyama, Y., Reniers, A., Roelvink, J. and Walstra, D. (2007). Modeling cross-shore sandbar behavior on the timescale of weeks. *J. Geophys. Res.* **112**, F03010.

Ruessink, B., Miles, J., Feddersen, F., Guza, R. and Elgar, S. (2001). Modeling the alongshore current on barred beaches, *J. Geophys. Res.* **106**, C10, pp. 22451–22463.

Ruessink, B. and van Rijn, L. (2011). Observations and empirical modelling of near-bed skewness and asymmetry, *in preparation* .

Ruessink, B., Walstra, D. and Southgate, H. (2003). Calibration and verification of a parametric wave model on barred beaches, *Coastal Eng.* **48**, pp. 139–149.

Sallenger, A. (2000). Storm impact scale for barrier islands, *J. of Coastal Research,* **16**, 3, pp. 890–895.

Sand, S. (1982). Long wave problems in laboratory models, *J. of Waterways, Port, Coastal and Ocean Eng.* **108**, pp. 492–503.

Sato, S. and Mitsunubo, N. (1991). A numerical model of beach profile change due to random waves, in *Proc. ASCE specialty Conf. Coastal Sediments* (Seattle, WA), pp. 674–687.

Schaffer, H. (1993). Infragravity waves induced by short-wave groups, *J. Fluid Mech.* **247**, pp. 551–588.

Schoonees, J. and Theron, A. (1995). Evaluation of 10 cross-shore sediment transport/morphological models, *Coastal Eng.* **25**, pp. 1–41.

Sha, L. and Van den Berg, J. (1993). Variation in ebb-tidal delta geometry along the coast of The Netherlands and the German Bight, *J. Coastal Res.* **9**, 3, pp. 730–746.

Sleath, J. (1999). Conditions for plug flow formation in oscillatory flow, *Continental Shelf Res.* **19**, pp. 1643–1664.

Slinn, D., Allen, J., Newberger, P. and Holman, R. (1998). Nonlinear shear instabilities of alongshore currents over barred beaches, *J. Geophys. Res.* **103**, C9, pp. 18357–18379.

Smit, M. (2010). *Formation and evolution of nearshore sandbar patterns*, Ph.D. thesis, Delft University of Technology, Delft, The Netherlands.

Smit, M., Reniers, A. and Stive, M. (2005). Nearshore bar response to time-varying conditions, in *Proc. Coastal Dynamics* (Barcelona, Spain).

Snodgrass, F., Groves, G., Hasselmann, K., Miller, G., Munk, W. and Powers, W. (1966). propagation of ocean swell across the pacific, *Phil. Trans. to Royal Soc. of London* **259**, 1103, pp. 431–497.

Soulsby, R. (1997). *Dynamics of Marine Sands* (Thomas Telford, London).

Soulsby, R., Hamm, L., Klopman, G., Myrhaug, D., Simons, R. and Thomas, G. (1993). Wave-current interaction within and outside the bottom boundary layer, *Coastal Eng.* **21**, pp. 41–69.

Speer, P. and Aubrey, D. (1985). A study of nonlinear tidal propagation in shallow inlet/estuarine systems, *Estuarine, Coastal and Shelf Science* **21**, pp. 207–224.

Spydell, M. and Feddersen, F. (2009). Lagrangian drifter dispersion in the surfzone: Directionally-spread normally incident waves, *J. Phys. Oceangr.* **39**, pp. 809–830.

Steetzel, H. (1987). A model for beach and dune profile changes near dune revetments, in *Proc. ASCE specialty Conf. Coastal Sediments* (New Orleans, LA), pp. 87–97.

Steetzel, H. (1990). Cross-shore transport during storm surges, in *Proc. 22nd Int. Conf. Coastal Eng.* (Delft, The Netherlands), pp. 1922–1934.

Steetzel, H. (1993). *Dune erosion*, Ph.D. thesis, Delft University of Technology, Delft, The Netherlands.

Stelling, G. (1984). On the construction of computational methods for shallow water flow problems, *Rijkswaterstaat Communications* **35**.

Stelling, G. and Duinmeijer, S. (2003). A staggered conservative scheme for every froude number in rapidly varied shallow water flows, *Int. J. Numer. Meth. Fluids* **43**, pp. 1329–1354.

Stelling, G. and Zijlema, M. (2003). An accurate and efficient finite-difference algorithm for non-hydrostatic free-surface flow with application to wave propagation, *Int. J. Numer. Meth. Fluids* **43**, pp. 1–23.

Stive, M. (1986). A model for cross-shore sediment transport, in *Proc. 20th Int. Conf. Coastal Eng.* (Taipei, Taiwan), pp. 1550–1564.

Stive, M. and de Vriend, H. (1990). Shear stress and mean flow in shoaling and breaking waves, in *Proc. 24nd Int. Conf. Coastal Eng.* (Kobe, Japan), pp. 594–608.

Stive, M. and Reniers, A. (2003). Sandbars in motion, *Science* **299**, pp. 1855–1856.

Stive, M. and Wang, Z. (2003). *Advances in coastal modeling*, chap. Morphodynamic modeling of tidal basins and coastal inlets (Elsevier), pp. 367–392.

Stive, M. J. B. (1984). A model for offshore sediment transport, in *Proc. 19th Int. Conf. Coastal Eng.* (New York, USA), pp. 1420–1436.

Stokes, C. (1847). On the theory of oscillatory waves, *Trans. Camb. Phil. Soc.* **8**, pp. 441–455.

Struiksma, N., Olesen, K., Flokstra, C. and De Vriend, H. (1985). Bed deformation in curved alluvial channels, *Journal of Hydraulic Res.* **23**, pp. 57–79.

Sutherland, J., Peet, A. and Soulsby, R. (2004a). Evaluating the performance of morphological models, *Coastal Eng.* **51**, 2, pp. 917–939.

Sutherland, J., Walstra, D., Chesher, T., Van Rijn, L. and Southgate, H. (2004b). Evaluation of coastal area modelling systems at an estuary mouth, *Coastal Eng.* **51**, 2, pp. 119–142.

Svendsen, I. (1984). Mass flux and undertow in a surf zone, *Coastal Eng.* **8**, pp. 347–365.

Svendsen, I. (2006). *Introduction to nearshore hydrodynamics* (World Scientific, Singapore).

Sverdrup, H. and Munk, W. (1947). Wind, sea and swell. theory of relations for forecasting, *Hydrographic Office Publication* **601**.

Swart, D. (1974). Offshore sediment transport and equilibrium beach profiles, Tech. Rep. 131, WL-Delft Hydraulics.

Szmytkiewicz, M., Biegowski, J., Kaczmarek, L. M., Okrj, T., Ostrowski, R., Pruszak, Z., Rzynsky, G. and Skaja, M. (2000). Coastline changes nearby harbour structures:

comparative analysis of one-line models versus field data, *Coastal Eng.* **40**, 2, pp. 119–139.

Tang, E. and Dalrymple, R. (1989). *Nearshore Sediment Transport Study*, chap. Nearshore circulation: rip currents and wave groups (Plenum Press), pp. 205–230.

Terrile, E., Reniers, A. and Stive, M. (2009). Acceleration and skewness effects on the instanteneous bed-shear stresses in shoaling waves, *J. of Waterway, Port, Coastal and Ocean Eng.* **228**, pp. 1–7.

Terrile, E., Reniers, A., Stive, M., Tromp, M. and Verhagen, H. (2006). Incipient motion of coarse particles under regular shoaling waves, *Coastal Eng.* **53**, pp. 81–92.

Thomson, J., Elgar, S., Raubenheimer, B., Herbers, T. and Guza, R. (2006). Tidal modulation of infragravity waves via non-linear energy losses in the surf zone, *Geophys. Res. Letters* **33**.

Thornton, E. and Guza, R. (1986). Surfzone longshore currents and random waves: Field data and models, *J. Phys. Oceanogr.* **16**, pp. 1165–1178.

Thornton, E., Humiston, R. and Birkemeier, W. (1996). Bar/trough generationon a natural beach, *J. Geophys. Res.* **101**, pp. 12097–12110.

Townend, I. (2005). An examination of empirical stability relationships for UK estuaries, *J. Coastal Res.* **21**, 5, pp. 1042–1053.

Trowbridge, J. and Madsen, O. (1984). Turbulent wave boundary layers:2. second-order theory and mass transport, *J. Geophys. Res.* **89**, pp. 7999–8077.

Tucker, M. (1952). Surf beats: sea waves of 1 to 5 min period, *Proc. Royal Soc. London* **A**, pp. 565–573.

Turner, I., Aarninkhof, S. and Holman, R. (2006). Coastal imaging applications and research in Australia, *J. Coastal Res.* **22**, 1, pp. 37–48.

Turner, I., Whyte, D., Ruessink, B. and Ranasinghe, R. (2007). Observations of rip spacing, persistence and mobility at a long straight coastline, *Marine Geology* **236**, pp. 209–221.

Ursell, F. (1952). Edge waves on a sloping beach, *Proc. Royal Soc. of London* **A**, pp. 79–97.

Van de Graaff, J. (1988). *Sediment concentration due to wave action*, Ph.D. thesis, Delft Univ. of Technology, Delft.

Van de Graaff, J. and Van Overeem, J. (1979). Evaluation of sediment transport formulae in coastal engineering practice, *Coastal Eng.* **3**, pp. 1–32.

Van de Kreeke, J. (1992). Stability of tidal inlets; Escoffiers analysis, *Shore and Beach* **60**, pp. 9–12.

Van de Kreeke, J. and Robaczewska, K. (1993). Tide-induced residual transport of coarse sediment: application to the Ems estuary, Netherlands, *J. Sea Res.* **31**, 3, pp. 209–220.

Van der Wegen, M. (2010). *Modeling morphodynamic evolution in alluvial estuaries*, Ph.D. thesis, Delft Univ. of Technology, Delft.

Van der Wegen, M., Dastgheib, A., Jaffe, B. and Roelvink, J. A. (2010a). Bed composition generation for morphodynamic modeling: case study of San Pablo Bay in California, U.S.A, *Ocean Dynamics* **61**, 2-3, pp. 173–186.

Van der Wegen, M., Dastgheib, A. and Roelvink, J. A. (2010b). Morphodynamic modeling of tidal channel evolution in comparison to empirical PA equilibrium relationship, *Coastal Eng.* **57**, pp. 827–837.

Van der Wegen, M., Roelvink, D., de Ronde, J. and van der Spek, A. (2008a). Long-term morphodynamic evolution of the Western Scheldt estuary, the Netherlands, using a process based model, in *Proc. COPEDEC VII Conference* (Dubai), pp. 367–368.

Van der Wegen, M. and Roelvink, J. A. (2008). Long-term morphodynamic evolution of a tidal embayment using a two-dimensional, process-based model, *J. Geophys. Res.* **113**, C03016.

Van der Wegen, M., Wang, Z. B., Savenije, H. and Roelvink, J. (2008b). Long-term morphodynamic evolution and energy dissipation in a coastal plain, tidal embayment, *J. Geophys. Res.* **113**, F03001.

van Dongeren, A. (1997). *Numerical modeling of quasi-3d nearshore hydrodynamics*, Ph.D. thesis, Univ. of Delaware, Newark, USA.

van Dongeren, A., Battjes, J., Janssen, T., van Noorloos, J., Steenhauer, K., Steenbergen, G. and Reniers, A. (2007). Shoaling and shoreline dissipation of low-frequency waves, *J. Geophys. Res.* **112**, C02011.

van Dongeren, A., Plant, N., Cohen, A., Roelvink, D., Haller, M. C. and Catalán, P. (2008). Beach Wizard: Nearshore bathymetry estimation through assimilation of model computations and remote observations, *Coastal Eng.* **55**, 12, pp. 1016–1027.

Van Dongeren, A., Wenneker, I., Roelvink, D. and Rusdin, A. (2006). A boussinesq-type wave driver for a morphodynamic model, in *Proc. 30th Int. Conf. Coastal Eng.* (San Diego, USA), pp. 3129–3141.

van Dongeren, A. R. and Svendsen, I. (1997). An absorbing-generating boundary condition for shallow water models, *J. of Waterways, Port, Coastal and Ocean Eng.* **123**, 6, pp. 303–313.

Van Enckevort, I. and Ruessink, B. (2003). Video observations of nearshore bar behaviour. Part 1: alongshore uniform variability. *Continental Shelf Res.* **23**, pp. 501–512.

van Gent, M. (2001). Wave runup on dikes with shallow foreshores, *J. Waterway, Ports, Coasts and Ocean Eng.* **127**, 5, pp. 254–262.

Van Goor, M., Zitman, T., Wang, Z. and Stive, M. (2001). Impact of sea level rise on the morphological equilibrium state of tidal inlets, *Marine Geology* **202**, 3-4, pp. 211–227.

van Rijn, L. (1984). Sediment pick-up functions, *J. Hydraulic Eng.* **110**, pp. 1494–1502.

Van Rijn, L. (1993). *Principles of sediment transport in rivers, estuaries and coastal seas* (AQUA Publications).

van Rijn, L. and Wijnberg, K. (1996). One-dimensional modelling of individual waves and wave-induced longshore currents in the surfzone, *Coastal Eng.* **28**, pp. 121–145.

Van Thiel de Vries, J. (2009). *Dune erosion during storm surges*, Ph.D. thesis, Delft University of Technology, Delft, The Netherlands.

van Thiel de Vries, J., van Gent, M., Walstra, D. and Reniers, A. (2008). Analysis of dune erosion processes in large-scale flume experiments, *Coastal Eng.* **55**, pp. 1028–1040.

van Veen, J. (1936). *Onderzoekingen in de Hoofden* (Algemeene Landsdrukkerij, 's Gravenhage).

Vellinga, P. (1986). *Beach and dune erosion during storm surges*, Ph.D. thesis, Delft University of Technology, Delft, The Netherlands.

Verboom, G. and Slob, A. (1984). Weakly reflective boundary conditions for two dimensional water flow problems, in *Proc. 5th International Conference on Finite elements in water resources*.

Verhagen, H. (1989). Sand waves along the ductch coast, *Coastal Eng.* **13**, pp. 129–147.

Walstra, D., Roelvink, J. and Groeneweg, J. (2000). Calculation of wave-driven currents in a mean flow model, in *Proc. 27th Int. Conf. Coastal Eng.* (Sidney, Australia), pp. 1050–1063.

Walstra, D. and Ruessink, B. (2009). Process-based modeling of cyclic bar behavior on yearly scales, in *Proc. Coastal Dynamics* (Lisbon, Portugal), pp. 1–12.

Walton, T. and Adams, W. (1976). Capacity of inlet outer bars to store sand, in *Proc. 15th Int. Conf. Coastal Eng.* (Honolulu, Hawaii), pp. 1919–1937.

Wang, P. and Horwitz, M. (2007). Erosional and depositional characteristics of regional overwash deposits caused by multiple hurricanes, *Sedimentology* **54**, pp. 545–564.

Wang, Z. (1992). Theoretical analysis on depth-integrated modelling of suspended sediment transport, *J. Hydraulic Res.* **30**, 3, pp. 403–421.

Wang, Z., Karssen, B., Fokkink, R. and Langerak, A. (1998). A dynamic/empirical model for long-term morphological development of estuaries, in *Physics of estuaries and coastal seas*, pp. 279–286.

Wang, Z., Louters, C. and De Vriend, H. (1995). Morphodynamic modelling for a tidal inlet in the Wadden Sea, *Marine Geology* **126**, pp. 289–300.

Watanabe, A. and Dibajnia, M. (1988). Numerical modeling of nearshore waves, cross-shore sediment transport and beach profile changes, in *Proc. IAHR Symp. on Math. Modeling of Sediment transport in the Coastal Zone*, pp. 166–174.

Wijnberg, K. (2002). Environmental controls on decadal morphologic behaviour of the holland coast, *Marine Geology* **189**, pp. 227–247.

Wright, L. and Short, A. (1984). Morphodynamic variability of surf zones and beaches: A synthesis, *Marine Geology* **56**, 1-4, pp. 93–118.

Xu, Z. and Bowen, A. (1993). Wave- and wind-driven flow in water of finite depth, *J. Phys. Oceanography* **24**, pp. 1850–1866.

Zou, Q., Bowen, A. and Hay, A. (2003). Vertical distribution of wave shear stress in variable water depth: Theory and field observations, *J. Geophys. Res.* **111**, pp. 1–17.

Zyserman, J. and Johnson, H. (2002). Modelling morphological processes in the vicinity of shore-parallel breakwaters. *Coastal Eng.* **45**, 3-4, pp. 261–284.